Microbiological methods

Alex M. Démlio

Microbiological methods

Fifth edition

C. H. Collins
MBE, MPhil, FRCPath, FIBiol, FIMLS
Senior Technical Officer, Public Health Laboratory Service, Dulwich, London

Patricia M. Lyne
MIBiol, AIFST

Butterworths
London Boston Durban Singapore Sydney Toronto Wellington

First published, 1964
Reprinted 1966
Second Edition 1967
Reprinted 1970
Third Edition 1970
Fourth Edition 1976
Reprinted with additions 1979
Fifth Edition 1984

British Library Cataloguing in Publication Data

Collins, C. H.
Microbiological methods.—5th ed.
1. Microbiology—Laboratory manuals
I. Title II. Lyne, Patricia M.
576′.028 QR65

ISBN 0-408-70957-X

Library of Congress Cataloging in Publication Data

Collins, C. H. (Christopher Herbert)
Microbiological methods.
Bibliography: p.
Includes index.
1. Microbiology—Technique. I. Lyne, Patricia M.
II. Title.
QR65.C6 1984 576′.028 83-15305

ISBN 0-408-70957-X

Filmset and printed in England by
Butler & Tanner Ltd Frome and London

Preface to the fifth edition

During the preparation of the fifth edition of this book we have frequently referred to earlier editions and revisions and it has been interesting to note how much of the basic technique and information of 20 years ago still holds good and how much has changed, not necessarily for the good of the art.

Once again, small changes in the method of presentation have been necessary, but we have held fast to the general format which, to judge from sales, has been successful in the past. We are, however, indebted to our publishers for giving this format a new and more attractive style.

As in previous editions, the names of some organisms have been changed to accommodate the whims of 'taxonomic tour operators', but for some microbes we have firmly retained the nomenclature in common use.

While we have incorporated much new material we are conscious that the book reflects our own interests, but we hope that these are also those of our colleagues and their students who work at the 'sharp' end in routine medical, public health and food control laboratories where unlimited money is not available for new and exciting equipment. We believe that 'manual' microbiology still has a long way to go before automation takes over completely.

We have also tried to increase the numbers of 'nuggets' and 'titbits' of useful information, acquired from other workers too numerous to name individually, and either unpublished or printed in journals which may not be available in many laboratories.

Hadlow

C. H. Collins
Patricia M. Lyne

Additional chapters by

A. CREMER, FIMLS, Senior Chief Medical Laboratory Scientific Officer, Bacteriology Department, University College Hospital, London.

T. J. DONOVAN, PhD, FIMLS, Senior Chief Medical Laboratory Scientific Officer, Public Health Laboratory, Maidstone.

A.L. FURNISS, MD, Dip Bact, formerly Director, Public Health Laboratory, Maidstone.

J.D. JARVIS, FIMLS, Senior Chief Medical Laboratory Scientific Officer, Microbiology Department, The London Hospital, London.

A.G.J. PROCTOR, TD, BD, MPhil, FIMLS, Senior Chief Medical Laboratory Scientific Officer, PHLS Mycology Reference Laboratory, London School of Hygiene and Tropical Medicine, London.

Contents

1 Safety in microbiology 1
2 Laboratory equipment 10
3 Sterilization, disinfection and the treatment of infected materials 32
4 Culture media 56
5 Cultural methods 89
6 Staining methods 97
7 Identification methods 102
8 Mycological methods 114
9 Counting micro-organisms 128
10 Agglutination tests 142
11 Clinical material 146
12 Fluorescent antibody techniques 160
13 Antibiotic sensitivity and assay tests 167
14 Bacterial food poisoning 182
15 Food—general principles 190
16 Meat 195
17 Fish 203
18 Milk and dairy products 207
19 Miscellaneous foods 223
20 Thermally processed foods 230
21 Soft drinks, fruit juices and alcoholic beverages 239
22 Sanitation control 241
23 Pharmaceuticals and cosmetics 247
24 Animal by-products, grass and silage 251
25 Water 253
26 Pseudomonas, acinetobacter, alkaligenes, achromobacter, flavobacterium, chromobacterium and acetobacter 261

27 Vibrios, aeromonas, plesiomonas and photobacterium 271

28 Escherichia, citrobacter, klebsiella, enterobacter, salmonella, shigella and proteus 280

29 Brucella, haemophilus, gardnerella, moraxella, bordetella, actinobacillus, yersinia, pasteurella, francisella, bacteroides, campylobacter and legionella 303

30 Neisseria, branhamella, gemella and veillonella 321

31 Staphylococcus, micrococcus, peptococcus and sarcina 325

32 Streptococcus, aerococcus, leuconostoc and pediococcus 331

33 Corynebacterium, microbacterium, brochothrix, propionibacterium, brevibacterium, erysipelothrix, listeria, lactobacillus and bifidobacterium 344

34 Bacillus 356

35 Clostridium 361

36 Mycobacterium 371

37 Actinomyces, nocardia and streptomyces 394

38 Spirochaetes 398

39 Yeasts 401

40 Moulds 406

41 Pathogenic fungi 413

Appendix 424

Index 435

Chapter 1

Safety in microbiology

Over 4000 laboratory-associated infections have been reported so far this century.[1,2] Many more have probably gone unreported. Most of the victims worked in research laboratories with organisms whose potential pathogenicity was unknown at the time, and/or used techniques now known to be hazardous. As a result of investigations over the last 25 years the causes of many of these infections have been discovered and methods for preventing them have been developed. There is now no need for microbiological laboratory workers (with certain notable exceptions which are outside the scope of this book) to feel at risk from the organisms they handle providing that they are aware of:

(1) the potential hazards of the organisms they handle,
(2) the routes by which these organisms may enter the body and cause infections,
(3) the correct methods of 'containing' these organisms so that they do not have access to those routes.

Classification of micro-organisms on the basis of risk

Micro-organisms vary in their ability to cause infections. Some are harmless; some may be responsible for diseases with mild symptoms; others can cause serious illnesses; a few have the potential for spreading in the community and causing serious epidemic disease. Experience and research into laboratory-acquired infections have enabled investigators to classify organisms and viruses into three or four groups. Three separate systems[3,4,5] have been published and are shown in *Tables 1.1* and *1.2*. The most recent, that of the Safety Measures in Microbiology Programme of the World Health Organization (*Table 1.2*)[5,6] is the best and is adopted in this book. This uses four Risk Groups (I–IV) in increasing order of hazard to the laboratory worker and the community. The Risk Group numbers also indicate the levels of 'containment', i.e. precautions and techniques necessary to prevent infections.

The World Health Organization published no lists of micro-organisms within each Risk Group. Member states are encouraged to prepare their own

TABLE 1.1. Summary of systems for classifying micro-organisms on the basis of hazards to laboratory workers and the community

	Hazard			
	Low			*High*
USPHS (1974)	Class 1 none or minimal	Class 2 ordinary potential	Class 3 special, to individual	Class 4 high, to individual and community
UK (1978)[a]		Category C no special potential	Category B special, to individual	Category A high, to individual and community
WHO (1979)	Risk Group 1 low individual low community	Risk Group II moderate individual limited community	Risk Group III high individual low community	Risk Group IV high, to individual and community
Appropriate laboratory	Basic	Basic	Containment[a]	Maximum containment[b]

[a] DHSS, 1978.
[b] Refers to precautions, equipment and design of laboratory.

TABLE 1.2 Classifications of micro-organisms on the basis of hazard. WHO Risk Group system[5, 6]

Risk Group I (low individual and community risk).
A micro-organism that is unlikely to cause human disease or animal disease of veterinary importance.

Risk Group II (moderate individual risk, limited community risk).
A pathogen that can cause human or animal disease but is unlikely to be a serious hazard to laboratory workers, the community, livestock or the environment. Laboratory exposures may cause serious infection, but effective treatment and preventive measures are available and the risk of spread is limited.

Risk Group III (high individual risk, low community risk).
A pathogen that usually produces serious human disease but does not ordinarily spread from one infected individual to another.

Risk Group IV (high individual and community risk).
A pathogen that usually produces serious human or animal disease and may be readily transmitted from one individual to another, directly or indirectly.

lists because organisms which may present a high risk in one country may offer few or none in another; there is also no point in taking elaborate and expensive precautions against an organism in the laboratory if the workers are exposed to it outside.

A few lists have been proposed, e.g. in the UK[4], and in the USA[3, 7] but there is still no general agreement within and between those countries. In *Tables 1.3* and *1.4* the organisms mentioned in this book have been classified into Risk Groups I, II and III, with reference to existing lists[4, 7], after seeking the advice of British, American and European microbiologists who actually work on the organisms and after considering the histories of infections which they have caused[1, 2]. As Risk Group IV contains only viruses no reference is made to it in this book.

TABLE 1.3 Risk Groups of bacteria mentioned in this book

Organism	*Risk Group*	*Organism*	*Risk Group*
Acetobacter	I	*Brevibacterium*	I
Achromobacter	I	*Brochothrix*	I
Acinetobacter	I	*Brucella*	III
Actinobacillus	I	*Campylobacter*	I
Actinomyces bovis	II	*Chromobacterium*	I
eriksonii	II	*Citrobacterium*	I
israelii	II	*Clostridium botulinum*	III
naeslundii		*difficile*	II
other spp.	I	*fallax*	II
Aerococcus	I	*novyi*	II
Aeromonas	I	*perfringens*	II
Alcalescens	I	*septicum*	II
Alkaligenes	I	*sordelli*	II
Arizona	I	*tetani*	II
Bacillus anthracis	II	other spp.	I
other spp.	I	*Corynebacterium diphtheriae*	II
Bacteroides	I	*equi*	I
Bifidobacterium	I	*pyogenes*	I
Bordetella	II	*renale*	I
Borrelia	II	*ulcerans*	II
Branhamella	I	other spp.	I

TABLE 1.3 *(continued)*

Organism	*Risk Group*	*Organism*	*Risk Group*
Edwardsiella	I	*Mycobacterium* (*continued*)	
Eikenella	I	*xenopi*	III
Enterobacter	I	other spp.	I
Erwinia	I	*Neisseria gonorrhoeae*	II
Erysipelothrix	II	*meningitidis*	II
Escherichia	I	other spp.	I
Flavobacterium meningo-		*Nocardia*	II
septicum	I	*Pasteurella*	II
other spp.	I	*Pediococcus*	I
Francisella	III	*Photobacterium*	I
Fusobacterium	I	*Plesiomonas*	I
Gardnerella	I	*Propionibacterium*	I
Gemella	I	*Proteus*	I
Gluconobacter	I	*Providence*	I
Klebsiella	I	*Pseudomonas mallei*	III
Lactobacillus	I	*pseudomallei*	III
Legionella	II	other spp.	I
Leptospira	II	*Salmonella paratyphi A*	III[a]
Leuconostoc	I	*typhi*	III[a]
Listeria	II	other serotypes	II
Microbacterium	I	*Serratia*	I[b]
Micrococcus	I	*Shigella*	II
Moraxella	I	*Staphylococcus aureus*	II
Mycobacterium		other spp.	I
africanum	III	*Streptobacillus*	II
avium	III	*Streptococcus* human and	
bovis	III	animal pathogens	II
chelonei	II	milk spp.	I
fortuitum	II	*Streptomyces madurae*	II
intracellulare	III	*pelleteri*	II
kansasii	III	*somaliensis*	II
leprae	II	*Treponema*	II
marinum	II	*Veillonella*	I
malmoense	III	*Vibrio cholerae*	II
scrofulaceum	III	*parahaemolyticus*	I
simiae	III	other spp.	I
szulgai	III	*Yersinia pestis*	III
tuberculosis	III	other spp.	I
ulcerans	II		

[a] Require a Containment laboratory but not a safety cabinet; airborne infection unlikely.
[b] Use a safety cabinet for experiments with aerosols.

TABLE 1.4 Risk Groups of fungi and yeasts mentioned in this book

Organism	*Risk Group*	*Organism*	*Risk Group*
Acremonium	I	*Epidermophyton*	III
Alternaria	I	*Fonseceae*	I
Aspergillus	I[a]	*Geotrichum*	I
Blastomyces	II	*Gliocladium*	I
Botrytis	I	*Hansenula*	I
Candida	I	*Helminthosporium*	I
Cladosporium	I	*Histoplasma capsulatum*	III
Coccidioides immitis	III	*Kloeckera*	I
Cryptococcus neoformans	II	*Madurella*	II
Debaromyces	I	*Microsporon*	II
Endomyces	I	*Neurospora*	I

TABLE 1.4 *(continued)*

Organism	*Risk group*	*Organism*	*Risk group*
Paecilomyces	I	*Saccharomyces*	I
Paracoccidioides braziliensis	II	*Scopulariopsis*	I
Penicillium	I	*Sporobolomyces*	I
Phialophora	I	*Sporothrix*	II
Pichia	I	*Torulopsis*	I
Pullularia	I	*Trichoderma*	II
Rhodotorula	I	*Trichophyton*	II

[a] Use a safety cabinet for experiments which generate spores. These may be allergenic.

Routes of infection

Micro-organisms can enter the body through the mouth (ingestion), through the lungs (inhalation), through the skin (injection) and through the eyes.

They may be ingested during mouth pipetting and may also enter the mouth from contaminated fingers and articles which have been contaminated from the laboratory benches, e.g. cigarettes, food, pencils. Such environmental contamination may be the result of spills and splashes, either unrecognized or inadequately disinfected.

Inhalation of infected airborne particles (aerosols) released during many common laboratory manipulations have probably caused the largest number of laboratory-associated diseases.[1, 2]

Injection may result from accidental stabbing with hypodermic needles, pasteur pipettes or broken and infected glassware. Organisms may also enter the body through cuts and abrasions in the skin, some of which may be too small to be obvious to the worker himself.

Splashing of infected liquids into the eyes has been the cause of a number of infections, some fatal.

The routes of infection in laboratory-acquired infections are not necessarily the same as those of 'naturally'-acquired infections. In addition, the infecting dose may be much greater and the symptoms may therefore be different.

Aerosols: infected airborne particles

These are likely to be released and may be inhaled or may contaminate the hands, bench, etc. during the following manipulations: work with loops and syringes; pipetting, centrifuging, blending and homogenizing, pouring; opening culture tubes and dishes. (For details *see* Collins[2].)

Prevention of laboratory-acquired infections: a code of practice

The principles of containment involve the provision of:

(1) primary barriers around the organisms to prevent their dispersal into the laboratory,
(2) secondary barriers around the worker to act as a safety net should the primary barriers be breached,

(3) tertiary barriers to prevent the escape into the community of any organisms which are not contained by the primary and secondary barriers.

This Code of Practice draws very largely on the requirements and recommendations published in other works.[2, 3, 4, 7, 8, 9]

Primary barriers

These are techniques and equipment designed to contain micro-organisms and prevent their direct access to the worker and their dispersal as aerosols.

(1) Mouth pipetting should be banned in all circumstances. Pipetting devices (p. 29) should be provided.
(2) No article should be placed in or at the mouth. This includes mouthpieces of rubber tubing attached to pipettes, pens, pencils, labels, smoking materials, fingers, food and drink.
(3) The use of hypodermic needles should be restricted. Cannulas are safer.
(4) Sharp glass pasteur pipettes should be replaced by soft plastic varieties (p. 28).
(5) Cracked and chipped glassware should be replaced.
(6) Centrifuge tubes should be filled only to within 2 cm of the lip. All Risk Group III materials should be centrifuged in sealed centrifuge buckets (p. 14).
(7) Bacteriological loops should be completely closed, with a diameter not more than 3 mm and a shank not more than 5 cm and be held in metal, not glass, handles.[2]
(8) Homogenizers should be inspected regularly for defects which might disperse aerosols (p. 15). Only the safest models should be used[2]. Glass Griffith's tubes and tissue homogenizers should be held in a pad of wadding in a gloved hand when operated.
(9) All Risk Group III materials should be processed in a Containment Laboratory (*see* Tertiary barriers *below*) and in a Microbiological Safety Cabinet (p. 16) unless exempted in *Table 1.3*.
(10) A supply of suitable disinfectant (p. 40) at use dilution should be available at every work station.
(11) Benches and work surfaces should be disinfected regularly and after spillage of potentially infected material.
(12) Discard jars for small objects and for re-usable pipettes should be provided at each work station. They should be emptied, decontaminated and refilled with freshly prepared disinfectant daily (p. 49).
(13) Discard bins or bags supported in bins should be placed near to each work station. They should be removed and autoclaved daily (p. 49).
(14) Broken culture vessels, e.g. after accidents, should be covered with a cloth. Suitable disinfectant should be poured over the cloth and the whole left for 30 min. The debris should then be cleared up into a suitable container (tray or dustpan) and autoclaved. The hands should be gloved and stiff cardboard used to sweep up.
(15) Infectious and pathological material to be sent through the post or by airmail should be packed in accordance with government, post office and

airline regulations, obtainable from those authorities. Full instructions are also given elsewhere.[2,4,6,8]

(16) Specimen containers should be robust and leak proof.

Secondary barriers

These are intended to protect the worker should the primary barriers fail. They should, however, be observed as strictly as the latter.

(1) Proper overalls should be worn at all times and fastened. They should be kept apart from outdoor and other clothing.
(2) Laboratory protective clothing should be removed when the worker leaves the laboratory and not worn in any other area such as canteens and rest rooms.
(3) Hands should be washed after handling infectious material and always before leaving the laboratory.
(4) Any obvious cuts, scratches and abrasions on exposed parts of the worker's body should be covered with waterproof dressings.
(5) In laboratories where pathogens are handled there should be medical supervision.
(6) All illnesses should be reported to the medical or occupational health supervisors. They may be laboratory associated. Pregnancy should also be reported. It is inadvisable to work with certain organisms during pregnancy.
(7) Any member of the staff who is receiving steroids or immunosuppressive drugs should report this to the medical supervisor.
(8) Workers should be immunized where possible and practicable against likely infections.[2,4]
(9) In laboratories where tuberculous materials are handled the staff should have received BCG or have evidence of a positive skin reaction before starting work. They should have annual chest X-rays thereafter[2,4].

Tertiary barriers

These are intended to offer additional protection to the worker and to prevent the escape into the community of micro-organisms that are under investigation in the laboratory.

They concern architectural and engineering design and the reader is referred to other publications.[2,6,10] The WHO Risk group system requires three levels of laboratory design, summarized below.[6]

The Basic laboratory

This is intended for work with organisms in Risk Groups I and II. Ample space should be provided. Walls, ceilings and floors should be smooth, nonabsorbent, easy to clean and disinfect and resistant to the chemicals that are likely to be used. Floors should also be slip resistant. Lighting and heating should be adequate. Hand basins, other than laboratory sinks, are essential.

Bench tops should be wide, the correct height for work at a comfortable sitting position, smooth, easy to clean and disinfect and resistant to chemicals likely to be used. Adequate storage facilities should be provided.

Access should be restricted to authorized persons.

The Containment laboratory

This is intended for work with organisms in Risk Group III. All the features of the Basic laboratory should be incorporated, plus the following.

The room should be physically separated from other rooms, with no communication (e.g. by pipe ducts, false ceilings) with other areas, apart from the door, which should be lockable, and transfer grilles for ventilation (*see below*).

Ventilation should be one way, achieved by having a lower pressure in the Containment laboratory than in other, adjacent rooms and areas. Air should be removed to atmosphere (total dump, not recirculated to other parts of the building) by an exhaust system coupled to a microbiological safety cabinet (*see* p. 19) so that during working hours air is continually extracted by one or the other and airborne particles cannot be moved around the building. Replacement air enters through transfer grilles.

Access should be strictly regulated. The international biohazard warning sign, with appropriate wording, should be fixed to the door (*Figure 1.1*).

Figure 1.1 International biohazard sign

The Maximum Containment laboratory

This is required for work with Risk Group IV materials and is outside the scope of this book. Construction and use generally requires government licence or supervision.

For further information about the safe design of Basic and Containment laboratories *see* Collins[2] and WHO[6].

Instruction in safety

This instruction should be part of the general training given to all microbiology laboratory workers and should be included in what is known as Good Laboratory Practice. In general, methods that protect cultures from contamination also protect workers from infection but this should not be taken for granted. Personal protection can be achieved only by good training and careful work. Programmes for training in safety in microbiology have been published.[2, 6]

For a review of the history, incidence, causes and prevention of laboratory-acquired infection *see* Collins[2].

References

1. PIKE, R. M. (1979) *Annual Review of Microbiology*, **33,** 41
2. COLLINS, C. H. (1983) *Laboratory-acquired Infections*. London: Butterworths
3. US PUBLIC HEALTH SERVICE (1974) *Classification of Etiologic Agents on the Basis of Hazard*. Atlanta: Centers for Disease Control
4. DEPARTMENT OF HEALTH AND SOCIAL SECURITY (1978) *Code of Practice for the Prevention of Infection in Clinical Laboratories*. London: HMSO
5. WORLD HEALTH ORGANIZATION (1979) *Weekly Epidemiological Records*, **No. 44,** p. 340.
6. WORLD HEALTH ORGANIZATION (1983) *Laboratory Biosafety Manual*. Geneva: WHO
7. US PUBLIC HEALTH SERVICE (1981) *Federal Register*, **46,** 59379
8. US PUBLIC HEALTH SERVICE (1974) *Lab Safety at the Centre for Disease Control*. Atlanta: Centers for Disease Control
9. MICROBIOLOGICAL CONSULTATIVE COMMITTEE (1982) *Guidelines for Microbiological Safety*. Reading: Society for General Microbiology
10. DEPARTMENT OF HEALTH AND SOCIAL SECURITY (1981) *Hospital Building Note No. 15. Pathology Department*. London: DHSS

Chapter 2

Laboratory equipment

It is important to choose the correct equipment for microbiological work. A wide range of laboratory apparatus, glassware and plastic goods is available from scientific equipment suppliers, but unfortunately not all of it is suitable for microbiological use. Before purchasing any new piece of equipment, it is best to obtain the personal advice of microbiologists who have had experience in its use. It is bad policy to rely entirely on advertisements, catalogues, extravagant claims of representatives and the opinions of purchasing officers who are mainly concerned with balancing budgets. The best is not always the most expensive, but it is rarely the cheapest. Few microbiologists require very expensive research-type microscopes, but centrifuges should be the very best available. Laboratory autoclaves should be designed for laboratory, not pharmacy or hospital, use. Thermostatic equipment designed for chemical work is usually suboptimal for microbiological use.

In this chapter, we have indicated the types of equipment that we and our associates and friends have found adequate for use in medical, veterinary and food microbiology (except virology) work and have given the names of some suppliers in the UK, and a few in the USA, with whose equipment we are familiar or which has been recommended to us by our colleagues. In the UK regular reviews of equipment and information about new apparatus are given in *Laboratory Practice*, *Medical Laboratory World* and *The Medical Technologist*, all of which appear monthly. Our experience of American equipment is limited but we are advised that two publications are likely to be very useful to American and Canadian microbiologists. These are *The Guide to Scientific Instruments*, published annually in *Science* (NY) by the American Association for the Advancement of Science, and the *Annual Directory of Medical Laboratory Product and Service Literature*, published in *Laboratory Management* by United Business Publications.

Failure to mention any other products or suppliers merely indicates that we have had no personal experience of them. We welcome information from other suppliers about their products.

The operation of some microbiological equipment may result in the release of infectious aerosols (*see* Chapter 1 and Collins[1]).

Microscopes

Although the microscope is an essential piece of microbiological apparatus, it is not used as often as might be expected and the purchase of elaborate and expensive (prestige) instruments is not justified. The medium priced instruments are adequate.

Laboratory workers should be familiar with the general mechanical and optical principles, but a detailed knowledge of either is unnecessary and, apart from superficial cleaning, maintenance should be left to the manufacturers, who will arrange for periodic visits by their technicians. Most manufacturers publish handbooks containing useful explanations and information.

Wide-field compensating eye-pieces should be fitted; they are less tiring than other eye-pieces and are more convenient for spectacle wearers. For low power work, ×5 and ×10 objectives are most useful and for high power (oil immersion) microscopy, ×50 and ×100 fluorite objectives are desirable. These fluorite lenses are much better than achromatics and much less expensive than the hardly justifiable apochromats.

Microscopes for fluorescence microscopy

Suitable equipment is described by Cremer in Chapter 12.

Low power microscopes

Although many observations on colony forms can be made with a hand-lens, a low power binocular 'dissecting' microscope is very useful for examining very small colonies and doing subcultures. An instrument that can be illuminated from above and below is desirable. Paired ×5 eye-pieces and 59-mm objectives give a satisfactory image.

Incubators and water-baths

Incubators

Incubators are available in various sizes. In general, it is best to obtain the largest possible model that can be accommodated, although this may create space problems in laboratories where several different incubation temperatures are employed. The small incubators suffer wider fluctuations in temperature when their doors are opened than do the larger models and most laboratory workers find that incubator space, like refrigerator space, is subject to 'Parkinson's law'.

Although incubators rarely develop faults, it is advisable, before choosing one, to ascertain that service facilities are available. The circuits are not complex but require expert technical knowledge to repair. Transporting incubators back to the manufacturers is most inconvenient.

In medical and veterinary laboratories, incubators are usually operated only at 35–37°C. The food or industrial laboratory usually requires machines operated at 55, 28–32 and 15–20°C. For temperatures above ambient there are no problems, but lower temperatures may need a cooling as well as a heating device.

In laboratories where capnophiles are cultured routinely, a carbon dioxide incubator is convenient. Several models are now available from specialist incubator manufacturers, e.g. LEEC, Lab Line, Grant. Automatic control of carbon dioxide and humidity is now possible in some of these.

Cooled incubators

For incubation at temperatures below the ambient incubators must be fitted with modified refrigeration systems with heating and cooling controls. These need to be correctly balanced (Jennings, LEEC).

Automatic temperature changes

These are necessary when cultures are to be incubated at different temperatures for varying lengths of time, as in the examination of water by membrane filtration, and when it is not convenient to move them from one instrument to another. Two thermostats are required, wired in parallel, and a time switch wired to control the thermostat set at the higher temperature.

Incubator rooms

A 'hot-room' or 'walk-in incubator' is not difficult to adapt from a small room or a corner of a large room. Windows must, of course, be blocked up.

The walls need two layers of building paper on which is glued 48-mm thick slabs of expanded polystyrene (cork is much more expensive) between battens at 600-mm centres. The inner lining can be ordinary plaster or insulation board. The ceiling must be lagged in the same way and in high rooms lower false ceilings are preferable. The floor can be covered with insulation board and hardboard and the doors lagged on their inner surfaces in the same way, and fitted in their jambs on sponge rubber draught-prevention strips.

Two methods of heating are possible. Tubular heaters around the walls are satisfactory and a power of 3 kW is more than adequate for a room of 5–7 m^3 that is to be maintained at 37°C. A large circulating fan, to avoid hot and cold spots, must be fitted on one wall and should operate constantly. An alternative arrangement is the greenhouse- or space-heater in which a 2.5–3.0 kW heater and a fan are built into a steel casing. The wiring must be altered so that the fan is always on and the heater connected to a sensitive thermostat (the thermostats built into these heaters are too coarse for this purpose).

These two methods of heating may be combined. Smaller tubular heaters, permanently switched on, will supply background heat and the space heater will maintain the required temperature.

Wooden shelving and racking is undesirable. If a high humidity is maintained, fungi may grow on the wood. Steel or aluminium racks are preferable and can be tailored by a racking specialist. We have used steel racking made by Denley to accommodate the usual aluminium culture bottle racks, thus avoiding solid shelving. Any shelves should be free, i.e. easily removed for cleaning, and there should be air spaces between the shelves and the walls so as to allow for circulation of air.

Complete incubator rooms can be purchased from several companies, e.g. Astell, Gallenkamp, Blickman.

Rooms in which a constant temperature at or below ambient has to be maintained need a refrigerator coil through which a fan blows cold air into the room. Such a device can be thermostatically operated.

Water-baths

The contents of a test-tube placed in a water-bath are raised to the required temperature much more rapidly than in an incubator. These instruments are therefore used for short-term incubation. If the level of water in the bath comes one-half to two-thirds of the way up that of the column of liquid in the tube, convection currents are caused which keep the contents of the tube well mixed, and hasten reactions such as agglutination.

All modern water-baths are equipped with electrical stirrers and in some the heater, thermostat and stirrer are in one unit, easily detached from the bath for use elsewhere or for servicing. Water-baths must also be lagged so as to prevent heat loss through the walls. A bath that has not been lagged by the manufacturers can be insulated with slabs of expanded polystyrene.

Water-baths should be fitted with lids in order to prevent heat loss and evaporation. These lids must slope so that condensation water does not drip on the contents. To avoid chalky deposits on tubes and internal surfaces, only distilled water should be used, except, of course, on those baths which operate at or around ambient temperature, which are connected to the cold water supply and fitted with a constant level and overflow device. This kind of bath offers the same problems as the low-temperature incubators unless the water arrives at a much lower temperature than that required in the water-bath. In premises that have an indirect cold water supply from a tank in the roof, the water temperature in summer may be near or even above the required water-bath temperature, giving the thermostat no leeway to operate. In this case, unless it is possible to make direct connection with the rising main, the water supply to the bath must be passed through a refrigerating coil. Alternatively, the water-bath can be placed in a refrigerator and its electricity supply adapted from the refrigerator light or brought in through a purpose-made hole in the cabinet. The problem is best overcome by fitting laboratory refrigeration units, designed for this purpose (Grant, Gilson, Techne).

Metal block heaters

These give better temperature control than water-baths and do not dry out, but they can be used only for tubes and thin glass bottles which must fit neatly into the holes in the steel or aluminium blocks (Grant, Camlab, Techne).

Centrifuges

The ordinary laboratory centrifuge is capable of exerting a force of up to 3000*g*, and this is the force necessary to deposit bacteria within a reasonable time.

For general microbiology, a centrifuge capable of holding 15- and 50-ml buckets and working at a maximum speed of 4000 rev/min is adequate. The swing-out head is safer than the angle head[1].

For maximum microbiological safety, sealed buckets should be fitted[1]. These confine aerosols if breakage occurs and are safer than sealed rotors. Sealed buckets ('safety cups') are supplied by several companies. Windshields offer no protection. There should be an electrically operated safety catch so that the lid cannot be opened when the rotors are spinning.

Centrifuge buckets are usually made of stainless steel and are fitted with rubber buffers. They are paired and their weight is engraved on them. They are always used in pairs, opposite to one another, and it is convenient to paint each pair with different coloured patches so as to facilitate recognition. If the buckets fit in the centrifuge head on trunnions, these are also paired.

Centrifuge tubes to fit the buckets are made of glass, plastic, nylon or spun aluminium. Conical and round-bottomed tubes are made. Although the conical tubes concentrate the deposit into a small button, they break much more readily than do the round-bottomed tubes. When the supernatant, not the deposit, is required, round-bottomed tubes should be used.

Instructions for using the centrifuge

(1) Select two centrifuge tubes of identical lengths and thicknesses. Place liquid to be centrifuged in one tube and water in the other to within about 2 cm of the top.
(2) Place the tubes in paired centrifuge buckets and place the buckets on the pans of the centrifuge balance. This can be made by boring holes large enough to take the buckets in small blocks of hard wood and fitting them on the pans of a simple balance.
(3) Balance the tubes and buckets by adding 70% alcohol to the lightest bucket. Use a pasteur pipette and allow the alcohol to run down between the tube and the bucket.
(4) Place the paired buckets and tubes in diametrically opposite positions in the centrifuge head.
(5) Close the centrifuge lid and make sure that the speed control is at zero before switching on the current. (Many machines are fitted with a 'no-volt' release to prevent the machine starting unless this is done.)
(6) Move the speed control slowly until the speed indicator shows the required number of revolutions per minute.

Precautions

(1) Make sure that the rubber buffers are in the buckets, otherwise the tubes will break.
(2) Check the balancing carefully. Improperly balanced tubes will cause 'head wobble', spin-off accidents and wear out the bearings.
(3) Check that the balanced tubes are really opposite one another in multi-bucket machines.
(4) Never start or stop the machine with a jerk.
(5) Observe the manufacturer's instructions about the speed limits for the various loads.
(6) Open sealed centrifuge buckets in a microbiological safety cabinet.

Maintenance

The manufacturers will, by contract, arrange periodic visits for inspection and maintenance in the interests of safety and efficiency.

Blenders, tissue grinders and shakers

These range from instruments suitable for grinding and emulsifying large samples, e.g. of food, to glass devices for homogenizing small pieces of tissue.

Most electrically driven machines, e.g. the Waring Blender, VirTis Homogenizer and MSE Atomix are efficient but suffer from the disadvantage of requiring a fresh, sterile cup for each sample. These cups are expensive and it takes time to clean and re-sterilize them.

Some of them may release aerosols during operation and when they are opened.[1] A heavy Perspex or metal cover should be placed over them during use. This should be decontaminated afterwards. All blenders, etc. should be opened in a microbiological safety cabinet because the contents will be warm and under pressure. Aerosols will be released.

The Stomacher Lab-Blender (Seward) overcomes these problems. Samples are emulsified in heavy-duty sterile plastic bags by the action of paddles. This is an efficient machine, capable of processing large numbers of samples in a short time without pauses for washing and re-sterilizing. There is little risk of aerosol dispersal. The bags are automatically sealed while the machine is working.

For smaller samples, the MSE Masticator is useful and inexpensive. A glass bottle of square section and holding about 25 ml of fluid (a 'square universal container') is fitted with a metal cap carrying a stuffing box through which passes a shaft provided with metal cutting blades inside the bottle and a knurled knob outside. This knob is held against a rotating wheel (e.g. of any laboratory electric motor), or is attached to the drive of the MSE Homogenizer, which is another useful piece of equipment.

Tissue grinders (Griffith's tubes), essential for emulsifying small pieces of tissue for microbiological examination, are available in several sizes. A heavy glass tube is constricted near to its closed end and a pestle, usually made of glass covered with PTFE or of stainless steel, is ground into the constriction. With the pestle in place, the tissue and some fluid are put into the tube. The pestle is rotated by hand or in the chuck of the MSE Homogenizer or Griffin Hollow Shaft Stirrer (Gallenkamp). The tissue is ground through the constriction and the emulsion collects in the bottom of the tube.

These grinders present some hazards. The tubes, even though made of borosilicate glass, may break and disperse infected material. They should be used inside a microbiological safety cabinet and be held in the gloved hand in a wad of absorbent material.

Tissue grinders are sold by several companies, including Payne, Camlab, Horwell, Jencon, Thomas and Beckman.

Small pieces of soft tissue, e.g. curettings, may be emulsified by shaking in a screw-capped bottle (Bijou) with a few glass beads and 1 or 2 ml of broth on a Vortex Mixer.

Shaking machines

Conventional shaking machines are sold by almost all laboratory suppliers. They are useful for mixing and shaking cultures and for serological tests, but should be fitted with racks, preferably made of polypropylene, which hold the bottles or tubes firmly, and covered with a stout Perspex box when in use so as to prevent the dispersal of aerosols.

The Vortex Mixer (Scientific Industries, Gallenkamp) is useful for mixing the contents of single bottles, e.g. emulsifying sputum, but should always be operated in a microbiological safety cabinet.

We believe that all bottles that contain infected material and which are shaken in any machine should be placed inside individual self-sealing or heat-sealed plastic bags in order to minimize the dispersal of infected airborne particles.

Microbiological safety cabinets

Microbiological safety cabinets are intended to capture and retain infected airborne particles released in the course of certain manipulations and to protect the laboratory worker from infection which may arise from inhaling them.

There are three kinds: Classes I, II and III. Class I and Class II cabinets are used in diagnostic and Containment laboratories for work with Risk Group III organisms. Class III cabinets are used exclusively for Risk Group IV viruses.

A Class I cabinet is shown in *Figure 2.1*. The operator sits at the cabinet, works with his hands inside and sees what he is doing through the glass screen. Any aerosols released from his cultures are retained because a current of air passes in at the front of the cabinet. This sweeps the aerosols up through the filters which remove all or most of the organisms. The clean air then

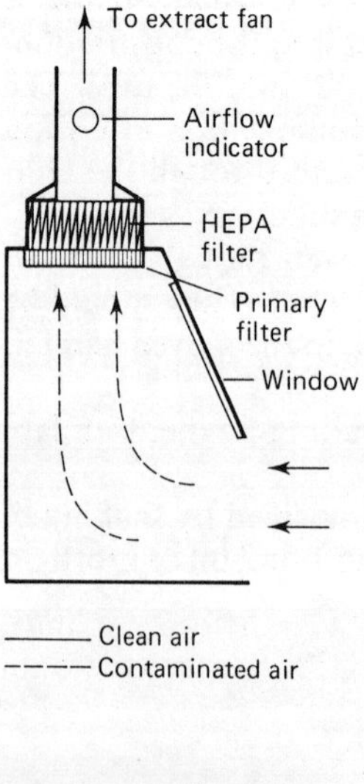

Figure 2.1 Class I microbiological safety cabinet

passes through the fan, which maintains the airflow, and is exhausted to atmosphere, where any particles or organisms that have not been retained on the filter are so diluted that they are no longer likely to cause infection if inhaled. A minimum airflow of 0.75 m/s must be maintained through the front of the cabinet and modern cabinets have airflow indicators and warning devices. The filters must be changed when the airflow falls below this level.

The Class II cabinet (*Figure 2.2*) is more complicated and is sometimes called a laminar flow cabinet but as this term is also used for clean air cabinets

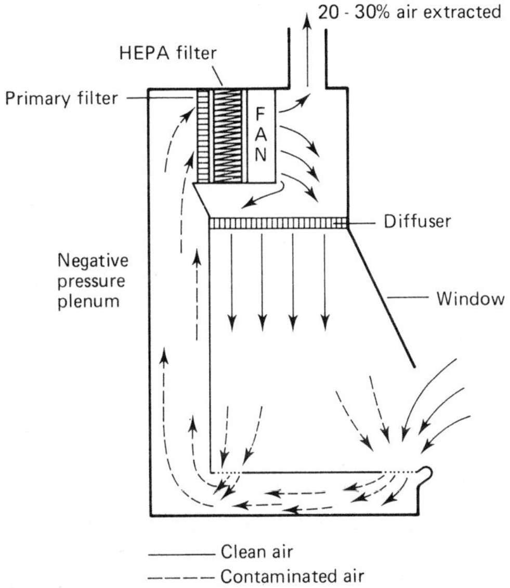

Figure 2.2 Class II microbiological safety cabinet

which do not protect the worker it should be avoided. In the Class II cabinet about 70% of the air is recirculated through filters so that the working area is bathed in clean (almost sterile) air. This entrains any aerosols produced in the course of the work and these are removed by the filters. Some of the air (about 30%) is exhausted to atmosphere and is replaced by a 'curtain' of room air which enters at the working face. This prevents the escape of any particles or aerosols released in the cabinet.

Class III cabinets are totally enclosed and are tested under pressure to ensure that no particles can leak from them into the room. The operator works with gloves which are an integral part of the cabinet. Air enters through a filter and is exhausted to atmosphere through one or two more filters (*Figure 2.3*).

Purchasing

Safety cabinets should comply with national standards, e.g. British Standards Institution[3], Standards Association of Australia[4], National Sanitation Foun-

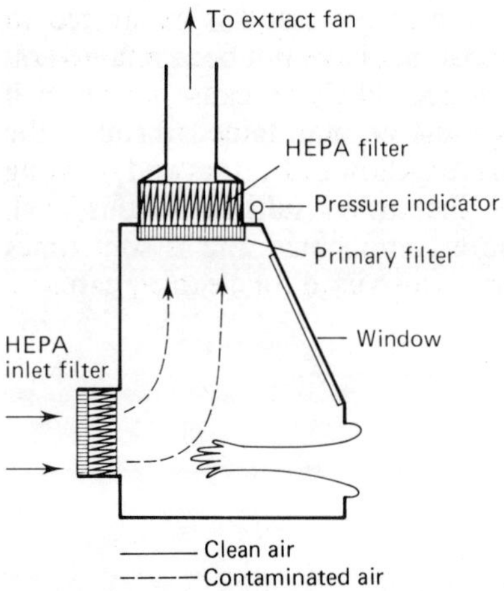

Figure 2.3 Class III microbiological safety cabinet

dation (USA)[5]. Suppliers include Baker, Bassaire, Envair, Foramaflow, Gelaire, Hepaire, Hi-Tech, How-Air, Howorth, LTA, LEEC, Microflow, Oliphant.

Use of safety cabinets

These cabinets are intended to protect the worker from airborne infection. They will not protect him from spillage and the consequences of poor techniques. The cabinet should not be loaded with unnecessary equipment or it will not do its job properly. Work should be done in the middle to the rear of the cabinet, not near the front and the worker should avoid bringing his hands and arms out of the cabinet while he is working. After each set of manipulations and before withdrawing his hands he should wait for 2–3 min to allow any aerosols to be swept into the filters. The hands and arms may be contaminated and should be washed immediately after ceasing work in a safety cabinet. Do not use bunsens, even micro-incinerators. They disturb the airflow. Use disposable plastic loops (p. 30).

Siting

The efficiency of a safety cabinet depends also on correct siting and proper maintenance. Possible sites for cabinets in a room are shown in *Figure 2.4*. A is a poor site, as it is near to the door and airflow into the cabinet will be disturbed every time the door is opened and someone walks past the cabinet into the room. Site B is not much better, as it is in almost a direct line between door and window, although no-one is likely to walk past it. Its left side is also close to the wall and airflow into it on that side may be affected by the 'skin effect', i.e. the slowing down of air when it passes parallel and close to a surface. Air passing across the window may be cooled, and will meet warm air from the rest of the room at the cabinet face, when turbulence may result.

Site C is better and site D is best of all. If two cabinets are required in the same room, sites C and D would be satisfactory, but they should not be too close together, or one may disturb the airflow of the other. We remember a

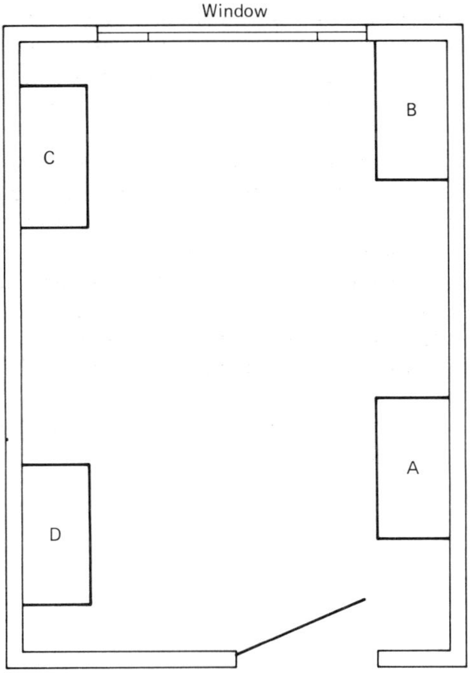

Figure 2.4 Possible sites for safety cabinets in relation to cross draughts from door and window and movement of staff. A is bad: B is poor: C is better: D is best

laboratory where a Class I cabinet was placed next to a Class II cabinet. When the former was in use it extracted air from the latter and rendered that cabinet quite ineffective.

Care must also be taken in siting any other equipment that might generate air currents, e.g. fans and heaters. Mechanical room ventilation may be a problem if (rarely) it is efficient, but this can be overcome by linking it to the electric circuits of the cabinets so that either, but not both are extracting air from the room at any one time. Alternatively baffles may be fitted to air inlets and outlets to avoid conflicting air currents near to the cabinet. Tests with smoke generators (*see below*) will establish the directions of air currents in rooms.

None of these considerations need apply to Class III cabinets which are in any case usually operated in more controlled environments.

Testing airflows

The presence and direction of air currents and draughts are determined with 'smoke'. This may indeed be from burning material in a device like that used by bee keepers, but is usually a chemical which produces a dense, visible vapour. Titanium tetrachloride is commonly used. If a cotton wool tipped stick, e.g. a throat swab, is dipped into this liquid and then waved in the air a white cloud is formed which responds to quite small air movements.

Commercial airflow testers are more convenient. They are small glass tubes, sealed at each end. Both ends are broken off with the gadget provided and a rubber bulb fitted to one end. Pressing the bulb to pass air through the tube causes it to emit white smoke. These methods are suitable for ascertaining air movements indoors.

To measure airflow an anemometer is required. Small, vane anemometers, timed by a stopwatch, are useful for occasional work but for serious activities electronic models, with a direct reading scale are essential (Electronic Instruments; Prossor). The electronic vane type has a diameter of about 10 cm. It has a satisfactory time constant and responds rapidly enough to show the changes in velocity that are constantly occurring when air is passing into a Class I safety cabinet. Hot wire or thermistor anemometers may be used but they may show very rapid fluctuations and need damping. Both can be connected to recorders.

To obtain a useful figure it is necessary to measure the airflow into a Class I cabinet in at least five places in the plane of the working face (*Figure 2.5*).

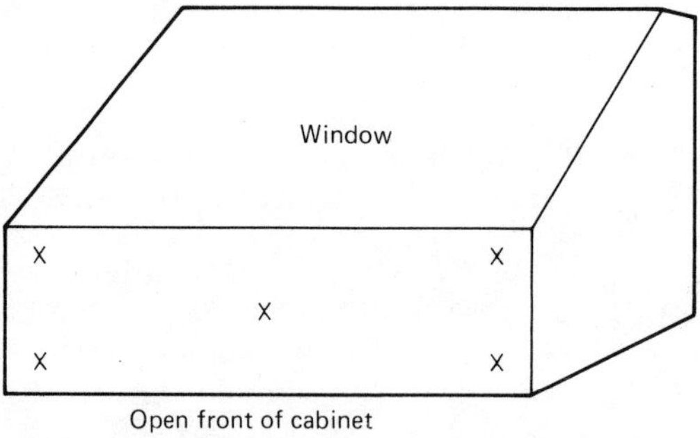

Figure 2.5 Testing airflow into a Class I safety cabinet. Anemometer readings should be taken at five places, marked with an X, in the plane of the working face with no-one working at the cabinet

An average is then calculated, but at no place should there be a reading that is 20 lfm (0.1 m/s) more or less than any of the others. If there is such a difference there will be turbulence within the cabinet. Usually some piece of equipment, inside or outside the cabinet, or the operator himself, is influencing the airflow.

Although airflows into Class I cabinets are usually much the same at all points on the working face this is not true for Class II cabinets, where the flow is greater at the bottom than at the top. The average inward flow can be calculated by measuring the velocity of air leaving the exhaust and the area of the exhaust vent. From this the volume per minute is found and this is also the amount entering the cabinet. Divided by the area of the working face it gives the average velocity. A rough and ready way, suitable for day-to-day use requires a sheet of plywood or metal which can be fitted over the working face. In the centre of this an aperture is made which is 2 × 2.5 cm. The inward velocity of air through this is measured and the average velocity over the whole face calculated from this figure and the area of the working face. Whichever way is used a check should also be made with smoke to ensure

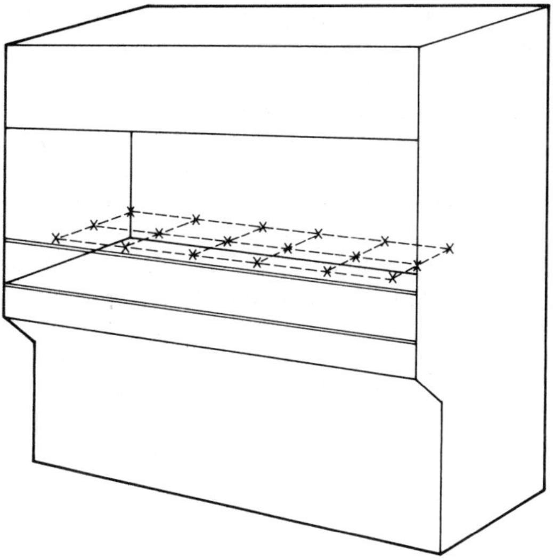

Figure 2.6 Testing the vertical airflow in a Class II safety cabinet. Anemometer readings should be taken at points marked X on an imaginary grid 6 inches within the cabinet walls and just above the level of the bottom of the glass window

that air is in fact entering the cabinet all the way round its perimeter and not just at the lower edge.

The downward velocity of air in a Class II cabinet should be measured with an anemometer at 18 points in the horizontal plane 10 cm above the top edge of the working face (*Figure 2.6*). No reading should differ from the mean by more than 20%.

It is usual, when testing airflows with an anemometer, to observe or record the readings at each position for several minutes, because of possible fluctuations.

To measure the airflow through a Class III cabinet the gloves should be removed and readings taken at each glove port. Measurements should also be taken at the inlet filter face when the gloves are attached.

Recommended airflows for each class of cabinet are shown in *Table 2.1*.

TABLE 2.1 Mean airflows (m/s) for microbiological safety cabinets according to British Standard 5726[3]

	Class I	*Class II*	*Class III*
Inflow at working face	0.7–1.0[a]	0.4	—
Downflow		0.25–0.5	
Inflow at glove ports			0.75
Inflow at filter			3

[a] With unused filters.
Data from British Standard 5726 (1979) by permission of the British Standards Institution.

Decontamination

As safety cabinets are used to contain aerosols which may be released during work with Risk Group III micro-organisms the inside surfaces and the filters will become contaminated. The working surface and the walls may be decontaminated on a day-to-day basis by swabbing them with disinfectant. Glutaraldehyde is probably the best disinfectant for this purpose as phenolics may leave sticky residues and hypochlorites may, in time, corrode the metal. For thorough decontamination, however, and before filters are changed formaldehyde should be used and precautions should be taken before use. The installation should be checked to ensure that none of the gas can escape to the room or to other rooms. The front closure (night door) of the cabinet should seal properly onto the carcase, or masking tape should be available to seal it. Any service holes in the carcase should be sealed and the HEPA filter seating examined to ensure that there are no leaks. If the filters are to be changed and the primary or roughing filter is accessible from inside the cabinet it should be removed and left in the working area. The supply filter on a Class III cabinet should be sealed with plastic film.

Formaldehyde is generated by:

(1) boiling formalin, which is a 40% solution of the gas in water;
(2) by heating paraformaldehyde, which is its solid polymer; or
(3) by mixing formalin or paraformaldehyde and water with potassium permanganate.

Boiling formalin

The volume used is important. Too little will be ineffective; too much leads to deposits of the polymer, which is persistent and which may contribute to the natural blocking of the filters. The British Standard 5726 specifies a concentration of 0.05 g/m^3 (0.0014 g/ft^3) which is approximately 25 ml of neat formalin for a Class I cabinet with a capacity of 0.34 m^3.

The formalin may be boiled in several ways. Some cabinets have built-in devices in which the liquid is dripped on to a hot-plate. Other manufacturers supply small boilers. A laboratory hot-plate, connected to a timing clock is adequate. The correct amount of formalin is placed in a flask. The formalin need not be diluted. There is enough water to achieve the humidity required already in the solution. The addition of water may diminish the efficiency of the gas and will result in a very wet cabinet.

Heating paraformaldehyde

Tablets (1 g) are available from chemical manufacturers. Three or four tablets are adequate for an average size cabinet. They may be heated on an electric frying pan connected to a time clock or by an electric hair dryer.

Oxidation with permanganate

Formaldehyde is released when potassium permanganate is mixed with formalin solution or with paraformaldehyde in the presence of water. The reaction may be violent but is not particularly dangerous unless too much permanganate is used. For most cabinets a mixture of 35 ml of formalin and 7.5 g of permanganate will suffice. The formalin is placed in a large beaker or bowl, in the cabinet, the permanganate, previously weighed out, is added and the front closure put in place immediately. The mixture boils almost at once. If paraformaldehyde is preferred 4 g of this and 8 g of permanganate are mixed immediately before they are required and placed in a large beaker or bowl in the cabinet and the 10 ml of water added. The front closure must be put on at once.

Exposure time

It is convenient to start decontamination in the late afternoon and let the gas act overnight. In the morning the fan should be switched on and the front closure 'cracked' open very slightly to allow air to enter and purge the cabinet. Some new cabinets have a hole in the front closure, fitted with a stopper which can be removed for this purpose. With Class III cabinets the plastic film is taken off the supply filter to allow air to enter the cabinet and purge it of formaldehyde. After several minutes the front closure may be removed. The fan is then allowed to run for about 30 min which should remove all formaldehyde.

Summary of decontamination procedure

The prefilter should be removed from inside the cabinet (where possible) and placed on the working surface. It is usually pushed or clipped into place and is easily removed, but as it is likely to be contaminated gloves should be worn. Neat formalin, 2 ml/ft^3 (70 ml/m^3), which is about 25 ml for a normal size Class I cabinet, should be placed in its container on the heater in the cabinet or in the reservoir. The front closure is then put in place and sealed if necessary. Then the heater is switched on and the formalin boiled away. After switching off the heater the cabinet is left closed overnight. The next morning the cabinet fan is switched on and then the front closure is opened very slightly to allow air to pass in and purge the cabinet of formaldehyde. After several minutes the front closure is removed and the cabinet fan allowed to run for about 30 min. Any obvious moisture remaining on the cabinet walls and floor may then be wiped away and the filters changed.

Changing the filters and maintenance

The cabinet must be decontaminated before filters are changed and any work is done on the motor and fans. If these are to be done by an outside contractor, e.g. by the manufacturer's service engineer the front closure should be sealed on again after the initial purging of the gas and a notice: *Cabinet decontaminated but not to be used* placed on it awaiting the engineer's arrival. He will also require a certificate stating that the cabinet has been decontaminated.

The primary, or prefilters, should be changed when the cabinet airflow approaches its agreed local minimum. Used filters should be placed in plastic or tough paper bags, which are then sealed and burned. If the airflow is not restored to at least the middle of the range then arrangments should be made to replace the HEPA filter. This is usually done by the service engineer, but may be done by laboratory staff if they have received instruction. Unskilled operators often place the new filter in upside down or fail to set it securely and evenly in its place. Used filters should be placed in plastic bags, which are then sealed for disposal. They are not combustible. Some manufacturers accept used filters and recover the cases, but this is not usually a commercial practice—no refund is given. When manufacturers replace HEPA filters they may offer a testing service.

Testing; further information

This is beyond the scope of this book and the reader is referred to other publications[1, 2] and national standards[3, 4, 5].

Preparation room equipment

Sterilizers are discussed in Chapter 3.

Inspissators

In the preparation of slopes of medium consisting largely of egg or serum, the amount of heating required to coagulate the protein must be carefully controlled. A steamer heats the medium too rapidly and raises the temperature too high. The modern inspissator is thermostatically controlled at 75–85 °C and fitted with a large internal circulating fan. The shelves on which the tubes are sloped are made of wire mesh so that circulation is not impeded. It is convenient to have wire mesh or aluminium racks made which hold tubes or bottles at the correct angle (5–10°) and which slide on the shelf brackets. These facilitate rapid loading and unloading while the instrument is hot.

An egg or serum medium is usually coagulated in 45–60 min at 80 °C. Whether the inspissator is loaded hot or cold is a matter of personal choice, but better control, required for media that contain drugs, e.g. mycobacterial sensitivity tests, is obtained by raising the temperature first and then putting in a standard load for a constant time.

Laminar flow clean air work stations

These cabinets are designed to protect the work from the environment and are most useful for aseptic distribution of certain media and plate pouring. A stream of sterile (filtered) air is directed over the working area, either horizontally into the room or vertically downwards when it is usually re-circulated. They are particularly useful for preventing contamination when distributing sterile fluids. They are not 'safety cabinets' and should not be used for handling bacteria or tissue cultures, but a Class II microbiology safety cabinet (p. 17) will also protect against contamination (Baker, Bassaire, Envair, Gelaire, Hepaire, Microflow).

Glassware-washing machines

These are useful in large laboratories which use enough of any one size of tube or bottle to give an economic load (Camlab; Gallay; Miele; Netzsch).

Glassware drying cabinets

A busy wash-up room requires a drying cabinet with wire-rack shelves and a 3-kW electric heater in the base, operated through a three-heat control. An extractor fan on the top is a refinement, otherwise the sides near the top should be louvred.

Media distributors

Although time-honoured methods such as the funnel, rubber tube and clip are still useful in laboratories where only a small amount of culture media is distributed or tubed, automatic machines are now available and are indispensable in the larger media room.

Two types of machines are available; one is based on a peristaltic pump and the other on a syringe pump.

The Jencons Accuramatic delivers a pre-determined volume from a reservoir (large media bottle) and can be operated manually, by a pedal switch or by a timing device that permits slow or rapid delivery to suit the operator. The fluid is transferred entirely through autoclavable plastic tubing by a peristaltic pump and the cycle ensures that some fluid is returned to the reservoir so as to ensure mixing. As some media stain the tubing, separate 'manifolds' can be kept for each medium to be sterilized.

We know of several syringe pump models. The Brewer (BBL), Oxford (Boehringer), Struers (Camlab), Multi-Dosamatt (Roth) and Filamatic (National Instruments) automatic pipetting machines deliver accurate, pre-determined volumes over a wide range automatically or by foot or hand control. Measurement is in glass or stainless steel syringes with stainless steel or nylon valve assemblies through plastic tubing. For sterile distribution, the syringe assemblies and the tubing may be autoclaved.

Useful fully automatic machines for pouring petri dishes are supplied by Oxoid, Camlab, New Brunswick, Turner, PBI and Horwell. Some of these will sterilize reconstituted media, pour and then stack the poured petri dishes.

The Zipettes (Jencons) are a series of useful dispensers for distributing liquids from screw-capped and other bottles.

Glassware, plastics and small equipment

For ordinary microbiological work, soda-glass tubes and bottles are satisfactory. In assay work, the more expensive resistance glass might be justified. An important consideration is whether glassware should be washed or discarded. Purchasing cheaper glassware in bulk and using plastic disposable petri dishes and culture tubes may, in some circumstances, be more economical than employing labour to clean them.

Plastics fall into two categories:

(1) disposable items, such as petri dishes, specimen containers and plastic loops, which are destroyed when autoclaved and cannot be sterilized by ordinary laboratory methods (but *see* Chapter 3), and
(2) recoverable material, which must be sterilized in the autoclave.

toclavable plastics

include polyethylene, styrene, acrylonitrile, polystyrene and rigid nyl chloride.

Autoclavable plastics

These withstand a temperature of 121 °C and include polypropylene, polycarbonate, nylon, PTFE (Teflon), polyallomer, TPX (methylpentene polymer), Viken and vinyl tubing.

It is best to purchase plastic apparatus from specialist firms, e.g. WCB, Azlon and Bel Art, or their agents, who will give advice on the suitability of their products for specific purposes. Most recoverable plastics used in microbiology can be washed in the same way as glassware.

Petri dishes

Disposable plastic petri dishes are used in most laboratories in developed countries. They are supplied already sterilized and packed in batches in polythene bags. They can be stored indefinitely, are cheap when purchased in bulk and are well made and easy to handle. Two kinds are available, vented and unvented. Vented dishes have one or two nibs that raise the top slightly from the bottom and are to be preferred for anaerobic and CO_2 cultures.

Glass petri dishes are, however, still popular in some areas.

In general, two qualities are obtainable. The thin, blown dishes, usually made of borosilicate glass, are pleasant to handle, their tops and bottoms are flat and they stack safely. They do not become etched through continued use but are fragile, must be washed with care and are expensive. The thick pressed glass dishes are often convex, cannot always be stacked safely and are easily etched and scratched with use and washing. On the other hand, they are cheap and not fragile. The 'life' of petri dishes can be doubled by purchasing aluminium lids (e.g. Oxoid) that are made to fit both the tops and bottoms of the dishes.

Glass petri dishes are sterilized by hot air in aluminium boxes obtainable from Denley, Luckham, Scientific Products and Thomas.

Test-tubes and bottles, plugs, caps and stoppers

Culture media can be tubed in either test-tubes or small bottles. Apart from personal choice, test-tubes are easier to handle in busy laboratories and take up less space in storage receptacles and incubators, but bottles are more convenient in the smaller workroom where media is kept for longer periods before use. Media in test-tubes may dry up during storage. Screw-capped test-tubes are available.

The most convenient sizes of test-tube are: 127 × 12.5 mm, holding 4 ml of medium; 152 × 16 mm, holding 5–10 ml; 152 × 19 mm, holding 10–15 ml; and 178 × 25 mm, holding 20 ml. Rimless test-tubes of heavy quality are made for bacteriological work. The lipped, thin glass chemical test-tubes are useless.

Cotton wool plugs have been used for many years to stopper test-tube cultures, but have largely been replaced by metal or plastic caps, or in some laboratories by soft, synthetic sponge bungs.

Aluminium test-tube caps were introduced some years ago but they have a limited tolerance and, in spite of alleged standard specifications of test-tubes, a laboratory very soon accumulates many tubes that will not fit the caps. Caps that are too loose are useless. Aluminium caps of a new design, Cap-O-Test closures (Payne) are better. They have a wider tolerance and fit most tubes of any normal size, being held in place with a small spring. These caps are cheap, last a long time, are available in many colours and save a great deal of time and labour. Stainless steel caps, held in place with small internal lugs, are made by Astell, and polypropylene caps in several sizes and colours and which stand up to repeated autoclaving are obtainable in several colours from Oxoid, Clark (Clark-Fin closures) and Nunc.

Temporary closures can be made from kitchen aluminium foil.

Aluminium capped test-tubes are sterilized in the hot air oven in baskets. Polypropylene capped tubes must be autoclaved.

Rubber stoppers of the orthodox shape are useless as they are blown out of the tubes in the autoclave. The Astell rubber seals are designed to avoid this and to allow steam to penetrate into the tubes as readily as with cotton wool plugs or aluminium or polypropylene closures. These stoppers fit 16-mm tubes and also the Astell culture bottles used for roll-tube work, etc. They have a very long laboratory life.

Several sizes of small culture bottles are made for microbiological work. Some of these bottles are of strong construction and are intended to be re-used. Others, although tough enough for safe handling, are disposable (Labco). The most useful sizes are: the 'bijou', holding 7 ml; the 'small McCartney' 14 ml; the 'McCartney', 28 ml; and the 'Universal Container', 28 ml, which has a larger neck than the others and is also used as a specimen container. These bottles usually have aluminium screw-caps with rubber liners. The liners should be made of black rubber; some red rubbers are thought to give off bactericidal substances. Polypropylene caps are also used; they need no liners but in our experience they may loosen spontaneously during long incubation or storage. The medium dries up. They are satisfactory in the short term. Astell culture bottles, holding about 20 ml and closed by Astell seals, are popular for roll-tube counting.

All of these bottles can be sterilized by autoclaving.

Media storage bottles

'Medical flats' or 'rounds' are made in sizes from 60 ml upwards. The most convenient sizes are 110 ml, holding 50–100 ml of medium, and 560 ml, for storing 250–500 ml. The flat bottles are easiest to handle and store but the round ones are more robust.

Specimen containers

We favour the robust glass 'Universal Container', which holds about 28 ml and has an aluminium screw-cap with a black rubber liner. There are too many different containers, many of plastic; some are satisfactory, others are not; some leak easily and others do not stand up to handling by patients. Only screw-capped containers should be used. Those with 'push-in' or 'pop-up' stoppers are dangerous. They generate aerosols when opened. Waxed paper pots should not be used as they invariably leak.

For larger specimens, there is a variety of strong screw-capped jars.

Before any containers are purchased in bulk, it is advisable to test samples by filling them with coloured water and standing them upside down on blotting paper for several days after screwing the caps on moderately well. Patients and nurses may not screw caps on as tightly as possible. Similar bottles should be sent through the post, wrapped in absorbent material in accordance with postal regulations. Leakage will be evident by the staining of the blotting paper or wrapping. More severe tests include filling the containers with coloured water and centrifuging them upside down on a wad of blotting paper. Specimen container problems and tests are reviewed by Collins[1].

Sample jars and containers

Containers for food samples, water, milk, etc., should conform to local or national requirements. In general, large screw-capped jars are suitable but are rather heavy if many samples are taken, and should not be used in food factories. Plastic containers may be used as leakage and spillage is not such a problem as with pathological material. Strong plastic bags are useful, particularly if they are of the self-sealing type. Otherwise 'quick-ties' may be used.

Pasteur pipettes

These pipettes are probably the most dangerous pieces of laboratory equipment in unskilled hands.[1] They are used, with rubber teats, to transfer liquid cultures, serum dilutions, etc.

Very few laboratories make their own pasteur pipettes nowadays; they are obtainable, plugged and unplugged, from most laboratory suppliers and are best purchased in bulk. Long and short forms are available and most are made of 6–7-mm diameter glass tubing.

After plugging, they can be sterilized in aluminium boxes made for this purpose by Denley, Luckham, Payne, Scientific Instruments, Thomas, etc.

New and safer pasteur pipettes with integral teats and made of low-density polypropylene are now available and are supplied ready sterilized. They are much safer than glass pipettes (Elkay).

Pasteur pipettes are used once only.

Graduated pipettes

Straight side blow out pipettes, 1–10 ml capacity are used. They must be plugged with cotton wool at the suction end to prevent bacteria entering from the pipettor or teat and contaminating the material in the pipette. These plugs must be tight enough to stay where they are during pipetting but not so tight that they cannot be removed during cleaning. About 25 mm of non-absorbent cotton wool is pushed into the end with a piece of wire. The ends are then passed through a bunsen flame to tidy them. Wisps of cotton wool which get between the glass and the pipettor or teat may permit air to enter and the contents to leak.

Pipettes are sterilized in the hot air oven in square section aluminium containers similar to those used for pasteur pipettes. A wad of glass wool at the bottom of the container prevents damage to the tips.

Disposable 1 ml and 10 ml pipettes are available. Some firms supply them already plugged and sterilized. Jobling, Sterilin, Horwell and Falcon (BBL) supply these.

Pipetting aids

Rubber teats and pipetting devices provide an alternative to the highly dangerous practice of mouth pipetting[1].

Rubber teats

Choose teats with a capacity greater than that of the pipettes for which they are intended, i.e. a 1-ml teat for pasteur pipettes, a 2-ml teat for 1-ml pipette, otherwise the teat must be used fully compressed, which is tiring. Most beginners compress the teat completely, then suck up the liquid and try to hold it at the mark while transferring it. This is unsatisfactory and leads to spilling and inaccuracy. Compress the teat just enough to suck the liquid a little way past the mark on the pipette. Withdraw the pipette from the liquid, press the teat slightly to bring the fluid to the mark and then release it. The correct volume is now held in the pipette without tiring the thumb and without risking loss. To discharge the pipette, press the teat slowly and gently and then release it in the same way. Violent operation usually fails to eject all the liquid; bubbles are sucked back and aerosols are formed.

'Pipettors'

A large number of devices that are more sophisticated than simple rubber teats are now available. Broadly speaking there are four kinds of these:

(1) rubber bulbs with valves that control suction and dispensing;
(2) syringe-like machines that hold pipettes more rigidly than rubber bulbs and have a plunger operated by a rack and pinion or a lever;
(3) electrically operated pumps fitted with flexible tubes in which pipettes can be inserted;
(4) mechanical plunger devices which take small plastic pipette tips and are capable of repeatedly delivering very small volumes with great accuracy.

It is extremely difficult to give advice on the relative merits of the various devices. That which suits one operator, or is best for one purpose may not be suitable for others. Choice should therefore be made by the operators, not by managers or administrators who will not use them. None of those in categories 1, 2 and 4 above are expensive, and it should be possible for several different models to be available. What is necessary is some system of instruction in their use and in their maintenance. None should be expected to last forever.

Micro-slides and cover-glasses

Unless permanent preparations are required, the cheaper cut-edge micro-slides are satisfactory. Slides with ground and polished edges are much more expensive. Most micro-slides are sold in boxes of 100 slides. They should not be washed and re-used but discarded.

Cover-glasses are sold in several thicknesses and sizes. Those of thickness grade No. 1, 16-mm square, are the most convenient. They are sold in boxes containing about 100 glasses. Plastic cover-glasses are available.

Durham's (fermentation) tubes

These are small glass tubes, usually 25–30 × 5–6 mm and closed at one end, which are placed inverted in tubes of culture media to detect gas formation. They are very cheap and are not worth washing.

Inoculating loops and wires

These are usually made of 25 SWG Nicrome wire, although this is more springy than platinum. They should be short (not more than 5 cm long) in order to minimize vibration and therefore involuntary discharge of contents[1]. Loops should be small (not more than 3 mm in diameter). Large loops are also inclined to empty spontaneously and scatter infected airborne particles. They should be completely closed, otherwise they will not hold fluid cultures. This can be carried out by twisting the end of the wire round the shank, or by taking a piece of wire 12 cm long, bending the centre round a nail or rod of appropriate diameter and twisting the ends together in a drill chuck. Ready-made loops of this kind are sold by Medical Wire.

Loops and wires should not be fused into glass rods. Aluminium holders are sold by most laboratory suppliers.

Disposable loops are excellent. Nunc make two sizes, 1 and 10 μl, and both are useful. Hughes & Hughes make 10 μl loops.

Spreaders

Cut a glass rod of 3–4 mm diameter into 180 mm lengths and round off the cut ends in a flame. Hold each length horizontally across a bat's wing flame so that it is heated and bends under its own weight approximately 36 mm from one end, and an L shape is obtained.

Racks and baskets

Test-tube and culture bottle racks should be made of polypropylene or of metal covered with polypropylene or nylon so that they can be autoclaved. These racks also minimize breakage, which is not uncommon when metal racks are used. Wooden racks are unhygienic. Suitable racks are supplied by Richardsons, Payne, Bel Art, Beckman and others.

Aluminium trays for holding from 10–100 bottles, according to size are widely employed in the UK. They can be autoclaved, and are easily taken apart for cleaning. Denley and Luckham supply standard sizes and make others to order.

The traditional wire baskets are unsafe for holding test-tubes. They contribute to breakage hazards and do not retain spilled fluids. For non-infective work, these baskets, covered with polypropylene or nylon, are satisfactory, but autoclavable plastic boxes of various sizes are safer for use with cultures. These are made by Azlon, WCB Plastics, Payne and Bel Art.

Bunsen burners

The usual bunsen, with a bypass is satisfactory for most work, but for material that may spatter or that is highly infected a hooded bunsen should be used. There are several versions of these, but we recommend the Kampff Micro-bunsen burner (Horwell) and the Bactiburner (Denley) which enclose the flame in a borosilicate tube.

Electric 'bunsen burners' are also available from most suppliers (Cherwell, Hoffman-Horo, Horwell). These are tubular micro-incinerators in which the loop or wire is inserted.

Other bench equipment

A hand magnifier, forceps and a knife or scalpel should be provided, and also a supply of tissues for mopping up spilled material and general cleaning. Swab sticks, usually made of wood and about 6 in long, are useful for handling some specimens, and wooden throat spatulae are useful for food samples. These can be sterilized in large test-tubes or in the aluminium boxes used for pasteur pipettes.

Some form of bench 'tidy' or rack is desirable to keep loops and other small articles together.

Discard jars and disinfectant pots are considered in Chapter 3.

References

1. COLLINS, C.H. (1983) *Laboratory-acquired Infections*, pp. 61–95, p. 141. London: Butterworths
2. WORLD HEALTH ORGANIZATION (1983) *Laboratory Biosafety Manual*. Geneva: WHO
3. BRITISH STANDARDS INSTITUTION (1979) *Specification for Microbiological Safety Cabinets*. BS 5726. London: BSI
4. STANDARDS ASSOCIATION OF AUSTRALIA (1980) *Biological Safety Cabinets*. Australian Standard 2251. Sydney, NSW
5. NATIONAL SANITATION FOUNDATION (1976) *Standard No. 49 Class II (Laminar Flow) Biohazard Cabinetry*. Michigan: Ann Arbor

Chapter 3

Sterilization, disinfection and the treatment of infected materials

'Sterilization' implies the complete destruction of all micro-organisms, including spores.

'Disinfection' implies the destruction of vegetative organisms which might cause disease, or, in the context of the food industries, which might cause spoilage. Disinfection does not necessarily kill spores.

The two terms are not synonymous.

Sterilization

The methods commonly used in microbiological laboratories are:

(1) red heat (flaming);
(2) dry heat (hot air);
(3) steam under pressure (autoclaving);
(4) steam not under pressure (Tyndallization); and
(5) filtration.

Incineration is also a method of sterilization but as it is applied outside the laboratory for the ultimate disposal of laboratory waste it is considered separately (p. 54).

Red heat

Instruments such as inoculating wires, loops and searing irons are sterilized by holding them in a bunsen flame until they are red hot. Micro-incinerators (p. 54) are recommended for sterilizing inoculating wires contaminated with highly infectious material (e.g. tuberculous sputum) to avoid the risk of spluttering contaminated particles over the surrounding areas.

Dry heat

This is applied in an electrically heated oven which is thermostatically controlled and is fitted with a large circulating fan to ensure even temperatures in all parts of the load. Modern equipment has electronic controls which can be set to raise the temperature to the required level, hold it there for a prearranged time and then switch off the current. A solenoid lock is incorporated in some models to prevent the oven being opened before the cycle is complete. This safeguards sterility and protects the staff from accidental burns.

Materials which can be sterilized by this method include glass petri dishes, flasks, pipettes and metal objects. Various metal canisters and cylinders which conveniently hold glassware during sterilization and keep it sterile during storage are available from laboratory suppliers.

Loading

Air is not a good conductor of heat so oven loads must be loosely arranged, with plenty of spaces to allow the hot air to circulate.

Holding times and temperatures

When calculating processing times for hot air sterilizing equipment, there are three time periods which must be considered:

(1) *the heating-up period*, which is the time taken for the entire load to reach the sterilization temperature; this may take about 1 h;
(2) *the holding periods* at different sterilization temperatures recommended by the Medical Research Council[1] which are 160 °C for 45 min, 170 °C for 18 min, 180 °C for 7½ min and 190 °C for 1½ min;
(3) *the cooling-down period*, which is carried out gradually to prevent glassware from cracking as a result of a too rapid fall in temperature; this period may take up to 2 h.

Control of hot air sterilizers

Tests for electrically operated fan ovens are described in British Standard 3421[2].

Hot air sterilization equipment should be calibrated with thermocouples when the apparatus is first installed and checked with thermocouples afterwards when necessary.

Ordinary routine control can be effected simply with the aid of Browne's tubes. The Browne's Type 3 tubes (Green Spot) are used for fan-operated ovens.

Steam under pressure

This is done by autoclaving. Bacteria are more readily killed by moist heat (saturated steam) than by dry heat. Steam kills bacteria by denaturing their protein. An agreed safe condition for sterilization is to use steam at 121 °C for 15 min[1]. This is suitable for culture media, aqueous solutions, treatment of discarded cultures and specimens, etc.

Air has an important influence on the efficiency of steam sterilization because its presence changes the pressure–temperature relationship. For example the temperature of saturated steam at 15 lb/in^2 is 121 °C, provided that all of the air is first removed from the vessel. With only half of the air removed, the temperature of the resulting air–steam mixture at the same pressure is only 112 °C. In addition, the presence of air in mixed loads will prevent penetration by steam.

All of the air that surrounds and permeates the load must first be removed before steam sterilization can commence.

Loads in autoclaves

As successful autoclaving depends on the removal of all the air from the chamber and the load, the materials to be sterilized should be packed loosely. 'Clean' articles may be placed in wire baskets, but contaminated material (e.g. discarded cultures) should be in solid bottomed containers not more than 8 in deep (*see* Disposal of Infected Waste, *below*). Large air spaces should be left around each container and none should be covered.

Types of autoclave

Only autoclaves designed for laboratory work and capable of dealing with a 'mixed load' should be used. 'Porous load' and 'bottled fluid sterilizers' are rarely satisfactory for laboratory work. There are two varieties of laboratory autoclave:

(1) pressure cooker types; and
(2) gravity displacement models with automatic air and condensate discharge.

'Pressure cooker' laboratory autoclaves

These are still in use in many parts of the world. The most common type is a device for boiling water under pressure. It has a vertical metal chamber with a strong metal lid which can be fastened down and sealed with a rubber gasket. An air and steam discharge tap, pressure gauge and safety valve are fitted in the lid (*Figure 3.1*). Water in the bottom of the autoclave is heated by external gas burners, an electric immersion heater or a steam coil.

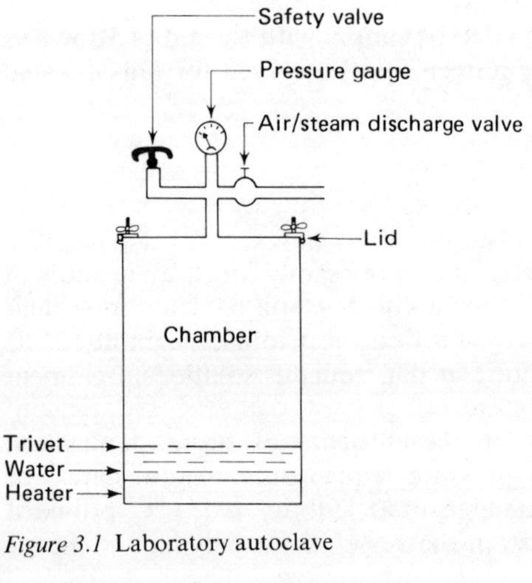

Figure 3.1 Laboratory autoclave

Operating instructions

There must be sufficient water inside the chamber. The autoclave is loaded and the lid is fastened down with the discharge tap open. The safety valve is then adjusted to the required temperature and the heat is turned on.

When the water boils, the steam will issue from the discharge tap and carry the air from the chamber with it. The steam and air should be allowed to escape freely until all of the air has been removed. This may be tested by attaching one end of a length or rubber tubing to the discharge tap and inserting the other end into a bucket or similar large container of water. Steam condenses in the water and the air rises as bubbles to the surface; when all of the air has been removed from the chamber, bubbling in the bucket will cease. When this stage has been reached, the air–steam discharge tap is closed and the rubber tubing removed. The steam pressure then rises in the chamber until the desired pressure, usually 15 lb/in^2, is reached and steam issues from the safety valve.

When the load has reached the required temperature (*see* Testing autoclaves, *below*), the pressure is held for 15 min.

At the end of the sterilizing period, the heater is turned off and the autoclave allowed to cool.

The air and steam discharge tap is opened very slowly after the pressure gauge has reached zero (atmospheric pressure). If the tap is opened too soon, while the autoclave is still under pressure, any fluid inside (liquid media, etc.) will boil explosively and bottles containing liquids may even burst. The contents are allowed to cool. Depending on the nature of the materials being sterilized, the cooling (or 'run-down') period needed may be several hours for large bottles of agar to cool to 80 °C, when they are safe to handle.

Autoclaves with air discharge by gravity displacement

These autoclaves are usually arranged horizontally and are rectangular in shape, thus making the chamber more convenient for loading. A palette and trolley system can be used.

Figure 3.2 shows in diagrammatic form a jacketed gravity displacement type of autoclave. Similar autoclaves can be constructed without jackets. The door should have a safety device to ensure that it cannot be opened while the chamber is under pressure.

The jacket surrounding the autoclave consists of an outer wall enclosing a narrow space around the chamber, which is filled with steam under pressure to keep the chamber wall warm. The steam enters the jacket from the mains supply, which is at high pressure, through a valve that reduces this pressure to the working level. The working pressure is measured on a separate pressure gauge fitted to the jacket. This jacket also has a separate drain for air and condensate to pass through.

The steam enters the chamber from the same source which supplies steam to the jacket. It is introduced in such a way that it is deflected upwards and fills the chamber from the top downwards, thus forcing the air and condensate to flow out of the drain at the base of the chamber by gravity displacement. The drain is fitted with strainers to prevent blockage by debris. The drain is usually fitted with a thermometer for registering the temperature of the issuing steam. The temperature recorded by the drain thermometer is often lower than that in the chamber. The difference should be found with thermocouple tests (*see below*). A 'near to steam' trap is also fitted.

The automatic steam trap or 'near-to-steam' trap is designed to ensure that only saturated steam is retained inside the chamber, and that air and condensate, which are at a lower temperature than saturated steam, are automatically discharged. It is called a 'near-to-steam' trap because it opens if the temperature falls to about 2 °C below that of saturated steam and closes within 2 °C or near to the saturated steam temperature. The trap operates by the expansion and contraction of a metal bellows, which open and close a valve. The drain discharges into a tundish in such a way that there is a complete air-break between the drain and the dish. This ensures that no contaminated water can flow back from the waste-pipe into the chamber.

Operation of a gravity displacement autoclave

If the autoclave is jacketed, the jacket must first be brought to the operating temperature. The chamber is loaded, the door is closed and the steam-valve is opened, allowing steam to enter the top of the chamber. Air and condensate flow out through the drain at the bottom (*Figure 3.2*). When the drain thermometer reaches the required temperature a further period must be allowed for the load to reach that temperature. This should be determined initially and periodically for each autoclave as described below. Unless this is done the load is unlikely to be sterilized. The autoclave cycle is then continued

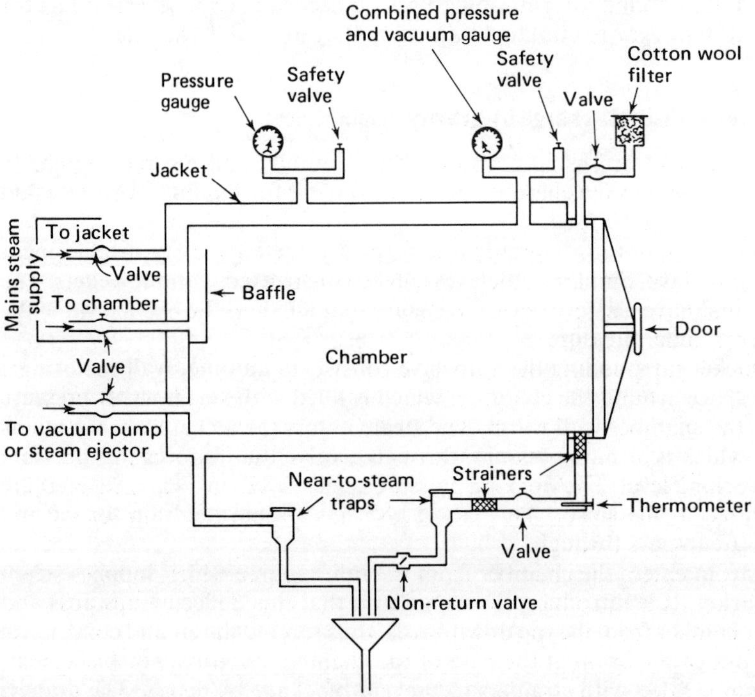

Figure 3.2 Gravity displacement autoclave

for the holding time. When it is completed the steam valves are closed and the autoclave allowed to cool until the temperature dial reads less than 80 °C. Not until then is the autoclave safe to open. It should first be 'cracked' or opened very slightly and left in that position for several minutes to allow steam to escape and the load to cool further (*see* Safety of the operator, *below*).

Tests for autoclave efficiency; determining time cycles

Steam sterilizers should be tested when purchased and at regular intervals thereafter with thermocouples.

Measurement of temperatures is by thermocouples, placed inside the autoclave. These are thin copper–constantan wires covered with PTFE and may be placed in the load. They are thin enough not to interfere with the door seal. The outside ends are connected to a digital or pen recorder (e.g. Comark). Four channels are usually sufficient to test laboratory autoclaves.

Autoclaves should be tested under the 'worst load' conditions. In most laboratories this would be a container full of $\frac{1}{2}$ oz (5 ml) screw-capped bottles. This should be placed in the centre of the autoclave, and if space is available other loaded containers placed around it. A thermocouple should be placed in a bottle in the middle of this load. Other thermocouples may be distributed in other parts of the autoclave. The sterilization cycle is then started and timed. The time taken for the temperature in the chamber drain to reach 121 °C is noted and then the time taken for the thermocouple in the load to register that temperature. This is when the sterilization time begins. After not less than 15 min the steam may be turned off when the cooling time begins. Thus there are four periods:

(1) *warming-up*, until chamber drain thermometer reaches 121 °C;
(2) *steam penetration into load*, until centre of load reaches 121 °C;
(3) *sterilization*, during which the load is maintained at 121 °C, usually 15 min;
(4) *cooling down time*, until the temperature in the load falls to 80 °C, when it may be removed.

As it is not practicable to use a thermocouple for each load (unless a sensor is built into the chamber), it is important to note these times for normal operation. The period of exposure *after* the temperature in the drain reaches 121 °C is therefore (2) + (3) min. Automatic cycle autoclaves, which depend on the temperature in the drain should be modified accordingly.

In some autoclaves the door cannot be opened until the temperature in the drain falls to 80 °C. This does not imply that the temperature in the load has also fallen to a safe level. That in large, sealed bottles may still be over 100 °C, when the contents will be at a high pressure. Sudden cooling may cause the bottles to explode. The autoclave should not be opened, therefore until the temperature in the load has fallen to 80 °C or below. This may take a very long time, and in some autoclaves there are locks which permit the door to be opened only fractionally to cool the load further before it is finally released. In others, complicated cooling arrangements with air blasts and cold water sprays are used.

For further information on autoclave problems *see* the PHLS publications[3, 4].

Protection of the operator

The only reliable way of testing autoclave efficiency is by instrumentation, but Browne's tubes may be used, type 2 (yellow spot), for temperatures above 126 °C. Browne's tubes should be stored in a cool place (below 20 °C) to prevent deterioration.

Another kind of sterilization indicator is made by Bennett. For some loads, e.g. glassware and culture media the wellknown autoclave tape, used in hospitals (3M) is useful.

In certain circumstances it may be desirable to test if bacterial spores are killed during the autoclave cycle. Spores of *Bacillus stearothermophilus* are used. The resistance to heat of the spores depends on many factors, including the medium in which they have been grown so it is best to use commercial preparations that have been quality controlled. Oxoid make spore strips, so do 3M (Attest) and BBL (Killit).

After exposure to the autoclave cycle the strips are removed from their carrier and placed in glucose tryptone broth recovery medium and incubated at 55–60 °C to test for viability.

Serious accidents, including burns and scalds to the face and hands have occurred when autoclaves have been opened, even when the temperature gauge read below 80 °C and the doors have been 'cracked'. Liquids in bottles may still be over 100 °C and under considerable pressure. The bottles may explode on contact with air at room temperature.

When autoclaves are being unloaded operators should wear full-face visors of the kind that cover the skin under the chin and throat. They should also wear thermal-protective gloves.

Tyndallization

This process, named after the scientist Tyndall employs a Koch (or Arnold steamer), which is a metal box in the bottom of which water is boiled by a gas burner, electric heater or steam coil. The articles to be processed rest on a perforated rack just above the water level. The lid is conical so that the condensation runs down the sides instead of dripping on the contents. A small hole in the top of the lid allows air and steam to escape. This method is used to sterilize culture media that might be spoiled by exposure to higher temperatures, e.g. media containing easily hydrolysed carbohydrates or gelatin. These are steamed for 30–45 min on each of three successive days. On the first occasion, vegetative bacteria are killed; any spores that survive will germinate in the nutrient medium overnight, producing vegetative forms that are killed by the second or third steaming.

Filtration

Bacteria can be removed from liquids by passing them through filters with very small pores that trap bacteria but, in general, not mycoplasmas or viruses. The method is used for sterilizing serum for laboratory use, antibiotic solutions and special culture media that would be damaged by heat. It is also used for separating the soluble products of bacterial growth (e.g. toxins) in fluid culture media.

Types of filters

The following types of filters are mainly of historic interest:

(1) earthenware candles such as the Berkefeld and Mandler type, which are made from Kieselghur, and the Chamberland filter made from unglazed porcelain;
(2) asbestos pads (Seitz filters);
(3) sintered-glass filters, which are made of finely ground glass, fused sufficiently to make the glass granules adhere to one another.

The disadvantages of earthenware candles and asbestos pads are that they are absorptive and have a comparatively slow filtration rate.

Sintered-glass filters have the advantage that little absorption takes place, but the rate of filtration is comparatively slow.

Membrane filters

These filters are made from cellulose esters (cellulose acetate, cellulose nitrate, collodion, etc.). They have many advantages over the early types of bacteriological filters because they are much less absorptive and have a high filtration rate. A range of pore sizes is available and some membranes can be used to separate virus particles by using different gradations of pore size. Bacterial filters have a pore size of less than 0.75 μm. The membranes can be sterilized by autoclaving.

For use, the membrane is mounted on a supporting platform, usually made of stainless steel, which is sealed together between the upper and lower funnels. Filtration is achieved by applying either a positive pressure to the entrance side of the filter or a negative pressure to the exit side.

Small filter units for filtering small volumes of fluid (e.g. 1–5 ml) are available, such as the Hemmings filters (Beaumaris; Colab). The fluid passes through the filter by gravitational force in a centrifuge. In the Swinney and Swinnex type (Millipore), the fluid is forced through a small filter from a syringe.

Membrane filters and filter holders to suit different purposes are obtainable from BBL, Gelman, Millipore and Oxoid, each of which publishes useful booklets or leaflets about their products.

Disinfection

Many different chemicals may be used and they are collectively described as disinfectants or biocides. The former term is used in this book. Some are ordinary reagents, others are special formulations, marketed under trade names. Microbiologists and laboratory managers are often under pressure from salesmen to buy products for which extravagant claims are made. There are usually marked differences between the activity of some disinfectants when tested under optimal conditions by the Rideal–Walker or less discredited techniques and when they are used in practice. The effects of time, temperature, pH, and the chemical and physical nature of the article to be disinfected and of the organic matter present are often not fully appreciated.

Types, and laboratory uses of disinfectants

There is a rough spectrum of susceptibility of micro-organisms to disinfectants. The most susceptible are vegetative bacteria, fungi and lipid-containing viruses. Mycobacteria and non lipid-containing viruses are less susceptible and spores are generally resistant.

In the choice of disinfectants, some consideration should be given to their toxicity and any harmful effects that they may have on the skin, eyes and respiratory tract.

Only those disinfectants which have a laboratory application and considered here. For other information and proprietary names *see* the standard works[5, 6, 7].

The most commonly used disinfectants in laboratory work are clear phenolics and hypochlorites. Aldehydes have a more limited application, and alcohol and alcohol mixtures are less popular but deserve greater attention. Iodophors and quaternary ammonium compounds (QAC) are more popular in the USA than in the UK, while mercurial compounds are the least used. The properties of these disinfectants are summarized below and in *Table 3.1*. Other substances, such as ethylene oxide and propiolactone, are used commercially in the preparation of sterile equipment for hospital and laboratory use but are not used for decontaminating laboratory waste and the other activities mentioned above. They are therefore excluded from this discussion.

Clear phenolics

These compounds are effective against vegetative bacteria (including mycobacteria) and fungi. They are inactive against spores and non lipid-containing viruses. Most phenolics are active in the presence of considerable amounts of protein but are inactivated to some extent by rubber, wood and plastics. They are not compatible with cationic detergents. Laboratory uses include discard jars and disinfection of surfaces. Clear phenolics should be used at the highest concentration recommended by the manufacturers for 'dirty situations', i.e. where they will encounter relatively large amounts of organic matter. This is usually 2–5%. Dilutions should be prepared daily and diluted phenolics should not be stored for laboratory use for more than 24 h, although many diluted clear phenolics may be effective for more than seven days.

Skin and eyes should be protected.

TABLE 3.1. Properties of some disinfectants

	Active against					*Inactivated by*					*Toxicity*		
	Fungi	*Bacteria G+*	*Bacteria G−*	*Myco-bacteria*	*Spores*	*Protein*	*Natural materials*	*Man-made materials*	*Hard water*	*Detergent*	*Skin*	*Eyes*	*Lungs*
Phenolics	+++	+++	+++	++	−	+	++	++	+	C	+	+	−
Hypochlorites	+	+++	+++	++	++	+++	+	+	+	C	+	+	+
Alcohols	−	+++	+++	+++	−	+	+	+	+	−	−	+	+
Formaldehyde	+++	+++	+++	+++	+++[a]	+	+	+	+	−	+	+	+
Glutaraldehyde	+++	+++	+++	++	+++[b]	NA	+	+	+	−	+	+	+
Iodophors	+++	+++	+++	++	+	+++	+	+	+	A	+	+	−
QAC	+	+++	++	−	−	+++	+++	+++	+++	A(C)	+	+	−

+++ good
++ fair
+ slight
− nil
[a] above 40 °C
[b] above 20 °C
From Collins[14].

G Gram
C Cationic
A Anionic

Hypochlorites

The activity is due to chlorine, which is very effective against vegetative bacteria (including mycobacteria), spores and fungi. Hypochlorites are considerably inactivated by protein and to some extent by natural non protein material and plastics and they are not compatible with cationic detergents. Their uses include discard jars and surface disinfection but as they corrode some metals care is necessary. They should not be used on the metal parts of centrifuges and other machines which are subjected to stress when in use. For general purposes and for pipette and other discard jars a chlorine concentration of 2500 ppm is recommended instead of 1000 ppm[8], but for spillage of blood and discard jars that may receive much protein a concentration of 10 000 ppm should be used[9]. The hypochlorites sold for industrial and laboratory application in the UK contain 100 000 ppm available chlorine and would be diluted either 1:40 or 1:10 for use. Some household hypochlorites (e.g. those used for babies' feeding bottles) contain 10 000 ppm and would be diluted 1:4 or used neat. Household 'bleaches' in the UK and USA contain 50 000 ppm available chlorine and dilutions of 1:20 and 1:5 are appropriate. Solid preparations including those that contain hypochlorites, used for domestic disinfection, and chlorinated isocyanurates, used in swimming pools, may have laboratory applications.

Hypochlorites decay rapidly in use, although the products as supplied are stable. Diluted solutions should be replaced after 24 h. The colouring matter added to some commercial hypochlorites is intended to identify them: it is not an indicator of activity.

Hypochlorites may cause irritation of skin, eyes and lungs.

Aldehydes

Formaldehyde (gas) and glutaraldehyde (liquid) are good disinfectants. They are active against vegetative bacterial (including mycobacteria), spores and fungi. They are active in the presence of protein and are not very much inactivated by natural or man-made materials, or detergents.

Formaldehyde is not very active at temperatures below 20 °C and requires a relative humidity of at least 70%. It is not supplied as a gas, but as a solid polymer, paraformaldehyde, and a liquid, formalin, which contains 37–40% of formaldehyde. Both forms are heated to liberate the gas, which is used for disinfecting enclosed spaces such as safety cabinets and rooms. Formalin, diluted 1:10 to give a solution containing 4% formaldehyde, is used for disinfecting surfaces and, in some circumstances, cultures. Solid, formaldehyde-releasing compounds are now on the market[10] and these may have laboratory applications but they have not yet been evaluated for this purpose. Formaldehyde is used mainly for decontaminating safety cabinets (p. 77) and rooms.

Glutaraldehyde usually needs an activator, such as sodium bicarbonate, which is supplied with the bulk liquid. Most activators contain a dye so the user can be sure that the disinfectant has been activated. Effectiveness and stability after activation varies with product and the manufacturers literature should be consulted. Kelsey *et al.*[11] found that the sporicidal activity of one brand was halved in seven days, and Coates[8] suggested that if activated glutaraldehydes are used up to this age then the exposure time should be doubled. They may be used in discard jars (but are expensive) and are particularly useful for disinfecting metal surfaces as they do not cause corrosion.

Aldehydes are toxic. Formaldehyde is particularly unpleasant as it affects the eyes and causes respiratory distress. Special precautions are required (*see below*). Glutaraldehyde is less harmful, but contact with skin and eyes should be avoided.

Alcohol and alcohol mixtures

Ethanol and propanol, at concentrations of about 70–80% in water are effective, albeit slowly, against vegetative bacteria. They are not effective against spores or fungi. They are not especially inactivated by protein and other material or detergents.

Effectiveness is enhanced by the addition of formaldehyde, e.g. a mixture of 10% formalin in 70% alcohol[12] or hypochlorite to give 2000 ppm of available chlorine[13].

Alcohols and alcohol mixtures are useful for disinfecting surfaces and, alcohol–hypochlorite mixtures excepted, for balancing centrifuge buckets.

They are relatively harmless to skin but may cause eye irritation.

Quaternary ammonium compounds

These are cationic detergents known as QACs or quats, and are effective against vegetative bacteria and some fungi but not against mycobacteria or spores. They are inactivated by protein and by a variety of natural and plastic materials and by nonionic detergents and soap. Their laboratory uses are therefore limited but they have the distinct advantages of being stable and of not corroding metals. They are usually employed at 1–2% dilution for cleaning surfaces, but are very popular in food hygiene laboratories because of their detergent nature.

QACs are not toxic and are harmless to the skin and eyes.

Iodophors

Like chlorine compounds these iodines are effective against vegetative bacteria (including mycobacteria), spores, fungi, and both lipid-containing and non lipid-containing viruses. They are rapidly inactivated by protein, and to a certain extent by natural and plastic substances and are not compatible with anionic detergents. For use in discard jars and for disinfecting surfaces they should be diluted to give 75–150 ppm iodine[15], but for hand-washing or as a sporicide, diluted in 50% alcohol to give 1600 ppm iodine[12]. As sold, iodophors usually contain a detergent and they have a built-in indicator: they are active as long as they remain brown or yellow. They stain the skin and surfaces but stains may be removed with sodium thiosulphate solution.

Iodophors are relatively harmless to skin but some eye irritation may be experienced.

Mercurial compounds

Activity against vegetative bacteria is poor and mercurials are not effective against spores. They do have an action on viruses at concentrations of 1:500 to 1:1000 and a limited use, as saturated solutions for safely making microscopic preparations of mycobacteria.

Their limited usefulness and highly poisonous nature make mercurials unsuitable for general laboratory use.

Precautions in the use of disinfectants

As indicated above, some disinfectants have undesirable effects on the skin, eyes and respiratory tract. Disposable gloves and safety spectacles, goggles or a visor should be worn by anyone who is handling strong disinfectants, e.g. when preparing dilutions for use.

The laboratory testing of disinfectants

The following are the main tests that have been devised for assessing the efficiency of disinfectants.

Manufacturers' tests

These tests are used to control the quality of batches during production. The Rideal-Walker[16] and Chick-Martin[17] tests were originally developed for testing the efficiency of phenolic-type disinfectants against a pure phenol standard. The Rideal-Walker test compares the performance of a disinfectant with that of phenol, and the results are calculated as a phenol coefficient, which is given as a number following the letters RW. Higher numbers indicate better disinfectant performances. This test was introduced during the early years of this century when phenol was much used as a disinfectant. A few years after the Rideal-Walker test had been described, Chick and Martin published their phenol coefficient test method. It is similar in design to the Rideal-Walker test in many ways but is done in the presence of organic material. As most disinfectants are inactivated by organic material, the Chick-Martin coefficient (CM) is therefore normally a lower number than the RW coefficient.

The Rideal-Walker and Chick-Martin tests are manufacturers' tests for phenolic disinfectants but the more recently introduced nonphenolic types of disinfectants have very different properties and vary greatly, and they are unsuitable for either the Rideal-Walker or the Chick-Martin test.

Other tests include AOAC[18], the German Society for Hygiene and Microbiology (DGHM)[19], the (Dutch) Phytopharmacy Commission[20] and the Kelsey-Sykes[21] tests. Much has been written about these tests which need not be reviewed here and the reader is referred to the work by Croshaw[22].

There now seems to be no justification for doing any of them in the clinical or research microbiology laboratory. They are best left to the reference and other specialized establishments. Results of 'one-off' tests may not be reliable except in very skilled hands. Tests may be wrong in principle, when like is not compared with like[22] and—a common mistake—neutralization procedures may be inadequate or incorrectly applied[23].

There is also another good reason why certain of these tests should not be done in clinical and teaching laboratories. Most of them retain *Salmonella typhi* as the test organism, although there is a great deal of scientific and commonsense objection to the use of this organism[22]. It is claimed that the strains of typhoid bacilli specified, e.g. *S.typhi* NCTC 786 is not virulent because it has no Vi antigen, an assumption now questioned by not a few bacteriologists.

A more serious objection, noted by ourselves is that clinical and teaching laboratories have used strains of *S.typhi*, recently isolated from cases of typhoid fever, for these tests. This observation, made earlier to an official working party, contributed to the decision to ban the use of *S.typhi* for testing disinfectants in clinical laboratories in the UK.

To meet a growing need for more realistic tests for hospital use, the Public Health Laboratory Service (PHLS) has been responsible for the development of use-dilution tests[7, 24]. These tests are employed to suggest a practical use-dilution for a particular brand of disinfectant which is to be used under hospital or laboratory conditions, e.g. for decontaminating dirty instruments, cleaning hospital walls and operating theatre trolleys and in the bacteriological laboratory for such uses as in laboratory discard jars. These tests are:

(1) the capacity test[25],
(2) the stability test[26], and
(3) the 'in-use' test[27] (described in detail below).

The capacity and stability tests are carried out mainly in manufacturers' and reference laboratories. The 'in-use' test should be employed by the users of disinfectants to check the efficiency of the disinfectants actually in use in their own hospitals or laboratories. This test is easy to perform and is strongly recommended for routine checks.

The Kelsey–Sykes capacity test[24]

The Kelsey–Sykes capacity test is used to estimate the concentration of disinfectants that can be recommended for use in hospitals under 'clean' and 'dirty' conditions, that is, in the absence or in the presence of organic material. It is suitable for testing all types of disinfectants.

The term 'clean condition' implies a situation in which organic material is absent, such as food preparation surfaces and trolley tops, which are first washed clean with hot water and detergent before the application of a disinfectant. The term 'dirty condition' is applied to those situations where organic material is present, e.g. the surfaces of bedpans and laboratory discard jars, and higher concentrations of disinfectants are needed to neutralize the inactivating effect of the organic material and still ensure disinfection.

Basically, the test consists in adding a known volume of a suspension of a test organism at set intervals to a standard volume of the particular concentration of the disinfectant under test, with or without added organic matter. The organic matter added is usually yeast. At regular time intervals after the addition of each of three volumes of the test suspension, a standard amount is removed from the reaction mixture and added to five tubes of recovery medium.

During the progress of the test, the disinfectant is increasingly diluted, but it is the particular concentration at the start which is said to pass or fail the test.

A starting concentration which, after the second addition of organism suspension, yields no growth in at least two out of the five tubes of recovery medium is said to pass the test and can be recommended for use under 'clean' conditions, or under 'dirty' conditions when yeast was included in the test.

The test is not intended for use by those employed in hospitals, but is for official laboratories or for manufacturers, as a guide for recommending the concentrations of disinfectants to be used in hospitals, catering establishments, etc.

The stability test[26]

This is a simplified test designed to check the stability and long-term effectiveness of disinfectants in concentrations recommended by the manufacturers for use in 'clean' and 'dirty' situations. The test is used to supplement the information given by the Kelsey–Sykes capacity test and is suitable only for specialist laboratories. A study of the paper by Maurer[26] shows evidence that bacteria can not only survive but also multiply in some disinfectant solutions.

The 'in-use' test[27]

Samples of liquid disinfectants are taken from such sources as laboratory discard jars, floor-mop buckets, mop wringings, disinfectant liquids in which cleaning materials or lavatory brushes are stored, disinfectants in Central Sterile Supply Departments, used instrument containers and stock solutions of diluted disinfectants. The object is to determine whether the fluids contain living bacteria, and in what numbers. The test is described here in detail for use in any laboratory, because meaningful results can be obtained only in the light of local circumstances.

Method (as described by Maurer)[28]

Stage 1 A 1-ml volume of disinfectant solution in use is taken from each pot or bucket with a separate sterile pipette.
Stage 2 The 1-ml sample is added to 9 ml of diluent in a sterile universal container or a 25-ml screw-capped bottle. The diluent should be selected according to the group to which the disinfectant belongs (*Table 3.2*).
Stage 3 The bottle of diluent is returned to the laboratory within 1 h of the addition of the disinfectant. A separate, sterile pasteur pipette or 50-drop pipette is used to withdraw a small volume of the disinfectant/diluent and to place ten drops, separately, on the surface of each of two well dried nutrient agar plates.
Stage 4 The two plates are incubated as follows:

(1) One plate is incubated for three days at 32 or 37 °C, whichever is most convenient. The optimum temperature for most pathogenic bacteria is

37 °C but those which have been damaged by disinfectants often recover more readily at 32 °C.

(2) The second plate is incubated for seven days at room temperature.

Stage 5 After incubation, the plates are examined and bacterial growth is recorded.

TABLE 3.2 'In-use' test diluents

Diluent	*Disinfectant group*
Nutrient broth	Alcohols Aldehydes Hypochlorites Phenolics
Nutrient broth + Tween 80, 3% w/v	Hypochlorites + detergent Iodophors Phenolics + detergent QACs

'In-use' test results

The growth of bacterial colonies on one or both of a pair of plates is evidence of the survival of bacteria in the particular pot from which the sample is taken. One or two colonies on a plate may be ignored. A disinfectant is not a sterilant and the presence of a few live bacteria in a pot is to be expected.

However, the growth of five or more colonies on one plate should arouse suspicion that all is not well. The relationship between the number of colonies on the plate and the number of live bacteria in the pot can easily be calculated, as the disinfectant sample is diluted 1 in 10 and the 50-drop pipette delivers 50 drops/ml. When five colonies are grown from ten drops of disinfectant/diluent, then five live bacteria were present in one drop of disinfectant and 250 live bacteria were present in 1 ml of disinfectant.

When only an occasional sample drop shows one or two colonies, the use-dilution of the disinfectant can be regarded as satisfactory. The disinfectant process cannot be considered satisfactory when five or more colonies are grown on one or both of a pair of plates.

Precautions when testing disinfectants

In spite of the misgivings expressed above and by others about the use of *S.typhi* in the testing of disinfectants it is apparent that the manufacturers wish to cling to the practice. They cannot be blamed entirely for this, as they are in business to sell disinfectants and their customers demand 'phenol coefficients', even when they do not understand them. Too little has been done in the past to educate the purchaser.

A Code of Practice embodying appropriate safety precautions was drawn up by the British Association of Chemical Specialities[29]. This Code was timely and necessary, because of misapplication, in certain areas, of the (Howie) Code of Practice to non-clinical and industrial laboratories. The Howie Code, as mentioned above, forbids the use of *S.typhi* for testing disinfectants in clinical laboratories. It also places *S.typhi* in Category B. Special accommodation and conditions for containment are specified for handling Category B micro-organisms and these include biological safety cabinets. This caused some confusion because the Howie Code did not specify which organisms in Category B are likely to cause infection by the airborne route and therefore require handling in safety cabinets, and which, like *S.typhi*, infect (usually) by the alimentary route and may be handled on the open bench but away from the mainstream of laboratory activities. The BACS Code of Practice, which has official approval, further clarifies this issue and it is this Code, not the Howie Code which should be observed in non clinical laboratories where disinfectants are tested.

The use of disinfectants in hospitals

It is advisable that hospitals work out a detailed code of practice on the use of disinfectants and antiseptics which applies to their own particular circumstances. Guides for this purpose have been published by the Public Health Laboratory Service in their Monograph No. 2[7], and also by a working party of the South Western Regional Health Authority[30].

For further information about disinfection and sterilization *see* the book by Russel *et al.*[31].

Treatment and disposal of infected materials

It is a cardinal rule that no infected material shall leave the laboratory.

In the context of the treatment and disposal of contaminated laboratory waste (including that contained in re-usable articles), this simple precept offers no problems to properly-equipped and well-managed laboratories.

It is clearly the responsibility of the laboratory management to ensure that no waste containing viable micro-organisms leaves the laboratory premises. The only possible exception would be when it is to be incinerated under the direct supervision of a member of the laboratory staff. Unfortunately, placing this responsibility firmly on the laboratory management may not solve the problem. In some establishments there are very sketchy ideas on freeing material from living organisms. There is a touching faith in the ability of disinfectants at varying concentrations and indefinite ages to kill microbes submerged in or even placed near to them.

We have always believed that disinfection is a first line defence, and for discarded bench equipment it is a temporary measure, to be followed, as soon as possible by autoclaving or incineration. Disinfectants should not be used as the sole method of treating bacterial cultures, even if they are completely submerged and all air bubbles are removed.

Containers for discarded infected material

In the laboratory there should be four important types of receptacles for discarded infected materials:

(1) discard bins or bags to receive cultures and specimens;
(2) discard jars to receive slides, pasteur pipettes and small disposable items;
(3) pipette jars for graduated (recoverable) pipettes; and

(4) plastic bags for combustibles such as specimen boxes and wrappers which might be contaminated.

All of these will go to the preparation room for final treatment and disposal (*see below*).

Discard bins and bags

Discard bins should have solid bottoms which should not leak; otherwise contaminated materials may escape[9]. To overcome steam penetration problems these containers should be shallow, not more than 8 in deep and as wide as will fit loosely into the autoclave. They should never be completely filled. Suitable plastic (polypropylene) containers are available commercially although not specifically designed for this purpose (WCB, Xlon).

Plastic bags are popular but only those made for this purpose should be used. They should be supported in buckets or discard bins. Even the most reliable may burst if roughly handled (Jencons, Sterilin).

For safe transmission to the preparation room the bins may require lids and the bags may be fastened with wire ties.

Discard bins and bags should be colour coded, i.e. marked in a distinctive way so that all workers recognize them as containing infected material.

Discard jars

The jars or pots of disinfectant that sit on the laboratory bench and into which used slides, pasteur pipettes and other rubbish are dumped have a long history of neglect and abuse. In all too many laboratories these jars are filled infrequently with unknown dilutions of disinfectants, are overloaded with protein and articles that float and are infrequently emptied. The contents are rarely disinfected.

Choice of container

Old jam jars and instant coffee jars are not suitable. Glass jars are easily broken and broken glass is an unnecessary laboratory hazard especially if it is likely to be contaminated. Discard jars should be robust and autoclavable and the most serviceable articles are 1-litre polypropylene beakers or screw-capped polypropylene jars. These are deep enough to hold submerged most of the things that are likely to be discarded, are quite unbreakable, and survive many autoclave cycles. They go dark brown in time, but this does not affect their use. Screw-capped polypropylene jars are better because they can be capped after use, inverted to ensure that the contents are all wetted by the disinfectant and air bubbles which might protect objects from the fluid are removed.

Correct dilution

A 1-litre discard jar should hold 750 ml of diluted disinfectant and leave space for displacement without overflow or the risk of spillage when it is moved. A mark should be made at 750 ml on each jar, preferably with paint (grease pencil and felt-pen marks are less permanent). The correct volume of neat disinfectant to be added to water to make up this volume for 'dirty situations' can be calculated from the manufacturers' instructions. This volume is then marked on a small measuring jug, e.g. of enamelled iron, or a plastic dispenser is locked to deliver it from a bulk container. The disinfectant is added to the beaker and water added to the 750 ml mark.

Sensible use

Laboratory supervisors should ensure that inappropriate articles are not placed in discard jars. There is a reasonable limit to the amount of paper or tissues that such a jar will hold, and articles that float are unsuitable for disinfectant jars, unless these can be capped and inverted from time to time to wet all the contents.

Large volumes of liquids should never be added to dilute disinfectants. Discard jars, containing the usual volume of neat disinfectant can be provided for fluids such as centrifuge supernatants, which should be poured in through a funnel that fits into the top of the beaker. This prevents splashing and aerosol dispersal[14]. At the end of the day water can be added to the 750 ml mark and the mixture left overnight. Material containing large amounts of protein should not be added to disinfectants but should be autoclaved or incinerated.

Regular emptying

No material should be left in disinfectant in discard jars for more than 24 h, or surviving bacteria may grow. All discard jars should therefore be emptied once daily, but whether this is at the end of the day or the following morning is a matter for local choice. Even jars that have received little or nothing during the time should be emptied.

'Dry discard jars'

Instead of jars containing disinfectants there is a place in some laboratories for the rigid card and metal containers used in hospitals for discarded disposable syringes and their needles, e.g. Cin Bins (Labco). These will accommodate pasteur pipettes, slides, etc. and can be autoclaved or incinerated.

Pipette jars

Jars for recoverable pipettes should be made of polypropylene or rubber. These are safer than glass. The jars should be tall enough to allow pipettes to be completely submerged without the disinfectant overflowing. A compatible detergent should be added to the disinfectant to facilitate cleaning the pipettes at a later stage. Tall jars are inconvenient for short people, who tend to place them on the floor. This is hazardous. The square based rubber jars (Camlab) may be inclined in a box or on a rack which is convenient and safer.

Plastic bags for combustibles

These bags are to be found in all hospitals and are colour coded. Clinical laboratories should follow the local code. Others should introduce one, e.g. red bags may be designated for this, and no other purpose.

Treatment and disposal procedures

There are three practical methods for the treatment of contaminated, discarded laboratory materials and waste:

(1) autoclaving;
(2) incineration; and
(3) chemical disinfection.

The choice is determined by the nature of the material: if it is disposable, or recoverable, and if the latter if it is affected by heat. With certain exceptions none of the methods excludes the others. It will be seen from *Figure 3.3* that incineration alone is advised only if the incinerator is under the control of the laboratory staff[9,14]. Disinfection alone is advisable only for graduated, recoverable pipettes.

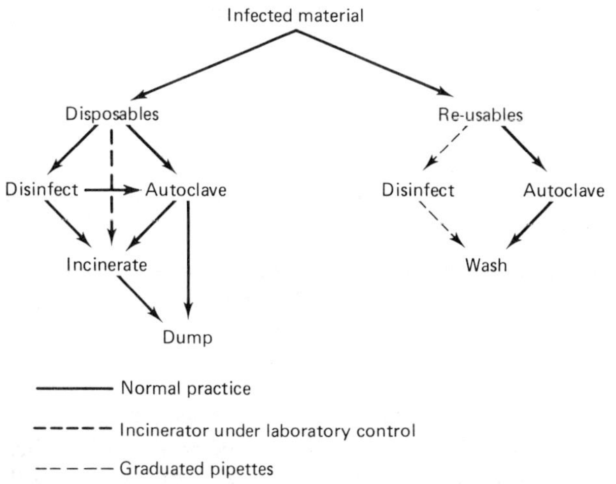

Figure 3.3 Flow chart for the treatment of infected material

Organization of treatment

The design features of a preparation room for dealing with discarded laboratory materials should include autoclaves, a sluice, a waste disposal unit plumbed to the public sewer, deep sinks, glassware washing machines, drying ovens, sterilizing ovens and large benches.

These should be arranged to preclude any possible mixing of contaminated and decontaminated materials. The designers should therefore work to a flow, or critical pathway chart provided by a professional microbiologist.

Such a chart is shown in *Figure 3.4*. The contaminated material arrives in

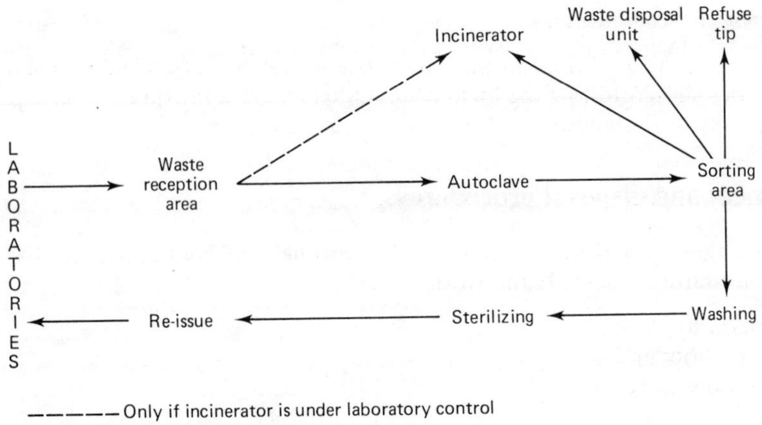

Figure 3.4 Design of preparation (utility) rooms. Flow chart for the disposal of infected laboratory waste and re-usable materials

its colour coded containers onto a bench or into an area designated and used for that purpose only. These are then sorted according to their colour codes and despatched to the incinerator or loaded into the autoclave. Nothing bypasses this area. After autoclaving the containers are taken to the sorting bench where the contents are separated into:

(1) waste for incineration, which is put into colour coded bags,
(2) waste for the rubbish tip, which is also put into different colour coded bags,
(3) waste suitable for the sluice or waste disposal unit,
(4) recoverable material which is passed to the next section or room for washing and re-sterilizing.

This section or room should have a separate autoclave. Contaminated waste and materials for re-use or re-issue should not be processed in the same autoclave.

Procedures for various items

Before being autoclaved the lids of discard bins should be removed and then included in the autoclave load in such a way that they do not interfere with steam penetration. Plastic bags should have the ties removed and the bags opened fully in the bins or buckets that support them.

Contaminated glassware

After autoclaving culture media may be poured away or scraped out and the tubes and bottles, etc., washed by hand or mechanically with a suitable detergent. The washing liquid or powder used will depend on the hardness of the water supply and the method of washing. The advice of several laboratory detergent manufacturers should be sought.

Busy laboratories require glassware washing machines. Before purchasing one of these machines, it is best to consider several and to ask other laboratories which models they have found satisfactory. Not all laboratory glassware washing machines are as good as the manufacturers claim. A prerequisite is a good supply of distilled or deionized water.

If hand-washing is practised, double sinks, for washing and then rinsing, are necessary, plus plastic or stainless steel bowls for final rinsing in distilled or deionized water. Distilled water from stripper stills, off the steam line, is rarely satisfactory for bacteriological work.

Rubber liners should be removed from screw-caps and the liners and caps washed separately and re-assembled. Colanders or sieves made of polypropylene are useful for this procedure.

New glassware, except that made of borosilicate or similar material, may require neutralization. When fluids are autoclaved in new soda-glass tubes or bottles, alkali may be released and alter the pH. Soaking for several hours in 2–3% hydrochloric acid is usually sufficient, but it is advisable to test a sample by filling with neutral water plus a few drops of suitable indicator and autoclaving.

Discard jars

After standing overnight to allow the disinfectant to act the contents of the jars should be poured carefully through a polypropylene colander. This, and its contents are then placed in a discard bin and autoclaved.

Rubber gloves should be worn for these operations.

The empty discard jars should be autoclaved before they are returned to the laboratory for further use. There may be residual contamination.[28]

Re-usable pipettes

After total immersion in disinfectant plus detergent overnight the pipettes should be removed with gloved hands.

Before the pipettes are washed, the cotton wool plugs must be removed. This can be carried out by inserting the tip into a piece of rubber tubing attached to a water tap. Difficult plugs can be removed with a small crochet hook. Several excellent pipette washing machines are manufactured that rely on water pressure and/or a siphoning action, but the final rinse should be in distilled or deionized water.

24-h urines

Although rare in microbiology laboratories other pathology departments may send them for disposal on the grounds that they may contain pathogens. Ideally they should be processed in the department concerned as follows.

Sufficient disinfectant, e.g. hypochlorite, should be added to the urine to give the use-dilution. After standing overnight then urine should be poured carefully down the sink or sluice to join similar material in the public sewer. The containers, which are usually plastic, may then be placed in colour coded bags for incineration.

Incineration

The problems with this method of disposal are in ensuring that the waste actually reaches the incinerator, and that if it does it is effectively sterilized: and that none escapes, either as unburned material or up the flue. Incinerators are rarely under the control of laboratory staff. Sometimes they are not even under the control of the staff of the hospital or institution, but are some distance away and contaminated and infectious material has to be sent to them on the public highway. It may never arrive or may not be incinerated.

Incinerators are not always efficient. Nor are incinerator operators. Unburned material may be found among the ash, and from its appearance it may be deduced that it may not have been heated enough to kill micro-organisms. We have recovered unconsumed animal debris, including fur and feathers and entrails, from a laboratory incinerator. The up-draught of air may carry micro-organisms up the flue and into the atmosphere if the load is too large or badly distributed. Darlow (personal communication) has recovered organisms on culture plates exposed over a flue. He recommends that incinerators used for infectious materials should have afterburners so that they consume their own smoke and render their effluent harmless. This is endorsed in the Code of Practice (DHSS, 1978) which also requires adequate supervision of the incineration of infected laboratory waste.[9]

In view of the problems and uncertainties associated with incineration it seems advisable to use this method only for material which has been autoclaved or disinfected and which, for aesthetic or other reasons, cannot be placed on a rubbish tip. There must, of course, be exceptions, and polythene 24-h urine containers are an example. Care must be taken when plastics are incinerated. A highly toxic smoke may be produced and it is usually recommended that not more than about 20% of any load is composed of plastic materials.

References

1. MEDICAL RESEARCH COUNCIL (1959) *Lancet*, **i**, 425
2. BRITISH STANDARDS INSTITUTION (1961) *Specification for Performance of Electrically Heated Sterilizing Ovens*. BS 3421. London: BSI
3. PUBLIC HEALTH LABORATORY SERVICE (1978) *Journal of Clinical Pathology*, **31**, 418
4. PUBLIC HEALTH LABORATORY SERVICE (1981) *Journal of Hospital Infection*, **2**, 377
5. RUBBO, S.D. and GARDNER, J.F. (1965) *A Review of Sterilization and Disinfection*. London: Lloyd-Luke
6. SYKES, G. (1965) *Disinfection and Sterilization*. 2nd edn. London: Spon
7. KELSEY, J.C. and MAURER, I.M. (1972) *The Use of Chemical Disinfectants in Hospitals*. Public Health Laboratory Service Monograph No. 2. London: HMSO
8. COATES, D.A. (1980) *Institute of Medical Laboratory Sciences Gazette*, **24**, 555
9. DEPARTMENT OF HEALTH AND SOCIAL SECURITY (1978) *Code of Practice for the Prevention of Infection in Clinical Laboratories*. London: HMSO
10. ALLWOOD, M.C. and MYERS, E.R. (1981) Formaldehyde releasing compounds. In *Disinfectants: Their Use and Evaluation of Effectiveness*. Edited by C.H. Collins *et al*. London: Academic Press
11. KELSEY, J.C., McKINNON, I.H. and MAURER, I.M. (1974) *Journal of Clinical Pathology*, **27**, 632
12. US PUBLIC HEALTH SERVICE (1978) *National Institutes of Health Laboratory Safety Monograph*. Bethesda, MD: US Department of Health and Human Services

13. COATES, D. and DEATH, J.E. (1978) *Journal of Clinical Pathology*, **31,** 148
14. COLLINS, C.H. (1983) *Laboratory-acquired Infections*. London: Butterworths
15. US PUBLIC HEALTH SERVICE (1974) *Laboratory Safety at the Center for Disease Control*. Atlanta: US Department of Health and Human Services
16. BRITISH STANDARDS INSTITUTION (1934) *Technique for Determining the Rideal-Walker Coefficient of Disinfectants*. BS 541. London: BSI
17. BRITISH STANDARDS INSTITUTION (1938) *Modified Technique of the Chick-Martin Test for Disinfectants*. London: BSI
18. ASSOCIATION OF ANALYTICAL CHEMISTS (1960) *Official Methods of Analysis of the Association of Analytical Chemists*. 9th edn. Washington, DC: AOAC
19. DEUTSCHE GESELLSCHAFT FÜR HYGIENE UND MIKROBIOLOGIE (1972) *Ritchlinien für die Prufung Chemischer Disinfektionsmittel*, **3,** DGHM. Stuttgart: Alfred G. Fischer
20. PHYTOPHARMACY COMMISSION (1974) *Appraisal of Disinfectants and Combined Disinfectants and Cleaning Agents Intended for use in the Food Industry*. Den Hague, Hogwen Phytopharmacy Commission: The Netherlands
21. KELSEY, J.C. and MAURER, I.M. (1974) *Pharmaceutical Journal*, **213,** 528
22. CROSHAW, B. (1981) Disinfectant testing, with particular reference to the Rideal-Walker and Kelsey-Sykes test. In *Disinfectants, Their Use and Evaluation of Effectiveness*. Edited by C.H. Collins *et al.* pp. 1-14. London: Academic Press
23. RUSSELL, A.D. (1981) Neutralization procedures in the evaluation of bacterial activity. In *Disinfectants. Their Use and Evaluation of Effectiveness*. Edited by C.H. Collins *etal.* pp. 45-57. London: Academic Press
24. KELSEY, J.C. (1969) *Journal of Medical Laboratory Technology*, **26,** 79
25. KELSEY, J.C. and SYKES, G. (1969) *Pharmaceutical Journal*, **202,** 67
26. MAURER, I.M. (1969) *Pharmaceutical Journal*, **203,** 529
27. KELSEY, J.C. and MAURER, I.M. (1966) *Monthly Bulletin of the Ministry of Health and Public Health Laboratory Service*, **25,** 180
28. MAURER, I.M. (1972) The management of laboratory discard jars. In *Safety in Microbiology*. Edited by D. Shapton and R.G. Board, pp. 53-59. London: Academic Press
29. BRITISH ASSOCIATION FOR CHEMICAL SPECIFICATIONS (1981) *Code of Practice for the Handling of Salmonella typhi*. NCTC 786. London: BACS
30. SOUTH WEST REGIONAL HEALTH AUTHORITY (1975) *Notes on Disinfection and Sterilization*. 3rd edn, Bristol
31. RUSSELL, A.D., HUGO, W.B. and AYLIFFE, G.A.J. (1982) *Principles and Practice of Disinfection, Preservation and Sterilization*. Oxford: Blackwell
32. COLLINS, C.H. (1974) Prevention of laboratory-acquired infection. In *Public Health Laboratory Service Monograph* No. 6. Edited by A.J. Willis and C.H. Collins, p. 35. London: HMSO

Chapter 4

Culture media

Many laboratories now purchase most of their culture media. Only very large laboratories can afford the time and labour necessary for the production and quality control of the wide range of media in current use.

Culture media may be purchased in dehydrated form or ready for use in tubes and petri dishes. When a proprietary medium is specified in this book it is given capital initials followed by an abbreviation of the manufacturer's name in brackets, as below. Some culture media, however, are made by several companies or may be made in the laboratory. The names of these media are given lower case initials.

The manufacturers known to us are: BBL (*B*), Difco (*D*), Gibco (*G*), lab m (*L*), Mast (*Ma*), Merck (*Me*), Oxoid (*O*) and the Pasteur Institute (*P*).

Poured plates are delivered (in the UK) by lab m, Merck and Tissue Culture Services. Redigel media, in a limited but useful number of formulations, are available from Cherwell. The cold liquid medium, plus additives if required, is poured into prepared petri dishes and sets almost immediately. A small, but again very useful range of ready-to-use media for mycology and industry, including contact plates, is marketed by Medical Wire.

There may be other media manufacturers and suppliers whose names we do not know. We do not claim that the lists above and below are complete.

The information given in this chapter is intended to guide users in their choice of media for particular organisms and purposes. Further guidance may be found in the publications of the various manufacturers. In particular, the Oxoid Manual[1], new editions of which appear regularly, is very useful.

We do not give the formulae for the preparation of most of the commercially available media in this book, but formulae are given for a number of media which are not on sale or which we have preferred to make ourselves.

General purpose media

These include a wide variety of 'nutrient' and 'infusion' broths and agars, made from meat, meat extracts or yeast extracts. Robertson's cooked meat medium is included, but *see* p. 80.

Media for fastidious organisms

These are enriched broths and agars which may be used alone or with the addition of supplements. They include blood agar bases, formulations containing brain, heart and veal extracts and various tryptic digests of meat and soybean. Some are marketed as Casman (*B*, *D*, *G*), Columbia (*B*, *D*, *L*, *G*, *O*), Eugon (*B*, *D*) agars, etc. For blood agar *see* p. 70.

Media for the selective isolation of bacteria

Although selective media may not be necessary for the isolation of some of the organisms named below it is advisable to use them if the material to be cultured is likely to contain many other organisms. It should be noted, however, that all selective media are, to some extent, inhibitory to the organisms they are intended to select. Growth promoting or inhibitory additives are listed beginning on p. 67.

Aeromonas

Blood agar; Furunculosis agar (*D*).

Bacillus cereus

B. Cereus Selective agar (*O*).

Bacteroides

Blood agar plus menadione–antibiotic supplements.

Bordetella

Bordet Gengou or Thayer Martin media plus antibiotic supplements.

Brucella

Blood agar or proprietary Brucella media plus antibiotic supplements.

Campylobacter

Columbia or Campylobacter agar (*O*) plus supplements.

Clostridia

Reinforced Clostridial agar; Tryptose–Sulphite Cycloserine agar, TSC (*Me*); Oleandomycin Polymyxin Sulphadiazine Perfringens agar, OPSP (*O*); Iron Sulphite agar (*O*); Clostrisel (*B*); Clostridium Difficile agar (*O*); Willis and Hobbs medium (p. 82).

Coliform bacilli

Broths Brilliant green deoxycholate; formate (or lactose) ricinoleate; lauryl sulphate, MacConkey; Mineral Modified Glutamate (*O*); Enterobacteria Enrichment EE (*B*).

Agars Cystine lactose electrolyte deficient (CLED); deoxycholate; Endo; eosin methylene blue (EMB), violet red bile.

Corynebacteria (*Diphtheria bacilli*)

Hoyle (*L*, *O*); proprietary tellurite media; Loeffler.

Enterobacteria

See under Coliform bacilli, Salmonellas, Shigellas.

Haemophilus

Blood agar plus yeast, haemin, Filde's extract or commercial supplements.

Lactobacilli

APT (*B*, *D*); MRS; Tomato Juice agar (*O*); Rogosa agar.

Legionella

Blood agar plus supplements. Proprietary media (*O*).

Mycobacteria

Lowenstein-Jensen (p. 77): Acid Egg (*O*); ATS (*D*); Kirchner (p. 76); Middlebrook 7H9, 7H10, 7H11 (*D*).

Neisseria

Thayer Martin; New York City medium (*O*); commercial GC media; all plus additives and antibiotics.

Pediococci

Wort agar.

Pseudomonas

Commercial selective media containing cetrimide (*B*, *O*, *D*).

Salmonellas

Enrichment Selenite broths; bismuth sulphite (Wilson and Blair); brilliant green broth (BEB); deoxycholate-citrate (DCA) (many formulations).

Plating Bismuth sulphite (Wilson and Blair); brilliant green; Endo; deoxycholate–citrate (many formulations).

Shigellas

Deoxycholate–citrate, especially that containing xylose and lactose (many formulations).

Staphylococci

Baird–Parker (*O*); Chapman–Stone; Mannitol-salt (*B*, *O*); milk salt agar (p. 78); phenolphthalein phosphate agar (p. 79); Staphylococcus 110; Tellurite Polymyxin Egg Yolk agar (*B*, *D*).

Streptococci

Azide media; BAGG broth; Barnes (*Me*); Edwards (*O*); ethyl violet azide; Mead's (*B*. p. 78); mitis salivarius agar; thallous acetate (Slanetz and Bartlet; Trypticase Yeast extract Cystine agar.

Vibrios

Alkaline peptone water; thiosulphate citrate bile sucrose agar (TCBS); Marine agar (*D*).

Yersinia enterocolitica

YE medium (*O*) plus supplements.

Growth-promoting or inhibiting supplements and additives

Single substances or mixtures of growth promoters and antibiotics, packed ready to add to culture media, are supplied by several companies. The most comprehensive selection is offered by Oxoid[1].

Purpose	*Supplement/additive*
Anaerobes	Cysteine – HCl – sodium dithiothreitol – Vitamin K
Bacteroides	Haemin – menadione – sodium pyruvate – nalidixic acid
Bacillus cereus	Polymyxin
Bordetella	Cephalexin selective supplement
Brucella	Polymyxin – bacitracin – cycloserine – nalidixic acid – nystatin – vancomycin
Campylobacter	Sodium pyruvate – metabisulphite – ferrous sulphate. Vancomycin – polymyxin – trimethoprim. Bacitracin – cycloheximide – colistin – cephazolin – novobiocin
Clostridium difficile	Cycloserine–cefoxitin
Corynebacteria	Potassium tellurite (3.5%)

Haemophilus	Haemoglobin. Filde's extract
Lactobacilli	Lactic acid
Legionella	Cysteine HCl - ferric pyrophosphate - sodium selenate. Colistin - vancomycin - trimethoprim - amphotericin (CVTA)
Mycobacteria	Bovine albumin. Oleic acid - albumin - dextrose - citrate (OADC)
Neisseria	Haemoglobin. Yeast autolysate. Vancomycin - colistin - nystatin (VCN). Vancomycin - colistin - nystatin - trimethoprim (VCNT) Vancomycin - colistin - ampherotericin - trimethoprim (VCAT) Lincomycin - colistin - ampherotericin - trimethoprim (LCAT)
Pasteurella	Haemoglobin
Salmonellas from sewage	Sulphacetamide - sodium mandelate
Streptococci (mitis, etc.)	Potassium tellurite (0.8%) Nalidixic acid - polymyxin - neomycin
Yersinia	Cefsulodin - Irgasan - novobiocin

Media for dermatophytes and pathogenic yeasts

General media

Czapek-Dox agar; wort agar; Sabouraud agar (various formulations).

Dermatophytes

Littman ox gall agar (*B*, *Me*); malt agar with and without chloramphenicol and cycloheximide.

Yeasts

BiGGY (Nickerson) medium (*O*, *Me*); corn meal agar; Niger seed agar (p. 79).

Transport media

Semisolid media with and without charcoal and antibiotics, intended to keep delicate microbes alive during transit to the laboratory are sold under the following names:

Amies; Cary-Blair; Stuart; Transgrow; Transport; Pertussis.

Media for food microbiology

Most general media (p. 56) are useful, especially glucose (dextrose) tryptone media.

Bacterial counts

Standard (plate) count agar; Standard methods agar; milk agar; yeast extract milk agar.

Counting yeasts and moulds

Buffered yeast agar (p. 69). OGYE agar (*O*); WL agar; Lysine agar (*O*); glucose salt agar; potato dextrose agar; Malt extract agar; Rose Bengal chloramphenicol agar (*O*).

Acidophilic yeasts

Malt agar.

Beer spoilage

Universal Beer agar (*O*).

Brochothrix

Gardner's STAA (p. 75).

Coliform bacilli

See p. 58.

Clostridia

See p. 57.

Enterobacteria

See coliform bacilli, p. 58.

Flat sour organisms

Glucose tryptone agar.

Hydrogen swell organisms

Glucose tryptone agar.

Lactobacilli

See p. 58.

Lipolytic organisms

Tributyrin agar (*O*).

Moulds and yeasts in dairy products and utensils

Malt agar plus chloramphenicol and cycloheximide; Buffered yeast agar (p. 69); Potato dextrose agar.

Moulds on meat

Potato glucose agar.

Staphylococci

See p. 59.

Streptococci

See p. 59.

Media for water bacteriology

Bacterial counts

Standard methods agar; standard plate count agar.

Coliform bacilli

See p. 58.

Clostridium perfringens

See p. 57.

Enterobacteria

See coliform bacilli, p. 58.

Salmonellas

See p. 58.

Membrane filtration media

Companies that sell these media place a suffix before the names of their products: BBL, *M*-; Difco, m-; Gibco, MF-; Merck, Membrane Filtration; Oxoid, M-. Membrane filtration media at present available include the following.

Counting bacteria

Glucose tryptone broths; standard methods broth; tryptone soya broths.

Coliform bacilli

Brilliant green broths; Endo media; eosin methylene blue broth; enrichment broth; lauryl sulphate media; MacConkey broths; resuscitation broths.

Flat sour organisms

Glucose tryptone broths.

Salmonellas, including S. typhi

Bismuth sulphite broth; tetrathionate broth.

Staphylococci

Staphylococcus 110 broth.

Streptococci

Azide broths; enterococcus media.

Moulds

Czapek-Dox broth.

Antibiotic sensitivity test media

Mueller Hinton medium is the best known. Most culture media manufacturers offer this or their own variations under trade names.

Sterility test media

For products not containing preservatives—any non-selective broth medium. For products containing preservatives—Clausen medium (*O*); thioglycollate media (various formulations); p. 81.

Identification media

These are used for various 'biochemical' tests. Most are available commercially; page numbers are given for those that are not. (*See also* Paper strip (p. 103) and 'Kit' bacteriology (p. 103).)

Aesculin hydrolysis

Aesculin medium, Edwards' medium.

Arginine hydrolysis

Arginine broth, combined arginine-ornithine media.

Casein digestion

Skim milk agar (p. 80).

Carbohydrate utilization (most available commercially)

These are usually a peptone broth base containing an indicator (phenol red or Andrade) to which the fermentable substance is added at 0.5–2% (1% for most bacteria). A Durham's tube is included if detection of gas is desired.

These simple media are not suitable for all purposes, however, and the following modifications are advisable for special purposes.

Baird–Parker medium

For staphylococci and micrococci (p. 59).

Ammonium salt basal synthetic medium

Add carbohydrate for aerobic spore bearers and pseudomonads (p. 59).

Gillies medium

Contains glucose, mannitol, sucrose and salicin for screening enterobacteria (*B*).

Kohn two-tube medium

This is a modification of Gillies medium (*O*).

Krumwiede double sugar agar

Contains lactose and sucrose (*B*).

MRS medium

Plus carbohydrates for lactobacilli (p. 78).

Robinson's serum water sugars

Preferred for pathogenic corynebacteria (p. 81).

Russell's double sugar agar

Contains glucose and lactose.

Triple sugar iron agar (TSI)

Contains glucose, lactose and sucrose and also indicator for H_2S production.

Citrate utilization

Koser; Simmons.

Decarboxylase tests

Falkow and Moeller lysine media, basal media to which arginine, lysine and ornithine must be added (*see also p.* 73).

Esculin hydrolysis

See Aesculin.

DNAse

DNAse agars with or without indicator.

Gelatin liquefaction

Nutrient gelatin, gelatin agar, various formulations (pp. 75, 79). Charcoal gelatin discs for use with nutrient broths.

Gluconate utilization

Gluconate broth (p. 75).

Hugh and Leifson test

Add glucose to commercial base (p. 76).

Hydrogen sulphide production

Lead acetate agar and various formulations of iron agar, iron plus carbohydrate and iron plus lysine medium.

Indole production

Any commercial peptone or tryptone water medium.

KCN test (potassium cyanide)

Add 0.5% KCN to commercial broth base.

Lecithinase

Egg yolk agar and broth (p. 74). Willis and Hobbs medium (p. 82).

Malonate utilization

Malonate and phenylalanine—malonate broths.

Milk reactions

Litmus milk, bromocresol purple milk, Crossley milk medium (*O*).

MR and VP tests

Glucose phosphate broth, MRVP media.

Motility

Semisolid media. *See also* Craigie method (p. 80).

Nitratase test

Nitrate broths, indole–nitrate broths and agars.

Nagler reaction

Egg yolk agar (p. 74). Willis and Hobbs medium (p. 82).

OF (Ox-ferm: Hugh and Leifson) test

Add glucose to commercial base (*see also* p. 76).

ONPG test

ONPG broth (p. 79).

Organic acid utilization

Add 1–2% organic acid to commercial organic acid base medium (*see also* p. 70).

Oxygen preference

Use sloppy agar (p. 80).

PPA (phenylalanine deamination)

Phenylalanine agar.

Phosphatase test

Phenolphthalein phosphate agar (p. 79).

Schleifer and Kloos test

For staphylococci, use Purple Agar (*D*).

Starch hydrolysis

Starch hydrolysis agar (p. 81).

Sulphatase test

Phenolphthalein sulphate agar and broth (p. 80).

Urease test

Usually sold as a base, with sterile (filtered) urea separately.

Tyrosine decomposition

Tyrosine agar (p. 82).

Xanthine decomposition

Xanthine agar (p. 82).

Supplements and additives for identification media

These are usually supplied weighed or measured and ready to add to the commercial base media. They include:

Carbohydrates (dulcitol, glucose, lactose, maltose, mannitol, salicin, sucrose, xylose)
Egg yolk emulsion for lecithinase test media
Lead acetate for hydrogen sulphide detection
Ox bile for bile solubility test and bile inhibition tests
Potassium lactate for lysine agar
Sodium nitrate (30%) for nitrate broths

Most commercial media companies supply some or all of these.

Laboratory prepared media

Acid broth

Used for high acid thermally processed foods

Proteose peptone	5 g
Yeast extract	5 g
Glucose	5 g
Dipotassium phosphate K_2HPO_4	4 g
Distilled water to	1000 ml

Dissolve and dispense in 12–15 ml amounts, sterilize at 121 °C for 15 min. Final pH should be 5.0.

Alkaline peptone water

This is used for isolating *V. cholerae*. Adjust the pH of peptone water to pH 8.6 with N NaOH.

Anaerobe agar[2]

For the isolation and enumeration of anaerobes from cosmetics.

Trypticase soy agar	40 g
Agar (bacteriological grade)	5 g
Yeast extract	5 g
Haemin solution (*see below*)	0.5 ml
Vitamin K (*see below*)	1.0 ml
L-Cysteine (dissolved in 5 ml N NaOH)	0.4 g
Distilled water	1000 ml

Dissolve all the ingredients by boiling. Adjust pH to 7.5; autoclave for 15 min at 121 °C. Pour into petri dishes. Reduce under anaerobic conditions for 24 h before use.

Stock haemin solution

Dissolve 1g haemin (Eastman Chemical No. 2203) in 100 ml distilled water. Autoclave for 15 min at 121 °C. Store under refrigeration.

Stock vitamin K solution

Dissolve 1 g vitamin K1 (Sigma Chemical Co. No. V-3501) in 100 ml 95% ethanol. Leave for 2–3 days, shaking at intervals. Refrigerate.

Antifungal drug sensitivity testing

(1)	Glucose monohydrate	20 g
	Casamino acids (*D*)	20 g
	Sodium glycerophosphate	5 g
	Yeast extract 5% (*D* or *O*)	2 ml
	Distilled water	1000 ml

Heat at 50–60 °C to dissolve and distribute in 100-ml lots.

(2)	Agar	3 g
	Distilled water	1000 ml

Dissolve by autoclaving and distribute in 100-ml lots.

For sensitivity to amphotericin and imidazole add 2 ml of 1% cytosine to medium (1). For the assay of 5-fluorocytosine mix 100 ml of (1) and (2) and distribute in 10-ml lots for tests on slopes or 20-ml lots for tests in petri dishes.

Antifungal drug assay medium

Yeast Nitrogen Base (*D*)	3.5 g
Glucose	10.0 g
KH_2PO_4	1.5 g
$Na_2HPO_4.2H_2O$	1.0 g
Distilled water	1000 ml

Dissolve, dispense in 110-ml lots. Add 1.2 g of agar (*O*) to each bottle and autoclave.

Ammonium salt sugars

See under Carbohydrate media.

Arginine broth

This is useful identifying some streptococci and Gram-negative rods.

Tryptone	5 g
Yeast extract	5 g
Dipotassium hydrogen phosphate	2 g
L-Arginine monohydrochloride	3 g
Glucose	0.5 g
Water	1000 ml

Dissolve by heating, adjust to pH 7.0, tube in 5–10-ml amounts and autoclave at 115 °C for 10 min.

For arginine breakdown of lactobacilli, use MRS broth in which the ammonium citrate is replaced with 0.3% arginine hydrochloride.

Baird–Parker's sugar media

See under Carbohydrate media.

Basal synthetic media

These are used when investigations are being made into the ability of bacteria to use various carbon sources or, with the addition of a suitable carbon source, of growth factor or vitamin requirements.

(*a*) *For carbohydrate utilization*

Ammonium dihydrogen phosphate	1.0 g
Potassium chloride	0.2 g
Magnesium sulphate	0.2 g
Agar	10.0 g
Water	1000 ml

Dissolve by heating, add 4 ml of 0.2% bromthymol blue and a final 1% of sterile (filtered) carbohydrate solution (0.1% aesculin; 0.2% starch; these are sterilized by steaming).

To investigate amino acid sources, add 1% glucose and 0.1 g of amino acid.

(*b*) *For organic acid utilization*

Magnesium sulphate	1.0 g
Sodium chloride	1.0 g
Diammonium hydrogen phosphate	1.0 g
Potassium dihydrogen phosphate	0.5 g
Agar	12.0 g
Water	1000 ml

Dissolve salts by heating in 200 ml of water. Dissolve agar in the remaining water. Mix and adjust to pH 6.8. Bottle in 100-ml amounts. Melt, add 0.2 g of organic acid (sodium salts of acetic, benzoic, citric, oxalic, propionic, pyruvic, succinic, tartaric acids; calcium salt of malic acid; mucic acid as free acid), re-adjust to pH 6.8 if necessary and add 0.4 ml of 0.2% phenol red. Tube and give a final steaming to sterilize.

To investigate amino acid sources, add 1% glucose and the indicator and 0.1 g of the amino acid.

Blood agar media

Blood agar

This is nutrient agar or one of the commercial dehydrated Blood Agar Bases to which sterile horse blood is added. It is used to detect haemolytic organisms and to encourage the growth of organisms that grow poorly or not at all on nutrient agar. Some specially enriched commercial blood agar bases are available for diagnostic and sensitivity tests.

Melt 100 ml of nutrient agar or prepare 100 ml of Blood Agar Base. Cool to 48 °C in a water-bath. Add 10 ml of sterile defibrinated or oxalated horse blood and mix gently so as to avoid bubbles. Pour into petri dishes by raising the lid only enough to introduce the neck of the bottle. This amount should give seven poured plates.

In pathology laboratories, to conserve blood and to see more easily the clear zone around colonies, a very thin layer of blood agar is poured on top of a thin layer of ordinary nutrient agar. Sometimes saline agar, a 1% agar containing 0.9% sodium chloride, at pH 7.4 is used.

Chloral hydrate agar

Used in the isolation of anthrax bacilli. Add 5% chloral hydrate to blood agar while still fluid.

Chocolate agar

Some fastidious bacteria grow best on a blood agar in which the cells have been disrupted and the haemoglobin altered by heat.

Make blood agar as described above but replace the bottle in the water-bath, raise the temperature of the water-bath slowly to 80 °C and leave for 5 min. Cool slightly and pour into petri dishes. The medium should be homogeneous and chocolate coloured, not flaky.

Chocolate cystine agar

For the cultivation of *Neisseria* and *Haemophilus*.

Solution A	
Cystine	0.1 g
Nutrient broth	50 ml
Solution B	
p-Aminobenzoic acid	0.1 g
Nutrient broth	50 ml

Make up when required and steam for 1 h.

To 950 ml of melted blood agar base add 25-ml amounts of each solution of *A* and *B*. Bottle and sterilize at 115 °C for 10 min. For use, melt, cool to 50 °C, add 10% blood and heat at 70–80 °C until reddish brown in colour. Pour plates and use as soon as possible.

Colistin-nalidixic acid agar

To isolate Gram-positive cocci from material heavily contaminated with *Pseudomonas*, *Proteus*, *Klebsiella*, etc., add 10 μg/ml of colistin and 15 μg/ml of nalidixic acid to blood agar base before adding the blood.

Crystal violet blood agar

The dye inhibits many organisms but not streptococci. Add 1 ml of a 0.001 % solution of crystal violet in water to 500 ml of blood agar while still melted.

Gentamicin blood agar

Add 50 μg of gentamicin to 100 ml of blood agar.

Lysed blood agar

This is used for sulphonamide sensitivity tests. Lysed blood is obtainable commercially (*see above*) or made by alternate freezing and thawing of whole blood. The medium is made as described under Blood Agar.

Neomycin blood agar

Add 75 μg of neomycin sulphate to 100 ml of blood agar.

PABA blood agar

To neutralize sulphonamides in an inoculum, add 10 mg % of *p*-aminobenzoic acid to the medium before sterilizing it.

Buffered yeast agar

This is useful for the detection of and counting yeasts and moulds in the food industry, utensils, washes, etc.

Yeast extract	5.0 g
Dextrose	20.0 g
Ammonium sulphate	0.72 g
Ammonium dihydrogen phosphate	0.26 g
Agar	12.0 g
Water	1000 ml

Dissolve by heating and adjust to pH 5.5, bottle and sterilize at 115 °C for 10 min. If required for bottle counts, increase agar to 20 g/l.

To adjust this medium to a more acidic pH, cool to 50 °C and add 1% citric acid monohydrate and 10% lactic acid to each 100 ml as follows (Davis[3]):

pH	*1% citric acid* (ml)	*10% lactic acid* (ml)
4.75	1.26	0.125
4.50	2.24	0.20
4.25	3.92	0.30
4.00	6.16	0.45
3.75	9.52	0.70
3.50	14.56	1.17

Carbohydrate fermentation and oxidation tests

Formerly known as 'peptone water sugars', some of these media are now more nutritious and do not require the addition of serum except for very fastidious organisms. Liquid, semisolid and solid basal media are available commercially and contain a suitable indicator, e.g. Andrade (pH 5–8), phenol red (pH 6.8–8.4) or bromthymol blue (pH 5.2–6.8).

Some sterile carbohydrate solutions (glucose, maltose, sucrose, lactose, dulcitol, mannitol, salicin) are available (*O*). Prepare others as a 10% solution in water and sterilize by filtration. Add 10 ml to each 100 ml of the reconstituted sterile basal medium and tube aseptically. Do not heat sugar carbohydrate media. Durham's (fermentation) tubes need be added to the glucose tubes only; gas from other substrates is not diagnostically significant.

These media are not suitable for some organisms; variations are given below.

Ammonium salt 'sugars'

Pseudomonads and spore bearers produce alkali from peptone water and the indicator may not change colour. Use this basal synthetic medium instead.

Ammonium dihydrogen phosphate	1.0 g
Potassium chloride	0.2 g
Magnesium sulphate	0.2 g
Agar	10.0 g
Water	1000 ml

Dissolve by heating, add 4 ml of 0.2% bromthymol blue and a final 1% of sterile (filtered) carbohydrate solution (0.1% aesculin; 0.2% starch; these are sterilized by steaming).

'Anaerobic sugars'

Peptone water or serum peptone water sugar media need to have a low oxygen tension for testing the reactions of anaerobes, even under anaerobic conditions. Add a clean wire nail to each tube before sterilization.

Baird–Parker's carbohydrate medium

For the sugar reactions of staphylococci and micrococci use Baird–Parker[4] formula.

Yeast extract	1 g
Ammonium dihydrogen phosphate	1 g
Potassium chloride	0.2 g
Magnesium sulphate	0.2 g
Agar	12 g
Water	1000 ml

Steam to dissolve and adjust to pH 7.0. Add 20 ml of 2% bromocresol purple and bottle in 95-ml amounts. Sterilize at 115 °C for 10 min. For use, melt and add 5 ml of 10% sterile carbohydrate.

Lactobacilli fermentation medium

For sugar reactions of lactobacilli make MRS base (p. 78) without Lemco and glucose and adjust to pH 6.2–6.5. Bottle in 100-ml amounts and autoclave at 115 °C for 15 min. Melt 100 ml, add 10 ml of 10% sterile (filtered) carbohydrate solution and 2 ml of 0.2% chlorophenol red. Tube aseptically.

Robinson's serum water sugars

Some organisms will not grow in peptone water sugars. Robinson's serum water medium is superior to that of Hiss.

Peptone	5 g
Disodium hydrogen phosphate	1 g
Water	1000 ml

Steam for 15 min, adjust to pH 7.4 and add 250 ml of horse serum. Steam for 20 min, add 10 ml of Andrade indicator and 1% of appropriate sugar (0.4% starch).

Unheated serum contains diastase and may contain a small amount of fermentable carbohydrate, so a buffer is desirable, particularly in starch fermentation tests.

Carbon assimilation media for yeasts

Ammonium sulphate	5 g
Magnesium sulphate	0.5 g
Potassium dihydrogen phosphate	1.0 g
Agar	1.5 g
Water	1000 ml

Dissolve by heating. Autoclave in 15-ml amounts at 115 °C for 15 min.

Decarboxylase medium (Moeller)[5]

Peptone (Evans)	5 g
Lab Lemco (*O*)	5 g
Pyridoxine HCl	5 mg
Glucose	0.5 g
Distilled water	990 ml

Dissolve and adjust to pH 6.0. Add 5 ml of 1:500 bromocresol purple and 2.5 ml of 1:500 cresol red. Make up to 1000 ml and divide into three batches. To one add 2 g% DL-lysine and adjust to pH 6.0 with 0.1 N NaOH. To another batch add 2 g% DL-ornithine and adjust to pH 6.0 with 0.1 N NaOH. The third is the control medium. Tube in 1.1-ml amounts in narrow tubes so that the column of medium is about 2 cm. Seal with a few millimetres of liquid paraffin, plug and steam for 1 h.

Medium to demonstrate arginine decarboxylase or dihydrolase is made in the same way.

Decarboxylase media (Falkow[6])

This formula is preferred for those organisms which require a less anaerobic medium than Moeller's medium.

Peptone	5 g
Yeast extract	3 g
Glucose	1 g
Water	1000 ml

Dissolve by heating, adjust to pH 6.7 and add 10 ml of 0.2% bromocresol purple. Bottle in 100-ml amounts and autoclave at 115 °C for 10 min.

To 100 ml add 0.5 g of the appropriate amino acid (lysine, ornithine, arginine) and tube. Ordinary-sized tubes or bottles can be used. Tube also one batch without amino acid as control. Sterilize by steaming.

Donovan's medium[7]

Originally devised to differentiate klebsiellas, this is useful for screening lactose fermenting Gram-negative rods.

Tryptone (*O*)	10 g
Sodium chloride	5 g
Triphenyltetrazolium chloride	0.5 g
Ferrous ammonium sulphate	0.2 g
1% aqueous bromothymol blue	3 ml
Inositol	10 g
Agar	3 g
Water	1000 ml

Dissolve by heating and adjust to pH 7.2. Distribute in 4-ml amounts and autoclave at 115 °C for 15 min. Allow to set as butts.

Dorset's egg medium

Wash six fresh eggs in soap and water and break into a sterilized basin. Beat with a sterile fork and strain through sterile cotton gauze into a sterile measuring cylinder. To 3 parts of egg add 1 part of nutrient broth. Mix, tube or bottle and inspissate in a sloped position for 45 min at 80–85 °C. Glycerol (5%) may be added if desired. (Griffith's egg medium uses saline instead of broth and is useful for storing stock cultures.)

Egg yolk agar (Lowbury and Lilly[8]**)**

This is selective for *Cl. perfringens* and shows the Nagler and pearly layer reactions.

Nutrient agar or blood agar, base, melted	100 ml
Fildes' extract	5 ml

Steam for 20 min to remove chloroform from Fildes' extract, cool to 55 °C, and add 10 ml of egg yolk suspension (*O*), 100–125 μg/ml of neomycin and pour plates.

Elek's medium[9] **(modified)**

Used for plate toxigenicity test on diphtheria bacilli.

Solution A

Evans peptone	20 g
Maltose	3 g

Lactic acid	0.7 g
Distilled water	500 ml

Heat to dissolve. Add 3.5 ml of 10 N NaOH. Mix well, heat to boiling, filter, adjust to pH 7–8.

Solution B

Agar	15 g
Sodium chloride	5 g
Distilled water	500 ml

Heat to dissolve. Adjust to pH 7.8.

Mix solutions *A* and *B* and bottle in 15-ml amounts. Autoclave at 115 °C for 10 min.

Ellner's medium[10]

This is used to persuade *Cl. perfringens* to form spores.

Peptone	10 g
Yeast extract	3 g
Starch, soluble	3 g
Magnesium sulphate ($MgSO_4.7H_2O$)	0.1 g
Disodium hydrogen phosphate ($Na_2HPO_4.7H_2O$)	50 g
Potassium dihydrogen phosphate	1.5 g
Water	1000 ml

Dissolve by heating, adjust to pH 7.8 and tube. Before use, steam to drive off oxygen, cool rapidly and inoculate heavily at the bottom of the tube with a pasteur pipette without introducing air.

Gardner's STAA medium[11]

For the isolation of *Brochothrix* from meat.

Peptone	20 g
Yeast extract (*O*)	2 g
Glycerol	15 g
K_2HPO_4	1 g
$MgSO_4.7H_2O$	1 g
Agar	13 g
Distilled water	1000 ml

Heat to dissolve, adjust to pH 6.0 and sterilize at 121 °C for 15 min. Tube in suitable amounts. Before use melt, cool to 50 °C and add these inhibitory agents to give the final concentrations stated: streptomycin sulphate to 500 μg/ml; actidione to 50 μg/ml; thallous acetate to 50 μg/ml.

Gelatin agar

Melt nutrient agar and add 5% bacteriological gelatin.

Gluconate broth

For differentiation among enterobacteria.

Yeast extract	1.0 g
Peptone	1.5 g
Dipotassium hydrogen phosphate	1.0 g
Potassium gluconate	40.0 g
(or sodium gluconate)	37.25 g
Water	1000 ml

Dissolve by heating, adjust to pH 7.0, filter if necessary, tube in 5–100-ml amounts and autoclave at 115 °C for 10 min.

Glucose phosphate medium

For MR and VP tests. Add 0.5 % each of glucose and K_2HPO_4 to commercial peptone broth.

Griffith's egg medium

For storing stock cultures, *see* Dorset's egg medium.

Hugh and Leifson's medium[12]

Peptone	2.0 g
Sodium chloride	5.0 g
Dipotassium hydrogen phosphate	0.3 g
Agar	3.0 g
Water	1000 ml

Heat to dissolve, adjust to pH 7.1 and add 15 ml of 0.2% bromthymol blue and sterile (filtered) dextrose to give a final 1% concentration. Tube aseptically in narrow (less than 1 cm) tubes in 8–10-ml amounts.

Baird-Parker[4] modification for staphylococci and micrococci:

Tryptone	10 g
Yeast extract	1 g
Glucose	10 g
Agar	2 g
Water	1000 ml

Steam to disolve, adjust to pH 7.2. Add 20 ml of 0.2% bromocresol purple, tube in 10-ml amounts in narrow (12 mm) tubes and sterilize at 115 °C for 10 min.

Kirchner's medium

For the culture and differentiation of mycobacteria.

$Na_2HPO_4.12H_2O$	19.0 g
KH_2PO_4	2.5 g
$MgSO_4.7H_2O$	0.6 g
Trisodium citrate	2.5 g
Asparagin	5.0 g
Glycerol	20.0 ml
Phenol red 0.4% aq.	3.0 ml
Distilled water to	1000.0 ml

Steam to dissolve. The pH should be 7.4–7.6. Bottle in 9-ml amounts and autoclave at 115 °C for 10 min. To each bottle at 1 ml of horse serum and antibiotics as desired.

Loeffler medium: inspissated serum

Mix 3 parts of horse serum with 1 part of glucose broth, tube and inspissate in slopes.

Lowbury and Lilly medium

See Egg yolk agar.

Lethheen agar (modified)[2]

For testing cosmetics.

Lethheen agar (*D*, *B*)	32 g
Trypticase peptone	5 g
Thiotone peptone	10 g
Yeast extract	2 g
NaCl	5 g
Sodium bisulphate	0.1 g
Distilled water	1000 ml

Dissolve the ingredients by boiling. Autoclave for 15 min at 121 °C and dispense into petri dishes.

Lethheen broth (modified)[2]

Prepare as above but use Lethheen broth (*D*, *B*). Dispense into dilution bottles.

Lowenstein–Jensen medium

This is used for the isolation of mycobacteria.

KH_2PO_4 (AnalaR)	4.0 g
$MgSO_4.7H_2O$ (AnalaR)	0.4 g
Magnesium citrate	1.0 g
Asparagin	6.0 g
Glycerol (AnalaR)	20.0 ml
Distilled water to	1000 ml

Dissolve in this order, steam for 2 h.

Clean fresh eggs with soap and water and break them into a sterile graduated cylinder. Take 1600 ml of egg fluid, mix in a screw-capped jar or polypropylene container with some large glass beads to break the yolks, filter through gauze and add to 1 litre of salt mixture. Add 50 ml of 1% aqueous malachite green, tube, slope and inspissate.

A 4-ml volume of pyruvic acid (neutralized with 3 N sodium hydroxide solution) can replace glycerol.

Malt agar modifications

To isolate acidophilic yeasts, cool the sterilized medium to 50 °C, add 1 ml of sterile 10% lactic acid and pour plates at once.

For dermatophytes or pathogenic yeasts add 10 ml of 5% cycloheximide in acetone and 10 ml of 0.5% chloramphenicol in alcohol. Tube in 10-ml amounts and autoclave at 115 °C for 10 min.

To suppress bacterial growth and isolate moulds, add only the chloramphenicol.

Mead's medium[13]

This medium allows *S. faecalis* to be recognized by its ability to ferment sorbitol, decompose tyrosine and reduce triphenyltetrazolium chloride. Cultures are incubated at 45 °C.

Peptone	10 g
Yeastrel	1 g
Sorbitol	2 g
Tyrosine	5 g
Agar	12 g
Water	1000 ml

Dissolve by autoclaving at 115 °C for 10 min and adjust to pH 6.2. Add 0.1 g of triphenyltetrazolium chloride and 1 g of thallous acetate. When these have dissolved, add a further 4 g of tyrosine to give a suspension. Cool to 50 °C, mix to give an even suspension of tyrosine and pour plates. Layered plates, the lowest layer of the same medium without tyrosine, are best.

Milk salt agar[14]

This is used for the recovery of *S. aureus* from food samples[14].

Peptone	5 g
Lab Lemco	3 g
Sodium chloride	6.5 g
Agar	15 g

Sterilize at 121 °C for 15 min. Add 10 % skim milk (*B*, *O*) and pour plates.

MRS (de Man *et al.*[15]) medium

Lab Lemco	10 g
Yeast extract	5 g
Peptone	10 g
Dipotassium hydrogen phosphate	2 g
Sodium acetate	5 g
Magnesium sulphate ($MgSO_4.7H_2O$)	0.2 g
Manganese sulphate ($MnSO_4.4H_2O$)	0.05 g
Tween 80	1 ml
Triammonium citrate	2 g
Water	1000 ml

Dissolve by heating, bottle and autoclave. For use, add 2% sterile (filtered) glucose or other carbohydrate solution.

N medium (Collins[16]**)**

This is used in the differentiation of mycobacteria.

NaCl	1.0 g
$MgSO_4.7H_2O$	0.2 g
KH_2PO_4	0.5 g
$Na_2HPO_4.12H_2O$	3.0 g
$(NH_4)_2SO_4$	10.0 g
Glucose	10.0 g
Water	1000 ml

Dissolve in that order, adjust to pH 6.8–7.0 and distribute. Autoclave at 115 °C for 10 min. This medium can be modified by the addition of 1.2% agar and 0.04% bromocresol purple.

Niger seed agar

For the differentiation of cryptococci.

Grind 70 g of niger seed (*Guizotia abyssinica*) in a blender and add to 350 ml of water. Extract by autoclaving at 115 °C for 15 min. Filter through cotton wool and then paper to remove oil and debris.

Extract	100 ml
Glucose	5 g
Creatinine	0.4 g
Chloramphenicol (0.5% in ethanol)	5 ml
Agar	7.5 g
Water	400 ml

Autoclave at 115 °C for 15 min, cool to 50 °C and add 50 ml of biphenyl dissolved in 5 ml of ethanol. Distribute in small bottles and store in a refrigerator.

Nitrogen assimilation media for yeasts

Glucose monohydrate	20 g
Magnesium sulphate	0.5 g
Potassium dihydrogen phosphate	1.0 g
Agar	1.5 g
Water	1000 ml

Dissolve by heating. Autoclave in 15-ml portions at 115 °C for 15 min.

Nutrient gelatin

Choose a high-quality gelatin and add 12–15% to a nutrient, Lemco or yeast extract broth. Heat to dissolve at 115 °C for 10 min.

ONPG broth

o-Nitrophenyl-D-galactopyranoside	6 g
0.001 M Na_2HPO_4	1000 ml

Tube in 2-ml amounts. Do not heat.

Phenolphthalein phosphate agar

Phenolphthalein phosphate agar for the phosphatase test is made by adding 1 ml of a 1% solution of phenolphthalein phosphate to 100 ml of nutrient agar or commercial blood agar base and pouring plates. Add 125 units of polymyxin B sulphate for the recovery of *S. aureus* from foods, etc.

Phenolphthalein sulphate agar

This medium for the aryl sulphatase test is made by dissolving 0.64 g of potassium phenolphthalein sulphate in 100 ml of water (0.01 M). Tube one of the Dubos or Middlebrook Broth media in 2.7-ml amounts and add 0.3 ml of the phenolphthalein sulphate solution to each tube.

Robertson's cooked meat medium

This is a useful general purpose medium and is used for the cultivation of anaerobes.

Mince 500 g of fresh lean beef and simmer for 1 h in 500 ml of boiling water containing 2 ml of 1 N NaOH. This neutralizes the acid in the medium; filter, dry the meat in a cloth and when dry put it into 28-ml screw-capped bottles so that there is about 25 mm of meat in each. Add 10–12 ml of infusion or Lemco broth to each tube, cap and autoclave at 115 °C for 10 min.

Salt agar

For cultivation of plague bacilli. Add 3% sodium chloride to nutrient agar.

Salt meat broth

Advantage is taken of the salt-tolerance (halophily) of *S. aureus* in investigating outbreaks of food poisoning due to this organism. Many other organisms are inhibited by high salt concentrations.

Prepare Robertson's cooked meat medium as above but add 10% sodium chloride to the liquid before tubing.

Semisolid oxygen preference medium

For identifying mycobacteria. Add 0.1% agar to one of the fluid Middlebrook liquid media (*D*). Autoclave at 115 °C for 10 min. Tube in 12-ml amounts in narrow bottles. Before use add 0.5 ml of OADC supplement (*D*).

Skim milk agar

Reconstitute skim milk powder (*B*, *O*) and mix equal volumes of this and double-strength nutrient agar melted and at 55 °C. Mix and pour plates.

'Sloppy agar' Craigie tubes

Broth containing 0.1–0.2% agar will permit the migration of motile organisms but will not allow convection currents. This permits motile organisms to be separated from non-motile bacteria in the Craigie tube.

Tube sloppy agar in 12-ml amounts in screw-capped bottles. To each bottle add a piece of glass tubing of dimensions 50 × 5 or 6 mm. There must be sufficient clearance between the top of the tube and the surface of the medium so that a meniscus bridge does not form and the only connection between the fluid and the inner tube and outside it is through the bottom.

Starch agar

Prepare a 10% solution of soluble starch in water and steam for 1 h. Add 20 ml of this solution to 100 ml of melted nutrient agar and pour plates.

Sucrose agar

To demonstrate, levan or dextran formation, add 5% sucrose to 100 ml of melted nutrient agar. Steam for 30 min, cool to 55 °C, add 5 ml of sterile horse serum and pour plates.

Tellurite blood agar media for corynebacteria

The addition of 0.04% potassium tellurite to blood agar medium inhibits many other organisms and permits differentiation of *gravis*, *mitis* and *intermedius* types of *C. diphtheriae* and other corynebacteria. The base medium should be an infusion, not a digest medium, and Hoyle's base (*O*) gives good results. For recognition and differentiation at 18–24 h, brain-veal infusion agar (*D*) is excellent. Some workers prefer lysed to fresh blood. Horse blood may be lysed by alternate freezing and thawing or by the addition of 5 ml of 2% saponin solution to 100 ml of blood.

Potassium tellurite is stored as a 2% solution.

Thioglycollate medium 135c[2]

For cosmetic testing.

Prepare commercial medium (BBL) according to the manufacturer's instructions. Before autoclaving add 0.5 ml of 1% haemin and 0.5 ml of 1% vitamin K/litre. Mix, dispense and autoclave at 121 °C for 15 min. Tighten screw caps after cooling and store in a dark room at room temperature. Use only freshly prepared medium.

Tributyrin agar

Fat-splitting organisms cause spoilage in butter, etc. Most of these bacteria split glyceryl tributyrate (tributyrin).

Peptone	5 g
Yeast extract	3 g
Tributyrin	10 g
Agar	12 g
Water	1000 ml

Heat to dissolve, adjust to pH 7.5, tube or bottle and autoclave at 115 °C for 15 min.

Colonies of lipolytic organisms clear the medium.

Tyrosine agar: xanthine agar

To 100 ml of melted nutrient agar, add 5 g of tyrosine or 4 g of xanthine and steam for 30 min. Mix well to suspend amino acid and pour plates. *See also* Mead's medium.

Willis and Hobbs medium[17]

This excellent medium for isolating and identifying clostridia sometimes needs minor modification to give the best results for particular species. Nagler reaction, 'pearly layer' proteolysis and lactose fermentation can be observed.

Nutrient broth	1000 ml
Lactose	12 g
Agar	12 g
Water	1000 ml

Heat to dissolve, adjust to pH 7.0, add 3.5 ml of 1% neutral red and 1 g of sodium thioglycollate, bottle in 100-ml portions and autoclave at 115 °C for 10 min.

To each 100 ml of melted base add 4 ml of egg yolk emulsion (*O*) and 15 ml of sterile whole milk. Pour plates.

Neomycin may be added to inhibit non-spore-bearing anaerobes and aerobic spore bearers but should be used with caution. It is desirable to experiment with varying concentrations up to 100 μg/ml.

Diluents

Saline and Ringer solutions

Physiological saline, which has the same osmotic pressure as micro-organisms, is a 0.85 % solution of sodium chloride in water. Ringer solution, which is ionically balanced so that the toxic effects of the metallic ions neutralize one another, is better than saline for making bacterial suspensions. It is used at one-quarter the original strength.

Sodium chloride (AnalaR)	2.15 g
Potassium chloride (AnalaR)	0.075 g
Calcium chloride, anhydrous (AnalaR)	0.12 g
Sodium thiosulphate pentahydrate	0.5 g
Distilled water	1000 ml

The pH should be 6.6. Tablets of sodium chloride and Ringer are useful (*O*).

Calgon Ringer

Alginate wool, used in surface swab counts, dissolves in this.

Sodium chloride	2.25 g
Potassium chloride	0.105 g
Calcium chloride	0.12 g
Sodium hydrogen carbonate	0.05 g
Sodium hexametaphosphate	10.00 g
Water	1000 ml

The pH is 7.0. It is conveniently purchased as tablets (*O*).

Thiosulphate Ringer

This may be preferred for dilutions of rinses where there may be residual chlorine. Conveniently manufactured in tablet form (*O*).

Peptone water diluent

The safest and least lethal of diluents is 0.1% peptone water.

Phosphate buffered saline

This is a useful general diluent at pH 7.3.

Sodium chloride	8.0 g
Potassium dihydrogen phosphate	0.34 g
Dipotassium hydrogen phosphate	1.21 g
Water	1000 ml

Indicators

Andrade

Dissolve 0.5 g of acid fuchsin in 100 ml of water. Add 1 N NaOH solution until the colour changes to yellow. The indicator is colourless at pH 7.2, pink in acidic solutions and yellow in alkaline solutions.

Bromocresol purple

Dibromo-*o*-cresolsulphonphthalein. Dissolve 0.1 g in 9.2 ml of 0.02 N NaOH and dilute to 250 ml with water. The indicator (0.4% solution) is yellow at pH 5.8 and blue at 6.8.

Bromothymol blue

Dibromothymolsulphonphthalein. Dissolve 0.1 g in 8 ml of 0.02 N NaOH and make up to 250 ml with water. This indicator (0.4% solution) is yellow at pH 6.0 and blue at pH 7.6.

Congo red

This indicator (0.1% solution in water) is blue at pH 3.0 and red at pH 5.2.

Methyl red

4′-Dimethylaminoazobenzene-4-sulphonate. Dissolve 0.1 g in 300 ml of ethanol and add 200 ml of water. The solution is red at pH 4.4 and yellow at pH 6.2.

Phenolphthalein

Dissolve 1 g in 100 ml of 95% ethanol. The solution is colourless at pH 8.3 and red at pH 10.

Phenol red

Phenolsulphonphthalein. Dissolve 0.1 g in 14.1 ml of 0.02 N NaOH and dilute to 250 ml with water. This indicator (0.4% solution) is yellow at pH 6.8 and red at pH 8.4.

Thymol blue

Thymolsulphonphthalein. Dissolve 0.1 g in 10.75 ml of 0.02 N NaOH and dilute to 250 ml with water. It has an acid range, red at pH 1.2 and yellow at 2.8, and an alkaline range, yellow at pH 8.0 and blue at pH 9.6.

Preparation of culture media

Utensils

Copper or zinc containers must not be used. Small amounts of these metals will dissolve in the culture media and are bactericidal. Large amounts of media can be made in the stainless steel buckets that are used in the dairy and food trades. Smaller amounts can be made in resistance-glass laboratory flasks. Large stainless steel funnels can be used for filtration, with folded filter-papers of the heavy Chardin type. Filtration and adjustment of pH should be completed before agar is added. The present-day agars give a clear gel but if it is necessary to filter agar medium, cut a sheet of heavy filter-paper to fit a large Büchner funnel and place wet filter-paper pulp on it to a depth of 5 mm. Place the funnel on a filter flask and suck the pulp dry. Heat the whole apparatus in a steamer in order to prevent gelling of the medium in the filter.

Inhibitory substances

Metals such as zinc and copper (*see above*) and other inhibitory substances such as fatty acids may be present in peptones, agar, carbohydrates and other chemicals which are not specifically made for culture media. It is inadvisable, for reasons of cost and convenience, to use any material that is not specified as being of 'bacteriological quality' and purchasing officers should be severely discouraged from seeking, for reasons of 'economy', suppliers of media ingredients, including chemicals and dyes, who are not known to bacteriologists.

Distilled water from commercial bulk supplies, 'factory' or 'battery' quality may contain substances that inhibit bacterial growth, e.g. oily materials from stripper stills attached to steam lines. Manesty-type distilled water, glass-distilled water and deionized water should be used, although it must be realized that some trace elements necessary for bacterial growth are in fact supplied by the distilled water, by glassware, or by recognized impurities in analytical quality reagents.

Glucose, when autoclaved with salts such as phosphate, may yield inhibitory substances. It is best to add carbohydrates as sterile solutions after the medium has been sterilized.

Some cotton wools, when sterilized in a hot-air oven, yield fatty materials.

Sometimes these materials can be seen as condensates on glass tubes. Again, cotton wool from a reliable source should be used.

Excessive heating and re-melting may destroy growth factors and gelling capacity and cause darkening or pH drift of culture media.

Adjustment of pH

The pH of reconstituted and laboratory-prepared media should be checked with the meter, but for practical purposes devices such as the BDH Lovibond Comparator is good enough.

Pipette 10 ml of the medium into each of two 152 × 16 mm test-tubes. To one add 0.5 ml of 0.04% phenol red solution. Place the tubes in a Lovibond Comparator with the blank tube (without indicator) behind the phenol red colour disc, which must be used with the appropriate screen. Rotate the disc until the colours seen through the apertures match. The pH of the medium can be read in the scale aperture. Rotate the disc until the required pH figure is seen, add 0.05 N sodium hydroxide solution or hydrochloric acid to the medium plus indicator tube and mix until the required pH is obtained.

From the amount added, calculate the volume of 1 or 5 N alkali or acid that must be added to the bulk of the medium. After adding, check the pH again.

Example 10 ml of medium at pH 6.4 requires 0.6 ml of 0.05 N sodium hydroxide solution to give the required pH of 7.2.

Then the bulk medium will require

$$\frac{0.6 \times 1000}{20 \times 10} \text{ ml of 1 N NaOH/litre} = 3.0 \text{ ml}$$

Discs of other pH ranges are available.

All final readings of pH must be made with the medium at room temperature because hot medium will give a false reaction with some indicators.

Identification and storage

Most laboratories employ a colour code to identify media that look alike. Coloured beads may be placed in bottles of bulk medium. Coloured cotton wool or coloured aluminium or polypropylene caps (*see* p. 27) can be used for tubed media, and the caps of bottled media can be dabbed with coloured enamel paint. This is particularly useful for carbohydrate fermentation test media.

The labelling machines used to place prices on goods in supermarkets are very useful for labelling and dating tubes, bottles and petri dishes of culture media.

Bulk bottled media in screw-capped bottles can be stored at room temperature.

Filling, tubing and pouring culture media

Liquid media can be tubed with the aid of a large funnel held in a retort stand and fitted with a short piece of rubber tubing and a spring clip. The semi-automatic commercial fillers (p. 25) are more satisfactory.

To make agar slopes, tube the medium in 3–7-ml amounts according to the size of the tube required. After sterilization and before the medium sets, slope the tubes individually on the bench by leaning them against a length of glass or metal rod 6 or 7 mm in diameter. When making large batches, slope the tubes or bottles on wire racks, e.g. old refrigerator shelves, sloped at a convenient angle.

When pouring plates, raise the lid only far enough to permit the mouth of the tube or bottle to enter. Pour about 12–15 ml in each plate. Dry the plates slightly open in an incubator, and store them medium side-up in a refrigerator. An automatic plate pouring machine (p. 25) is useful in laboratories where many plates are used. This reduces the number of plates contaminated with airborne organisms to almost nil.

Contamination of sterile media during tubing and pouring may be a serious problem in some buildings. It can be minimized by using clean air cabinets of the simple, horizontal outflow variety, or (more expensive) a Class II microbiological safety cabinet.

Quality control and performance tests

Manufacturers of certain media and their ingredients impose stringent controls on the quality and performance of their products. Reconstitution and handling of these media in the laboratory (e.g. in overheating) may reduce its efficiency. Laboratory-prepared media are rarely tested.

Some kind of quality control should therefore be imposed on all materials used. The national system of quality control (in the UK) and performance tests (in the USA) operate satisfactorily only in clinical laboratories and problems arise mainly in the smaller industrial and food control establishments. A full scale system such as that described by Anhalt[20] is expensive in staff and materials and entails the maintenance of a large number of stock cultures. Two simpler systems are described here, one for isolation and selective media, the other for identification media.

Isolation and selective media: efficiency of plating technique (EOP)

New batches of culture media may vary considerably and should be tested in the following way.

Prepare serial tenfold dilutions, for example, 10^{-2} to 10^{-7} of cultures of various organisms which will grow on or be inhibited by the medium. For example, when testing DCA use *S. sonnei*, *S. typhi*, *S. typhimurium*, several other salmonellas and *E. coli*. Use at least four well dried plates of the test medium for each organism and at least two plates each of a known satisfactory control medium, and of a general medium, e.g. nutrient or MacConkey agar. Do Miles and Misra drop counts with the serial dilutions of the organisms on these plates so that each plate is used for several dilutions. (*Figure 4.1*) (*see* p. 133).

Count the colonies, tabulate the results and compare the performances of the various media. These may suggest that in a new batch of a special medium it is necessary to alter the proportion of certain ingredients.

Stock cultures, however, will frequently grow on media which will not

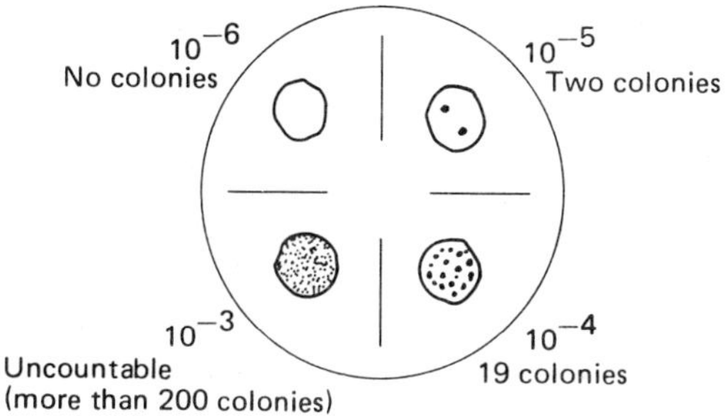

Figure 4.1 Miles and Misra count

support the growth of 'damaged' organisms from natural materials. It is best, therefore, to use dilutions or suspensions of such material.

With some culture media, these drop counts do not give satisfactory results because of reduced surface tension. An alternative procedure is recommended.

Make cardboard masks to fit over the tops of petri dishes and squares 25 × 25 mm in the centre of each. Place a mask over a test plate and drop one drop of the suspension through the square on the medium. Spread the drop over the area limited by the mask. This method has the additional advantage of making colony counts easier to perform but more plates must, of course, be tested.

In addition to these EOP tests, ordinary plating methods should be used to compare colony size and appearance.

Identification media

Maintain stock cultures (p. 94) of organisms known to give positive or negative reactions with the identification tests in routine use. The same organism may often be used for a number of tests. When a new batch of any identification medium is anticipated, subculture the appropriate stock strain so that a young, active culture is available. Test the new medium with this culture.

References

1. *Oxoid Manual* (1982) 5th edn. Basingstoke: Oxoid Ltd
2. FOOD AND DRUGS ADMINISTRATION (1978, 1979) *Bacteriological Analytical Manual.* 5th edn. Washington DC; Association of Official Analytical Chemists
3. DAVIS, J.G. (1931) *Journal of Dairy Research*, **3,** 133
4. BAIRD-PARKER, A.C. (1966) Methods for classifying staphylococci and streptococci. In *Identification Methods for Microbiologists*, Part A. Edited by B.M. Gibbs and F.A. Skinner. p. 59. London: Academic Press
5. MOELLER, V. (1955) *Acta pathologia et microbiologia scandinavica*, **36,** 161
6. FALKOW, S. (1958) *American Journal of Clinical Pathology*, **29,** 598
7. DONOVAN. T.J. (1966) *Journal of Medical Laboratory Technology*, **23,** 194
8. LOWBURY, E.J.L. and LILLY, H.A. (1955) *Journal of Pathology and Bacteriology*, **70,** 105
9. DAVIES, J.R. (1974) Elek's test for the toxogenicity of Corynebacteria. In *Laboratory Methods 1.* Public Health Laboratory Service Monograph No. 5. London: HMSO

10. ELLNER, P.D. (1956) *Journal of Bacteriology*, **71,** 495
11. GARDNER, G.A. (1966) *Journal of Applied Bacteriology*, **29,** 455
12. HUGH, R. and LEIFSON, F. (1953) *Journal of Bacteriology*, **66,** 24
13. MEAD, G.C. (1963) *Nature, London*, **197,** 1323
14. GILBERT, R.J., KENDALL, M. and HOBBS, B.C. (1969) Media for the isolation and enumeration of coagulase-positive staphylococci from foods. In *Isolation Methods for Microbiologists.* Edited by D.A. Shapton and G.W. Gould, pp. 9–14. London: Academic Press
15. DE MAN, J.C., ROGOSA, M. and SHARPE, M.E. (1960) *Journal of Applied Bacteriology*, **23,** 130
16. COLLINS, C.H. (1962) *Tubercle, London*, **43,** 292
17. WILLIS, A.T. and HOBBS, G. (1959) *Journal of Pathology and Bacteriology*, **77,** 511
18. ANHALT, J.P. (1981) Quality control. In *Laboratory Procedures in Clinical Microbiology.* Edited by J.A. Washington. New York: Springer-Verlag

Chapter 5

Cultural methods

General bacterial flora

To observe the general bacterial flora of a sample of any material, use non-selective media. No one medium nor any one temperature will support the growth of all possible organisms. It is best to use several different media, each favouring a different group of organisms, and to incubate at various temperatures aerobically and anaerobically. The pH of the medium used should approximate that of the material under examination.

Nutrient agar, one of the media for fastidious organisms with or without the addition of blood, glucose tryptone agar, skim milk agar and Rogosa agar are suitable media.

Methods for estimating bacterial numbers are given in Chapter 9, and the dilution methods employed can be used to make inocula for the following techniques. Three methods are given here for plating material. They depend on separating viable units (*see* p. 128) of bacteria so that each will grow into a separate colony. These colonies can be subcultured or picked for further examination.

Pour plate method

Melt several tubes containing 15 ml of medium. Place in water-bath at 45–50 °C to cool. Emulsify the material to be examined in 0.1% peptone water and prepare 1:10, 1:100 and 1:1000 dilutions in the same solution (*see* p. 131). To 15 ml of melted medium at 45 °C add 1 ml of a dilution, mix by rotating the tube between the palms of the hands and pour into a petri dish. Make replicate pour plates of each dilution for incubation at suitable temperatures. Allow the medium to set, invert the plate and incubate. This method gives a better distribution of colonies than that used for the plate count (p. 131), but cannot be used for that purpose because some of the inoculum, diluted in the agar, remains in the test-tube.

Spreader method

Dry plates of a suitable medium. Make dilutions of the emulsified material as described above. Place about 0.05 ml (1 drop) of dilution in the centre of a plate and spread it over the medium by pushing the glass spreader backward and forward while rotating the plate. Replace the petri dish lid and leave for 1–2 h to dry before inverting and incubating as above.

Looping-out method

Make dilutions of material as described above. Usually the neat emulsion and a 1:10 dilution will suffice. Place one loopful of material on the medium near the rim of the plate and spread it over the segment (*Figure 5.1*). Flame the loop and spread from area *A* over area *B* with parallel streaks, taking care not to let the streaks overlap. Flame the loop and repeat with area *C*, and so on. Each looping-out dilutes the inoculum. Invert the plates and incubate.

When looping-out methods are used, two loops are useful; one is cooling while the other is being used.

These surface plate methods are said to give better isolation of anaerobes than pour plates.

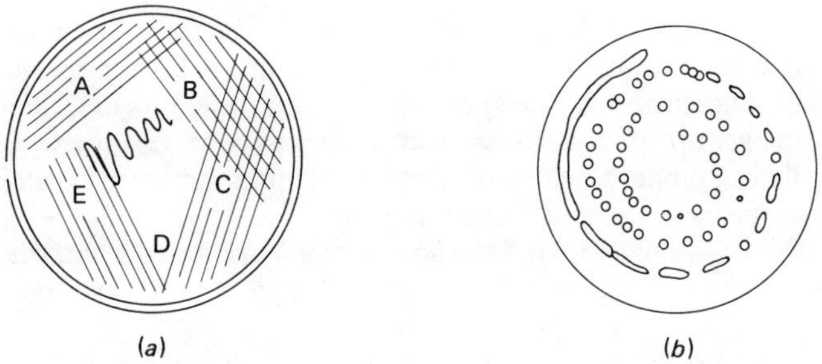

Figure 5.1 (a) 'Spreading' or 'looping-out' a plate; (b) rotary plating

Spiral or rotary plating[1,2,3]

This mechanical method is invaluable for plating large numbers of specimens or cultures and for doing sensitivity tests by Stokes method (p. 171). Hand-operated (with electrically-driven turntable) and fully-automated models are available. With the latter, exact volumes may be plated. A laser colony counter is on the market (Whitley, Horwell).

To use the hand-operated machine place the open plate on the turntable. Touch the rotating plate with the charged loop at the circumference of the medium and draw it slowly towards the centre. A spiral of bacterial growth will be obtained after incubation which will show discrete colonies (*Figure 5.1*).

Multipoint inoculators

These are convenient tools when many replicate cultures are needed[3]. They are fully or semi-automatic and spot-inoculate large numbers of petri dish cultures each with 10–20 cultures. They are particularly useful for antibiotic sensitivity testing and screeening (Mast, Oxoid).

Shake tube cultures

These are useful for observing colony formation in deep agar cultures, especially of anaerobic or microaerophilic organisms.

Tube media, glucose agar or thioglycollate agar, in 15–20-ml amounts in bottles or tubes 20–25 mm in diameter. Melt and cool to approximately 45 °C. Add about 0.1 ml of inoculum to one tube, mix by rotating between the palms, remove one loopful to inoculate a second tube, and so on. Allow to set and incubate. Submerged colonies will develop and will be distributed as follows: obligate aerobes grow only at the top of the medium, obligate anaerobes only near the bottom; microaerophiles grow near but not at the top; facultative organisms grow uniformly throughout the medium.

Stab cultures

These can be used to observe motility and gelatin liquefaction. The medium is tubed in 5-ml amounts in tubes or bottles of 12.5 mm diameter and inoculated by stabbing the wire down the centre of the agar.

Subcultures

For further examination, pure cultures are prepared from the various types of colonies in mixed cultures.

Place the plate culture under a low-power binocular microscope and select the colony to be examined. With a flamed straight wire (*not* a loop) touch the colony. There is no need to dig or scrape; too vigorous handling may result in contamination with adjacent colonies or with those beneath the selected colony.

With the charged wire, touch the medium on another plate, flame the wire and then use a loop to loop-out the culture to obtain individual colonies. This should give a pure culture but the procedure may be repeated.

To inoculate tubes or slopes, follow the same procedure. Hold the charged wire in one hand, the tube in the other in a nearly horizontal position. Remove the cap or plug of the tube by grasping it with the little finger of the hand holding the wire. Pass the mouth of the tube through a bunsen flame to kill any organisms that might fall into the culture and inoculate the medium by drawing the wire along the surface of a slope or touching the surface of a liquid. Replace the cap or plug and flame the wire. Several tubes can be inoculated in succession without recharging the wire.

Anaerobic culture

Anaerobic jars

Modern anaerobic jars (Oxoid, Whitley) are made of metal or plastic, are vented by Schrader valves and use 'cold' catalysts. They are therefore safer than earlier models: internal temperatures and pressures are lower. More catalyst is used and the escape of small particles of catalyst, which can ignite hydrogen–air mixtures is prevented. Nevertheless the catalyst should be kept dry, dried after use and replaced frequently.

For ordinary clinical laboratory work the commercial sachets are a convenient source of hydrogen and hydrogen–carbon dioxide mixtures (Gas Kit, Oxoid; Gas Pak, BBL).

Research laboratories prefer to use compressed hydrogen. The cylinder must be fitted with a suitable reducing valve and a pressure gauge. Mixtures of hydrogen and carbon dioxide or these gases mixed with nitrogen are safer but the cylinders containing carbon dioxide must be used horizontally, not vertically. A football bladder is filled from the cylinder and connected to the anaerobic vessel via a wash-bottle so that the passage of hydrogen can be observed.

Place the plates 'upside down', i.e. with the media at the bottom in the vessel. If they are incubated the right way up, the medium sometimes falls into the lid when the pressure is reduced. Dry the catalyst capsule by flaming and replace the lid. Connect to a vacuum pump across a manometer and remove air from the vessel until there is a negative pressure of about −300 mmHg. Close the tap on the vessel and connect it to the hydrogen bladder via a wash-bottle. Open the tap slowly and observe hydrogen bubbling in. After the pressure is equalized, residual oxygen will be removed by the catalyst and hydrogen and the gas will bubble very slowly through the bottle, usually at about one bubble per second. At this stage, if time is short, the tap can be closed and the apparatus disconnected and incubated. The reaction will continue.

Indicators of anaerobiosis

Indicators are provided with the commercial systems, and may be purchased separately. In clinical laboratories, the blood agar cultures themselves are often regarded as indicators of anaerobiosis: the haemoglobin is reduced and the plates have a purple-red instead of bright red appearance. This is not a very reliable method; the haemoglobin may appear to be reduced in conditions not anaerobic enough for some bacteria.

Anaerobic cabinets

These are highly desirable in laboratories where much of the work is done with anaerobes. The culture media is pre-reduced, inoculated and incubated in an inert, oxygen-free atmosphere. Most cabinets resemble bacteriologists' safety cabinets or glove boxes and are engineered to a high standard (Whitley, Horwell, Raven, Heinecke).

Culture under carbon dioxide

Some microaerophiles grow best when the oxygen tension is reduced. Capnophiles require carbon dioxide. For some purposes, candle jars are adequate. Place the cultures in a biscuit or dried-milk tin with a lighted candle or night-light. Replace the lid. The light will go out when most of the oxygen has been removed and replaced with carbon dioxide.

Alternatively, place 25 ml of 0.1 N HCl in a 100-ml beaker in the container and add a few marble chips before replacing lid.

Commercial carbon dioxide sachets or the hydrogen–carbon dioxide sachets may be used or carbon dioxide from a cylinder and an anaerobic jar. Remove air from the jar down to −76 mmHg and replace with carbon dioxide from a football bladder. This is the best method for capnophiles, which require 10–20% of carbon dioxide to initiate growth, but if extensive work of this nature is carried out, a carbon dioxide incubator is invaluable. A nitrogen–carbon dioxide mixture can be used instead of pure carbon dioxide.

Incubation of cultures

Incubators are discussed in Chapter 2. Instead of using an incubator for culturing at 20–22 °C, plates and tubes may be left on the bench, but a cupboard is better as draughts can be avoided. A maximum and minimum thermometer is necessary. Ambient temperatures often fluctuate widely when windows are open or heating is switched off at night.

Psychrophiles are usually incubated at 4–7 °C. A domestic refrigerator can be used provided that it is not opened too often. Temperatures of 20–25 °C are widely used for food bacteria and 35–37 °C for medical bacteria. Thermophiles require 55–60 °C.

Petri dish cultures are incubated upside down, i.e. medium half uppermost (except in anaerobic vessels), otherwise condensation water collects on the surface of the medium and prevents the formation of isolated colonies.

If it is necessary to incubate petri dish cultures for several days, they must be sealed in order to prevent the medium drying up. Plastic bags can be used.

Colony and cultural appearances

These appearances and the terms used to describe them are illustrated in *Figure 5.2*.

Examine colonies and growths on slopes with a hand-lens or plate microscope. Observe pigment formation both on top of and under the surface of the colony and note if pigment diffuses into the medium. In broth culture, surface growth may occur in the form of a pellicle. Some cultures develop characteristic odours.

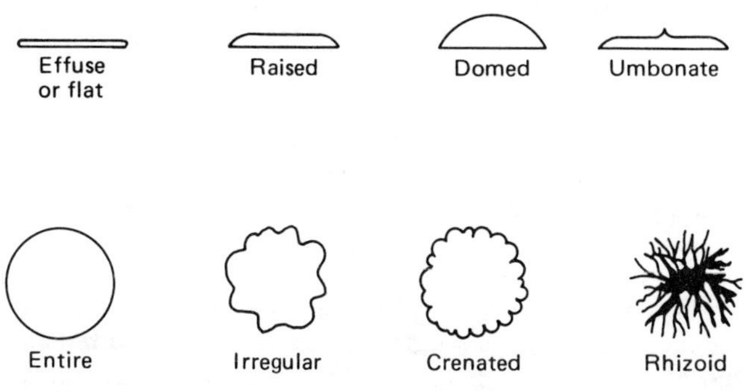

Figure 5.2 Description of colonies and cultures

Auxanograms

These are used to investigate the nutritional requirements of bacteria and fungi, for example, which amino acid or vitamin substance is required by 'exacting strains' or which carbohydrates the organisms can utilize as carbon sources.

Make a dense suspension of the organism to be tested in phosphate-buffered saline. Centrifuge, re-suspend in fresh phosphate-buffered saline. Wash again in the same way and add 0.1 ml to several 15-ml amounts of basal medium (p. 69) with indicator. Pour into plates. On each plate place a filter-paper disc of 5-mm diameter and on the filter-paper place one loopful of a 5% solution of the appropriate carbohydrate. Incubate and observe growth. Prepared discs are sold by BBL and Difco.

When testing for amino acid or vitamin requirements in this way, use the basal medium plus glucose. Several discs, each containing a different amino acid, can be used on a single plate.

Stock cultures

Most laboratories need stock cultures of a few 'standard strains' of micro-organisms that are used for testing culture media, controlling certain tests or for demonstration purposes. These must be subcultured periodically to keep them alive, with the consequent risk of contamination, and, more important in laboratories engaged on assay work, the involuntary selection of mutants. It is because of mutations in bacterial cultures that the so-called 'standard strains' or 'type cultures' so often vary in behaviour from one laboratory to another.

To overcome these difficulties, Type Culture Collections are maintained by government laboratories. In these establishments, the cultures are preserved in a freeze-dried state, and in the course of maintenance of the stocks they are tested to ensure that they conform to the official descriptions of the standard or type strains.

Cultures of micro-organisms may be obtained from: the National Collection of Type Cultures (NCTC), Central Public Health Laboratory, Colindale Avenue, London NW9 5HT; the National Collection of Industrial and Marine Bacteria (NCIMB), Torry Research Station, PO Box 13, Abbey Road, Aberdeen AB9 8DG; the American Type Culture Collection (ATCC), 12301 Parklawn Drive, Rockville, Maryland 20852; the National Collection of Dairy Organisms (NCDO), National Institute for Research in Dairying, Shinfield, near Reading, Berkshire; the National Collection of Pathogenic Fungi (NCPF), PHLS Mycology Reference Laboratory, London School of Hygiene and Tropical Medicine, Gower Street, London WC1E 7HT; the Commonwealth Mycological Collection (CMC), Ferry Lane, Kew, Surrey; the National Collection of Yeast Cultures (NCYC), Food Research Institute, Colney Lane, Norwich, Norfolk NR4 7UA. A fee is usually charged. These institutions also keep lists of private and commercial collections. Strains that are to be used occasionally, or rarely, should be obtained when required and not maintained in the laboratory. Strains that are in constant use may be kept for short periods by serial subculture, for longer periods by one of the drying methods and indefinitely by freeze-drying. When stocks of dried or freeze-dried cultures need renewing, however, it is advisable to start again with a culture from one of the National Collections rather than perpetuate one's own stock.

Serial subculture

The principle is to subculture as infrequently as possible in order to keep the culture alive and to arrest growth as early as possible in the logarithmic phase so as to avoid the appearance and possible selection of mutants. Useful media for keeping stock cultures of bacteria are Griffith's egg, Robertson's cooked meat and litmus milk. Fungi can be maintained on Sabouraud medium. These media must be in screw-capped bottles. It is difficult to keep cultures of some of the very fastidious organisms, for example gonococci, and it is advisable to have these freeze-dried.

Inoculate two tubes, incubate until growth is just obvious and active, for example, 12–18 h for enterobacteria, other Gram-negative rods and micrococci, two or three days for lactobacilli. Cool the cultures and store them in a refrigerator for 1–2 months.

Keep one of these cultures, *A*, as the stock, and use the other, *B*, for laboratory purposes.

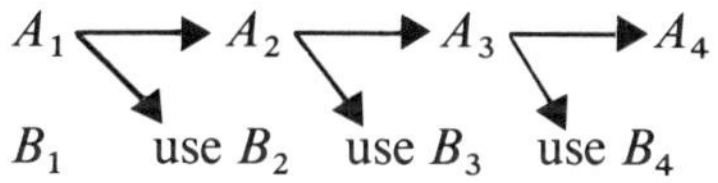

Drying cultures

Two methods are satisfactory. To dry by Stamp's method, make a small amount of 10% gelatin in nutrient broth and add 0.25% of ascorbic acid. Add a thick suspension of the organisms before the medium gels (to obtain a final density of about 10^{10} organisms/ml). Immerse waxed paper sheets in wax, drain and allow to set. Drop the suspension on the waxed surface. Dry in a vacuum desiccator over phosphorus pentoxide. Store the discs that form in screw-capped bottles in a refrigerator. Petri dishes smeared with a 20% solution of silicone in light petroleum and dried in a hot-air oven can be used instead of waxed paper.

Rayson's method is slightly different. Use 10% serum broth or gelatin broth and drop on discs of about 10 mm diameter cut from Cellophane and sterilized in petri dishes. Dry in a vacuum desiccator over phosphorus pentoxide. Store in screw-capped bottles in a refrigerator.

To reconstitute, remove one disc with sterile forceps into a tube of broth and incubate.

Freezing

Make thick suspensions of bacteria in 0.5 ml of sterile tap water or skim milk in screw-capped bottles and place in a deep freeze at −20 °C. Many organisms remain viable for long periods in this way. A method of freezing suspensions on glass beads at −76 °C is described by Feltham *et al.*[4].

Freeze-drying

Suspensions of bacteria in a mixture of 1 part of 30% glucose in nutrient broth and 3 parts of commercial horse serum are frozen and then dried by evaporation under vacuum. This preserves the bacteria indefinitely and is known as lyophilization. Yeasts, cryptococci, nocardias and streptomycetes can be lyophilized by making a thick suspension in sterile skimmed milk fortified with 5% of sucrose. This simple protective medium can be sterilized by autoclaving in convenient portions, e.g. 3 ml in bijou bottles.

Several machines are available. The simplest consists of a metal chamber in which solid carbon dioxide and ethanol are mixed. The bacterial suspension is placed in 0.2-ml amounts in ampoules or small tubes together with a label and these are held in the freezing mixture, which is at about −78 °C, until they freeze. They are then attached to the manifold of a vacuum pump capable of reducing the pressure to less than 0.01 mmHg. Between the pump and the manifold there is a metal container, usually part of the freezing container, in which a mixture of solid carbon dioxide and phosphorus pentoxide condenses the moisture withdrawn from the tubes. The tubes or ampoules are then sealed with a small blowpipe.

Simultaneous freezing and drying is carried out in centrifugal freeze-driers in which bubbling of the suspension is prevented by centrifugal action. Information about these machines can be obtained from Edwards (Pumps) Ltd.

For further information on the preservation of bacteria by freeze-drying and other methods *see* Lapage and Redway[5].

References

1. JARVIS, B., LACH, V.H. and WOOD, J.M. (1977) *Journal of Applied Bacteriology*, **43,** 149
2. KRAMER, J.M., KENDALL, M. and GILBERT, R.J. (1979) *European Journal of Applied Microbiology and Biotechnology*, **6,** 289
3. TROTMAN, R.E. (1978) *Technological Aids to Microbiology*. pp. 17–21. London: Edward Arnold
4. FELTHAM, R.K.A., POWER, A.K., PELL, P.A. and SNEATH, P.H.A. (1978) *Journal of Applied Bacteriology*, **44,** 313
5. LAPAGE, S.P. and REDWAY, M. (1975) *The Preservation of Bacteria and other Organisms*. Public Health Laboratory Service Monograph No. 6. London: HMSO

Chapter 6

Staining methods

The morphology of bacteria is difficult to observe in wet, unstained preparations and these are not permanent. It is usual to stain thin films of organisms prepared as follows.

Place a very small drop of saline or water in the centre of a 76×25 mm glass microslide. Remove a small amount of bacterial growth with an inoculating wire or loop, emulsify the organisms in the liquid, spreading it to occupy about 1 or 2 cm^2. Allow to dry in air or by waving high over a bunsen, taking care that the slide becomes no warmer than can be borne on the back of the hand. Pass the slide, film side down, once only through the bunsen flame to 'fix' it. This coagulates bacterial protein and makes the film less likely to float off during staining. It may not kill all the organisms, however, and the film should still be regarded as a source of infection. After staining, drain, blot with filter paper and dry by gentle heat over a bunsen.

Most laboratories now buy stains in solution ready for use, but some formulae are given below.

In the USA only stains certified by the Biological Stains Commission should be used. In the UK there is no comparable system and stains prepared by one of the specialist organizations should be used, e.g. Gurr, Lamb, BDH.

A large number of stains, mostly basic aniline dyes, have been used and described. For practical purposes one or two simple stains, the Gram method and an acid-fast stain are all that are required, plus one stain for fungi. A few other stains are described here including methods for staining capsules, spores and flagella.

Methylene blue stain

Stain for 1 min with a 0.5% aqueous solution of methylene blue.

Fuchsin stain

Stain for 30 s with the following: Dissolve 1 g basic fuchsin in 100 ml 95% ethanol. Stand for 24 h. Filter and add 900 ml water. *Or* use carbol fuchsin (*see* 'Acid-fast stain') diluted 1:10 with distilled water.

Gram stain

There are many modifications of this. This is Jensen's version:

(1) Dissolve 0.5 g methyl violet in 100 ml distilled water.
(2) Dissolve 2 g potassium iodide and 1 g iodine (ground together in a mortar to a fine powder) in not more than 20 ml distilled water. The iodine will not dissolve in a weaker solution of potassium iodide. Make up to 100 ml when dissolved.
(3) Basic fuchsin 0.1% (as above) or 1% aqueous safranin.

Stain with methyl violet solution for 20 s. Wash off and replace with iodine. Leave for 1 min. Wash off iodine with 95% alcohol or acetone, leaving on for a few seconds only. Wash with water. Counterstain with fuchsin or safranin.

Some practice is required with this stain to achieve the correct degree of decolorization. Acetone decolorizes much more rapidly than alcohol.

The Hucker method is commonly used in US laboratories:

(1) Dissolve 2 g of crystal violet in 20 ml of 95% ethanol. Dissolve 0.8 g of ammonium oxalate in 80 ml of distilled water. Mix these two solutions, stand for 24 h and then filter.
(2) Dissolve 2 g of potassium iodide and 1 g of iodine in 300 ml of distilled water using the method described for Jensen's Gram stain.
(3) Grind 0.25 g of safranin in a mortar with 10 ml of 95% ethanol. Wash into a flask and make up to 100 ml with distilled water.

Stain with the crystal violet solution for 1 min. Wash with tap water. Stain with the iodine solution for 1 min. Decolorize with 95% ethanol until no more stain comes away. Wash with tap water. Stain with the safranin solution for 2 min.

Acid-fast stain

The Ziehl–Neelsen method employs carbol fuchsin, acid alcohol and a blue or green counterstain. Colour blind technicians should use picric acid counterstain.

(1)

Basic fuchsin	5 g
Crystalline phenol	25 g
95% alcohol	50 ml
Distilled water	500 ml

Dissolve the fuchsin and phenol in the alcohol over a warm water-bath, then add the water. Filter before use.

(2)

95% ethyl alcohol	970 ml
Conc. hydrochloric acid	30 ml

(3)

0.5% methylene blue or malachite green, or 0.75% picric acid in distilled water.

Pour carbol-fuchsin on the slide and heat carefully until steam rises. Stain for 3–5 min but do not allow to dry. Wash well with water, decolorize with acid alcohol for 10–20 s, changing twice and counterstain for a few minutes with methylene blue or malachite green.

Some workers prefer the original decolorizing procedure with 20% sulphuric acid followed by 95% alcohol.

When staining nocardias, or leprosy bacilli use 1% sulphuric acid and no alcohol.

Cold staining methods are popular with some workers. For the Muller-Chermack method add 1 drop of Tergitol 7 (sodium heptadecyl sulphate) to 25 ml of carbol fuchsin and proceed as for Ziehl–Neelsen stain.

For Kinyoun's method make the carbol-fuchsin as follows:

Basic fuchsin	4 g
Melted phenol	8 g
95% ethanol	20 ml
Distilled water	100 ml

Dissolve the fuchsin in the alcohol. Shake gently while adding the water. Then add the phenol. Proceed as for Ziehl–Neelsen stain but do not heat. Decolorize with 1% sulphuric acid in water.

Fluorescent stain for acid-fast bacilli

There are several of these. We have found this auramine-phenol stain adequate.

(1)

Phenol crystals	3 g
Auramine	0.3 g
Distilled water	100 ml

(2)

Conc. HCl	0.5 ml
NaCl	0.5 g
Ethanol	75 ml

(3)

Distilled water	25 ml

Potassium permanganate 0.1% in distilled water.

Prepare and fix films in the usual way. Stain with auramine phenol for 4 min. Wash in water. Decolorize with acid alcohol for 4 min. Wash with potassium permanganate solution.

For microscopy use one of the commercial fluorescent microscopes (*see* Chapter 12).

Staining corynebacteria

To show the barred or beaded appearance and the metachromatic granules of these organisms Laybourn's modification of Albert's stain may be used.

(1) Dissolve 0.2 g malachite green and 0.15 g toluidine blue in a mixture of 100 ml water, 1 ml glacial acetic acid and 2 ml 95% alcohol.
(2) Grind 2 g iodine and 3 g potassium iodide in a mortar and dissolve in 40–50 ml water. Make up to 300 ml.

Stain with Solution (1) for 4 min. Wash, blot and dry and stain with Solution (2) for 1 min. The granules stain black and the barred cytoplasm light and dark green.

Spore staining

Make a thick film, stain for 3 min with hot carbol-fuchsin (Ziehl-Neelsen). Wash, flood with 30% aqueous ferric chloride for 2 min. Decolorize with 5% sodium sulphite solution. Wash and counterstain with 1% malachite green.

Flemming's technique substitutes nigrosin for the counterstain. Spread nigrosin over the film with the edge of another slide.

Spores are stained red. Cells are green, or with Flemming's method are transparent on a grey background.

Capsule staining

Place a small drop of indian ink on a slide. Mix into it a small loopful of bacterial culture or suspension. Drop a cover-glass on, avoiding air bubbles and press firmly between blotting paper. Examine with high-power lens.

For dry preparations mix one loopful of indian ink with one loopful of a suspension of organisms in 5% glucose solution at one end of a slide. Spread the mixture with the end of another slide, allow to dry and pour a few drops of methyl alcohol over the film to fix. Stain for a few seconds with methyl violet (Jensen's Gram solution No. 1). The organisms appear stained blue with capsules showing as haloes.

Churchman's method uses Wright's stain (used for staining blood films). Allow stain almost to dry on the slide and wash well with water. Capsules stain pink, bacteria blue.

Staining flagella

The method of Walc-Pokrzywnicki[1] gives good results.

Fixative

Ethanol	6 parts
Chloroform	3 parts
Formalin	1 part

Mordant

20% tannic acid	3 parts
5% ferric chloride	1 part

Silver stain

Dissolve a few grammes of silver nitrate in about 100 ml of distilled water. The exact amounts are not important. Keep about 10 ml in a test-tube and add strong ammonia drop by drop to the remainder until the heavy precipitate formed at the beginning just dissolves, leaving a faint opalescence. If too much ammonia is added and the opalescence disappears, add a little of the reserve silver nitrate. The final solution should be clear by transmitted light but faintly opalescent against a black background. Leave for 10 min and correct if necessary. This solution keeps indefinitely.

Grow swarming organisms on blood or nutrient agar and other organisms on a membrane filter on suitable media. Pour about 5 ml of distilled water on to the culture or on the membrane filter removed to another dish. Leave for 10–20 min. Remove about 3 ml to a test-tube with a wide bore pasteur pipette, add 2–3 drops of formalin and leave for about a week.

to the culture or on the membrane filter removed to another dish. Leave for 10–20 min. Remove about 3 ml to a test-tube with a wide bore pasteur pipette, add 2–3 drops of formalin and leave for about a week.

Use new slides, washed and stored in dichromate-sulphuric acid cleaning fluid. Wash with water and dry with a clean cloth.

Add 3–4 drops of the formalized suspension to 1 ml of distilled water and place one drop of the mixture on a clean slide. Allow to dry in air. Place in fixative for 3 min, wash with tap water and drain. Flood slide with mordant for 5–6 min. Wash with tap water, drain and dry the back of the slide with a clean cloth.

Pour the silver stain over the wet film through a Whatman No. 1 filter paper and leave for 15–20 min. Observe the development of a rusty brown colour in the thicker parts of the film by holding it against a white background. Wash with water and examine while wet with a 4-mm lens. If flagella are not seen, continue staining. If they are present, cover with a 5% solution of sodium thiosulphate for 2 min. Wash with tap water, distilled water and dry in air. When dry, cover with a cover-glass and suitable mounting medium. The organisms are stained black and the flagella golden brown.

Difco sell a flagella stain. It is a modification of the Leifson method. The technique is described in the Difco Supplementary Literature.

Lactophenol

Phenol, melted	20 g
Lactic acid	20 ml
Glycerol	40 ml
Water	20 ml

Warm the water and add the melted phenol, followed by the lactic acid and glycerol.

Lactophenol—cotton blue

Add 0.05% cotton blue (methyl blue) to lactophenol.

Dimethyl sulphoxide for clearing tissue for fungi

Mix 40 ml of dimethyl sulphoxide with 60 ml of water. Then dissolve 10 g of potassium hydroxide pellets in the mixture.

Reference

1. WALC-POKRZYWNICKI, A. (1969) *Journal of Medical Laboratory Technology*, **26,** 218

Chapter 7

Identification methods

Many conventional methods for the identification of micro-organisms are being replaced by paper strip and disc methods and by batteries of tests in 'kits'. These new methods offer certain advantages, e.g. the saving of time and labour but before an arbitrary choice is made between conventional tests and the newer methods, and between the several products available, the following factors should be considered.

(1) Conventional, paper strip and disc methods enable the user to make his own choice of tests. Kit methods do not; they give 10–20 test results whether the user needs them or not.
(2) Identification of organisms by conventional, paper strip or disc methods usually requires the judgement of the user. With kit methods the results are interpreted by numeric charts or computers.
(3) If final identification depends on serology then a few screening tests are usually adequate. If identification depends on biochemical tests then the more tests used the more reliable will be the results. Kits may then be the methods of choice, especially with unusual organisms and in epidemiological investigations when the kit manufacturers' services are particularly useful.
(4) Some techniques may be more hazardous than others. Those that involve pipetting increase the risk of dispersing aerosols and of environmental contamination. Those employing syringe and needle work increase the risk of self-inoculation.
(5) Conventional, strip and disc methods are usually based on the identification methods of Cowan and Steel[1]. The kit identifications are based on those of Edwards and Ewing[2].
(6) Some methods are much more expensive per identification than others.

It follows that choices should be made according to the nature of the investigations, the professional knowledge and skill of the workers and the size of the laboratory budget. The relative merits of the individual products mentioned below are not assessed here but some references to comparative investigations are given. *See also* the review by Goldschmit and Fung.[3]

As the paper and kit tests are unlikely to replace conventional tests in all laboratories the latter are also described below.

Paper strip tests

In these the strip is placed in a tube containing a small amount of culture, or a loopful of the organisms is rubbed on to the paper.

The *Pathotec* system of General Diagnostics includes strips for these tests: oxidase, nitratase, phenylalanine deamination, urease, indole, H_2S, VP, malonate, aesculin, lysine decarboxylase and ONPG.[4, 5]

Mast, Medical Wire and Oxoid sell β-lactamase detection papers.

Mast, Medical Wire and Merck make oxidase test strips.

Medical Wire sell urease test papers.

Paper discs

These are placed on seeded petri dish cultures and zones of inhibition or colour changes are noted after incubation. Oxoid and Mast make the special sets of antibiotic discs for the identification of Gram-negative anaerobes. Difco and Oxoid sell discs containing bacitracin (for Group A streptococci), optochin (for pneumococci) and X and V factors (for haemophilus).

Difco supply discs containing a wide range of fermentable substances, and also for the ONPG and urease tests.

Kit identification systems

These range from rows of small cells containing dry substrates arranged in plastic strips, through discs dispensed into special trays or dishes, to sets of substrates in plastic tubes or test-tubes.

API systems

API systems are available for

(1) enterobacteria, non-fermenting and oxidase positive fermentative bacilli (API 20, API 20E);
(2) yeasts (API 20C); staphylococci (20 STAPH);
(3) streptococci (API STREP); and
(4) enzyme tests (API ZYM).

Suspensions of the organisms are added to cupules in plastic galleries. These are incubated overnight. Some tests then require reagents to be added. Results are read from codes and a Profile Register. There is a computer identification service.[4-10]

Enterotubes (Roche)

These are designed for the identification of enterobacteria. A plastic tube contains the substrates for 14 tests. A wire needle passes through each substrate, from end to end. After uncapping, the wire is touched on the culture and then drawn through the tube and discarded. The tubes are recapped and incubated. One reagent must be added, using a hypodermic needle and syringe. The results are read from a set of codes and keys (ENCISE). There is also a computer consultation service.[4, 5, 6, 9]

Oxi-Ferm (Roche)

This is a similar system but extends identification to a number of other Gram-negative organisms.[7, 11]

Flow Laboratories (Diagnostic Research)

These systems use tubes and petri dishes of a special design[4]. The *r/b* set is for oxidase negative organisms. Four tubes allow enterobacteria to be identified. Results are interpreted from a computer code book after overnight incubation. The *N/F* set screens for some *Pseudomonas* species in 24 h and identifies other non-enterobacteria in a further 24 h. Results are interpreted by a 'logic' scheme or by computerized numerical coding. Two other sets are for identifying yeasts: Uni-Yeast-Tek identifies *Candida albicans* and *C. stellatoidea*; the *C/N* single tube identifies *Cryptococcus neoformans*. There is also a single tube, *Uni-OF*, for the Hugh and Leifson test.

Microbact (Disposable Products)

This provides three systems, for enterobacteria, enterobacteria and other Gram-negative rods and anaerobic bacteria. Suspensions of organisms are added to wells in prepared trays. Some wells must be covered with mineral oil. Some reagents have to be added after incubation. The results are read from charts.

The Micro ID system (General Diagnostics)

This identifies 33 species of enterobacteria with 15 tests. Suspensions of the organisms are added to impregnated paper discs in plastic trays and incubated for 4 h. Results are interpreted with the aid of a coded manual.[9]

Minitek systems (Bioquest)

These allow the identification of enterobacteria, neisseria, non-sporing anaerobes and yeasts. Discs, impregnated with substrates, are dispensed into specially prepared plates, inoculated with suspensions of the organisms and incubated overnight. Results are read from colour charts.

Enteroset (Fisher)

This is for identifying enterobacteria and consists of porous paper strips containing substrates sandwiched between plastic films. Suspensions of organisms are absorbed on the strips and results are read from charts after overnight incubation.

Auxotab (Wilson Diagnostics)

This consists of cards carrying reagent filled capillary chambers to which the bacterial suspensions are added. Enterobacteria are identified after 7-h incubation.[4]

AutoMicrobic (Vitek)

This is an automated system incorporating a dilution dispenser and filling module for dispensing the suspensions into prepared wells in trays. A computer module provides identifications after 8-h incubation.[9]

Conventional tests

Aesculin hydrolysis

Inoculate aesculin medium or Edwards' medium and incubate overnight. Organisms that hydrolyse aesculin blacken the medium.

Ammonia test

Incubate culture in nutrient or peptone broth for five days. Wet a small piece of filter-paper with Nessler reagent and place it in the upper part of the culture tube. Warm the tube in a water-bath at 50–60 °C. The filter-paper turns brown or black if ammonia is present.

Arginine hydrolysis

Incubate the culture in arginine broth (p. 69) for 24–48 h and add a few drops of Nessler reagent. A brown colour indicates hydrolysis. (But for lactobacilli, *see* p. 352.)

Carbohydrate fermentation and oxidation

Liquid, semisolid and solid media are described on p. 72. They contain a fermentable carbohydrate, alcohol or glucoside and an indicator to show the production of acid. To demonstrate gas production, a small inverted tube (Durham's tube; gas tube) is placed in the fluid media. In solid media (stab tube), gas production is obvious from the bubbles and disruption of the medium. Normally, gas formation is recorded only in the glucose tube and any that occurs in tubes of other carbohydrates results from the fermentation of glucose formed during the first part of the reaction.

For most bacteria, use peptone or broth-based media.

For *Lactobacillus* spp., use MRS base (p. 78) without glucose or meat extract and adjust to pH 6.2–6.5. Add chlorphenol red indicator.

For *Bacillus* spp. and *Pseudomonas* spp., use ammonium salt 'sugar' media (p. 72). These organisms produce ammonia from peptones and this may mask acid production.

For *Neisseria* spp. and other fastidious organisms use solid or semisolid media. Neisserias do not like liquid media. Enrich with Fildes' extract (5%) or with rabbit serum (10%). Horse serum may give false results as it contains fermentable carbohydrates. Robinson's buffered serum sugar medium (p. 73) overcomes this problem and is the medium of choice for *Corynebacterium* spp.

Treat all anaerobes as fastidious organisms. The indicator may be decolorized during incubation, so add more, after incubation. A sterile iron nail added to liquid media may improve the anaerobic conditions.

Tests may be carried out on solid medium containing indicator. Inoculate the medium heavily, spreading it all over the surface and place carbohydrate discs (p. 103) on the surface. Acid production is indicated by a change of colour in the medium around the disc. We do not recommend placing more than four discs on one plate.

Caesin hydrolysis

Use skim milk agar (p. 80) and observe clearing around colonies of casein-hydrolysing organisms. The milk should be dialysed so that acid production by lactose fermenters does not interfere. To detect false clearing due to this, pour a 10% solution of mercuric chloride in 20% hydrochloric acid over this medium. If the cleared area disappears, casein was not hydrolysed.

Catalase test

(1) Emulsify some of the culture in 0.5 ml of a 1% solution of Tween 80 in a screw-capped bottle. Add 0.5 ml of 20-vol hydrogen peroxide and replace the cap. Effervescence of oxygen indicates the presence of catalase. Do not do this test on an uncovered slide as the effervescence creates aerosols. Cultures on low-carbohydrate medium give the most reliable results.
(2) Add a mixture of equal volumes of 1% Tween 80 and 20-vol hydrogen peroxide to the growth on an agar slope. Observe effervescence after 5 min.
(3) Test mycobacteria by Wayne's method (p. 382).
(4) Test minute colonies which grow on nutrient agar by the Blue Slide Test[12]. Place one drop of a mixture of equal parts of methylene blue stain and 20-vol hydrogen peroxide on a slide. Place a cover-slip over the colonies to be tested and press down firmly to make an impression smear. Remove the cover-slip and place it on the methylene blue–peroxide mixture. Clear bubbles, appearing within 30 s, indicate catalase activity.

Citrate utilization

Inoculate solid (Simmons' or Christensens') medium or fluid medium (Koser) with a straight wire. Heavy inocula may give false-positive results. Incubate at 30–35 °C and observe growth.

Coagulase test

Staphylococcus aureus and a few other organisms coagulate plasma (*see* p. 326).

Decarboxylase tests

Falkow's method can be used for most Gram-negative rods but Moeller's method gives the best results with *Klebsiella* spp. and *Enterobacter* spp. Commercial Falkow media allow only lysine decarboxylase tests but commercial Moeller media permit lysine, ornithine and arginine tests to be used.

Inoculate Falkow medium (p. 73) and incubate for 24 h at 37 °C. The indicator (bromocresol purple) changes from blue to yellow due to fermentation of glucose. If it remains yellow, the test is negative; if it then changes to purple, the test is positive.

When using Moeller medium (p. 73), include a control tube that contains no amino acid. After inoculation, seal with paraffin to ensure anaerobic conditions and incubate at 37 °C for 3–5 days. There are two indicators, bromothymol blue and cresol red. The colour changes to yellow if glucose is fermented. Decarboxylation is indicated by a purple colour. The control tube should remain yellow.

DNAse test

Streak or spot the organisms heavily on DNAse agar. Incubate overnight and flood the plate with 1 N hydrochloric acid, which precipitates unchanged nucleic acid. A clear halo around the inoculum indicates a positive reaction.

Eijkman test

Inoculate suspected *E. coli* into brilliant green broth or MacConkey broth with a gas tube. Incubate at 44 ± 0.2 °C for 24 h. *E. coli* is one of the few organisms that produce gas at this temperature.

Gelatin liquefaction

(1) Inoculate nutrient gelatin stabs, incubate at room temperature for seven days and observe digestion. For organisms that grow only at temperatures when gelatin is fluid, include an uninoculated control and after incubation place both tubes in a refrigerator overnight. The control tube should solidify. This method is not entirely reliable.
(2) Inoculate nutrient broth. Place a denatured gelatin charcoal disc (Oxoid) in the medium and incubate. Gelatin liquefaction is indicated by the release of charcoal granules, which fall to the bottom of the tube.
(3) For a rapid test, use 1 ml of broth or 0.01 M calcium chloride in saline. Inoculate heavily (whole growth from slope or petri dish). Add a gelatin charcoal disc and place in a water-bath at 37 °C. Examine at 15-min intervals for 3 h.
(4) Inoculate gelatin agar medium. Incubate overnight at 37 °C and then flood the plates with saturated ammonium sulphate solution. Haloes appear around colonies of organisms producing gelatinase.

Gluconate oxidation

Inoculate gluconate broth (p. 75) and incubate for 48 h. Add an equal volume of Benedict's reagent and place in a boiling water-bath for 10 min. An orange or brown precipitate indicates gluconate oxidation.

Hugh and Leifson (HL) test. Oxidation fermentation test

This is also known as the 'oxferm' test. Some organisms metabolize glucose oxidatively, i.e. oxygen is the ultimate hydrogen acceptor and culture must therefore be aerobic. Others ferment glucose, when the hydrogen acceptor is another substance: this is independent of oxygen and cultures may be aerobic or anaerobic.

Heat two tubes of medium (p. 76) in boiling water for 10 min to drive off oxygen, cool and inoculate; incubate one aerobically and the other either anaerobically or seal the surface of the medium with 2 cm of melted Vaseline or agar to give anaerobic conditions.

Oxidative metabolism: acid in aerobic tube only.

Fermentative metabolism: acid in both tubes.

For testing staphylococci and micrococci, use the Baird–Parker modification (p. 76).

Hydrogen sulphide production

(1) Inoculate a tube of nutrient broth. Place a strip of filter-paper impregnated with lead acetate (Indicator Paper—Johnson's; BDH; Difco) in the top of the tube, holding it in place with a cotton wool plug. Incubate and examine for blackening of the paper.

(2) Inoculate one of the iron or lead acetate media. Incubate and observe blackening.

TSI (Triple Sugar Iron Agar) and similar media do not give satisfactory results with sucrose-fermenting organisms. The indicator paper (lead acetate) is the most sensitive method and the ferrous chloride media the least sensitive method, but the latter is probably the method of choice for identifying salmonellas and for differentiation in other groups where the amount of hydrogen sulphide produced varies with the species.

Indole formation

Grow the organisms for 2–5 days in peptone or tryptone broth.

(1) *Ehrlich's method* Dissolve 4 g of *p*-dimethylaminobenzaldehyde in a mixture of 80 ml of concentrated hydrochloric acid and 380 ml of ethanol (do not use industrial spirit; this gives a brown instead of a yellow solution). To the broth culture add a few drops of xylene and shake gently. Add a few drops of the reagent. A rose pink colour indicates indole.

(2) *Kovac's method* Dissolve 5 g of *p*-dimethylaminobenzaldehyde in a mixture of 75 ml of amyl alcohol and 25 ml of concentrated sulphuric acid. Add a few drops of this reagent to the broth culture. A rose pink colour indicates indole.

KCN test

Grow the test culture in tryptone broth. Inoculate KCN broth from this and stopper tube with a rubber bung. Incubate for 48 h and examine for growth.

Lecithinase activity

(1) Inoculate egg yolk agar (p. 74) or Willis and Hobbs' medium (p. 82) and incubate for three days. Lecithinase-producing colonies are surrounded by zones of opacity.

(2) Inoculate egg yolk salt broth and incubate for three days. Lecithinase producers make the broth opalescent. Some organisms (e.g. *B. cereus*) give a thick turbidity in 13 h.

Levan production

Inoculate nutrient agar containing 5% of sucrose. Levan producers give large mucoid colonies after incubation for 24–48 h.

Lipolytic activity

Inoculate tributyrin agar (p. 81). Incubate for 48 h at 25–30 °C. A clear zone develops around colonies of fat-splitting organisms. These media can be used for counting lipolytic bacteria in dairy products.

Malonate test

Inoculate one of the malonate broth media and incubate overnight. Growth and a deep blue colour indicate malonate utilization.

Methyl red test

Inoculate glucose phosphate broth (p. 76) and incubate for five days at 30 °C. Add a few drops of methyl red solution, prepared by dissolving 0.1 g of the dye in 300 ml of ethanol and making up to 500 ml with distilled water. A red colour indicates that the pH has been reduced to 4.5 or less. A yellow colour indicates a negative reaction

Motility

Hanging drop method

Place a very small drop of liquid bacterial culture in the centre of a 16 mm square No. 1 cover-glass, with the aid of a small (2 mm) inoculating loop. Place a small drop of water at each corner of the cover-glass. Invert over the cover-glass a microslide with a central depression—a 'well-slide'. The cover-glass will adhere to the slide and when the slide is inverted the hanging drop is suspended in the well.

Instead of using 'well-slides' a ring of Vaseline may be made on a slide. This is supplied in collapsible tubes or squeezed from a hypodermic syringe.

Bring the edge of the hanging drop, or the air bubbles in the water seal into focus with the 16-mm lens before turning to the high-power dry objective to observe motility.

Bacterial motility must be distinguished from Brownian movement. There is usually little difficulty with actively motile organisms, but feebly motile bacteria may require prolonged observation of individual cells.

Careful examination of hanging drops may indicate whether a motile organism has polar flagellation—a darting zig-zag movement—or peritrichate flagellation—a less vigorous and more vibratory movement.

To examine anaerobic organisms for motility grow them in a suitable liquid medium. Touch the culture with a capillary tube 60–70-mm long and about 0.5–1-mm bore. Some culture will enter the tube. Seal both ends of the tube in the bunsen flame and mount on Plasticine on the microscope stage. Examine as for hanging drops.

Craigie tube method

Craigie tubes can be used instead of hanging drop cultures (p. 80). Inoculate the medium (p. 80) in the inner tube. Incubate and subculture daily from the outer tube. Only motile organisms can grow through the sloppy agar.

Nagler test

This tests for lecithinase activity (*see above*) but the word has come to mean the half antitoxin plate test for *Cl. perfringens*. Egg yolk agar containing Fildes enrichment or Willis and Hobbs' medium are satisfactory.

Niacin test

This test is used exclusively for typing mycobacteria and is described on p. 378.

Nitratase test (nitrate reduction)

(1) Inoculate nitrate broth medium and incubate overnight.
(2) Make a dense suspension of the test organisms in 0.01 M sodium nitrate in M/45 phosphate buffer of pH 7.0 and incubate at 37 °C for 24 h.
(3) Grow the organisms in a suitable broth. Add a few drops of 1% sodium nitrate solution and incubate for 4 h.

Acidify with a few drops of 1 N hydrochloric acid and add 0.5 ml each of a 0.2% solution of sulphanilamide and 0.1% N-naphthylethylenediamine hydrochloride (these two reagents should be kept in a refrigerator and prepared fresh monthly). A pink colour denotes nitratase activity. But as some organisms further reduce nitrite, if no colour is produced add a very small amount of zinc dust. Any nitrate present will be reduced to nitrite and produce a pink colour, i.e. a pink colour in this part of the test indicates no nitratase activity and no colour indicates that nitrates have been completely reduced.

ONPG test

Lactose is fermented only when both β-galactosidase and permease are present. Deficiency of the latter gives late fermentation. True non-lactose fermenters do not possess β-galactosidase.

Inoculate ONPG broth (p. 79) and incubate overnight. If β-galactosidase is present a yellow colour due to *o*-nitrophenol is formed.

Optochin test

Pneumococci are sensitive but streptococci are resistant to optochin (ethylhydrocupreine hydrochloride).

Streak the organisms on blood agar and place an optochin disc on the surface. Incubate overnight and examine the zone around the disc.

Oxidase test (cytochrome oxidase test)

(1) Soak small pieces of filter-paper in 1% aqueous tetramethyl-*p*-phenylenediamine dihydrochloride or oxalate (which keeps better). Some filter-papers give a blue colour and these must not be used. Dry or use wet. Scrape some of a fresh young culture with a clean platinum wire or a glass rod (dirty or Nichrome wire gives false positives) and rub on the filter-paper. A blue colour within 10 s is a positive oxidase test. Old cultures are unreliable. Tellurite inhibits oxidase.

(2) Incubate cultures on nutrient agar slopes for 24–48 h at the optimum temperature for the strain concerned. Add a few drops each of freshly prepared 1% aqueous *p*-aminodimethylaniline oxalate and 1% α-naphthol in ethanol. Allow the mixture to run over the growth. A deep blue colour is a positive reaction.

Oxidative or fermentative metabolism of glucose

See Hugh and Leifson test *above*.

Phenylalanine test (PPA or PPD test)

Inoculate phenylalanine agar and incubate overnight. Pour a few drops of 10% ferric chloride solution over the growth. A green colour indicates deamination of phenylalanine to phenylpyruvic acid. Among the enterobacteria only proteus and providence strains have this property.

Phosphatase test

Some bacteria, e.g. *S. aureus*, can split ester phosphates. Inoculate a plate of phenolphthalein phosphate agar (p. 79) and incubate overnight. Expose the culture to ammonia vapour (caution). Colonies of phosphatase producers turn pink.

Proteolysis

Inoculate cooked meat medium (p. 80) and incubate for 7–10 days. Proteolysis is indicated by blackening of the meat or digestion (volume diminishes). Tyrosine crystals may appear.

Starch hydrolysis

Inoculate starch agar (p. 81). Incubate for 3–5 days then flood with dilute iodine solution. Hydrolysis is indicated by clear zones around the growth. Unchanged starch gives a blue colour.

Sulphatase test

Some organisms (e.g. mycobacteria) can split ester sulphates. Grow the organisms in media (Middlebrook 7H9 for mycobacteria) containing 0.001 M potassium phenolphthalein disulphate for 14 days. Add a few drops of ammonia solution. A pink colour indicates the presence of free phenolphthalein.

Tellurite reduction

Some mycobacteria reduce tellurite to tellurium metal (*see* p. 385).

Tween hydrolysis

Some mycobacteria can hydrolyse Tween (polysorbate) 80, releasing fatty acids that change the colour of an indicator. This test is used mostly with mycobacteria (*see* p. 383).

Tyrosine decomposition

Inoculate parallel streaks on a plate of tyrosine agar (p. 383) and incubate for 3–4 weeks. Examine periodically under a low-power microscope for the disappearance of crystals around the bacterial growth.

Urease test

Inoculate one of the urea media heavily and incubate for 3–12 h. The fluid media give more rapid results if incubated in a water-bath. If urease is present, the urea is split to form ammonia, which changes the colour of the indicator from yellow to pink.

Voges Proskauer Reaction (VP test)

This tests for the formation of acetyl methyl carbinol (acetoin) from glucose. This is oxidized by the reagent to diacetyl, which produces a red colour with guanidine residues in the media.

Inoculate glucose phosphate broth (p. 76) and incubate for five days at 30 °C. A very heavy inoculum and 6 h may suffice. Test by one of the following methods.

(1) Add 3 ml of 5% alcoholic α-naphthol solution and 3 ml of 40% potassium hydroxide solution (Barritt's method).
(2) Add a trace amount of creatinine and 5 ml of 40% potassium hydroxide solution (O'Meara's method).
(3) Add 5 ml of a mixture of 1 g of copper sulphate (blue) dissolved in 40 ml of saturated sodium hydroxide solution plus 960 ml of 10% potassium hydroxide solution (APHA method).

A bright pink or eosin red colour appearing in 5 min is a positive reaction. For testing *Bacillus* spp., add 1% sodium chloride to the medium.

Xanthine decomposition

Inoculate parallel streaks on a plate of xanthine agar (p. 82) and incubate for 3–4 weeks. Examine periodically under a low-power microscope for the disappearance of xanthine crystals around the bacterial growth.

References

1. COWAN, S. T. (1974) *Manual for the Identification of Medical Bacteria*. 2nd edn. Cambridge: Cambridge University Press
2. EDWARDS, P. R. and EWING, W.H. (1972) *Identification of Enterobacteriaceae*. 3rd edn. Minneapolis: Burgen Publishing Co.
3. GOLDSCHMIDT, M.C. and FUNG, D.Y.C. (1978) *Journal of Food Protection*, **41,** 201
4. NORD, C.E., LINDBERG, A.A. and DAHLBACK, A. (1974) *Medical Microbiology and Immunity*, **159,** 211
5. HOLMES, B., WILLCOX, W.R., LAPAGE, S.P. and MALNICK, H. (1977) *Journal of Clinical Pathology*, **30,** 381
6. HAYEK, L.J. and WILLIS, G.W. (1976) *Journal of Clinical Pathology*, **29,** 158
7. DOWDA, H. (1977) *Journal of Clinical Microbiology*, **6,** 605
8. OBERHOFER, T.R. (1979) *Journal of Clinical Microbiology*, **9,** 220
9. KELLEY, M.T. and LATIMER, J.M. (1980) *Journal of Clinical Microbiology*, **12,** 659
10. SHAYEGANI, M., MUAPIN, P.S. and McGLYNN, D. (1978) *Journal of Clinical Microbiology*, **7,** 539
11. OBERHOFER, T.R., ROWEN, J.W., CUNNINGHAM, G.F. and HIGBEE, J.W. (1977) *Journal of Clinical Microbiology*, **6,** 559
12. THOMAS, M. (1963) *Monthly Bulletin of the Ministry of Health and Public Health Laboratory Service*, **22,** 124

Chapter 8

Mycological methods

A.C.J. Proctor

Direct examination

Mount hair, skin or nail fragments on a slide in a drop of 10% potassium hydroxide in 40% dimethyl sulphoxide. Apply a cover-glass and then leave for a few minutes till the preparation 'clears'. Wet films of pus may be made with an equal volume of this mountant or an equal volume of glycerol if a more permanent preparation is required. Mix cerebrospinal fluid deposit with an equal volume of 2.5% nigrosin in 50% glycerol. Glycerol prevents mould growth in the solution, delineates capsules more sharply and prevents preparations, which should be thin, from drying out.

Cultural examination for moulds

Morbid material

Inoculate slopes or plates of chloramphenicol malt agar. If dermatophytes are sought, use a medium containing cycloheximide, either malt agar or one of the proprietary media containing an indicator. Push pieces of hair or skin into the medium at several points. Scrape nail clippings to a powder with a scalpel; do not put them on in lumps. Pieces of skin, hair or nail powder can be picked up more easily if the heated inoculating wire is first pushed into the medium to coat it with agar.

Spread pus, CSF deposit or biopsy material from cases of the deeper mycoses on slopes of glucose peptone agar with chloramphenicol. Cut tissues into small pieces or grind them in a Griffith's tube. Place in the bottom of a petri dish and pour on chloramphenicol malt agar (moulds) or glucose peptone agar (actinomycetes), melted and at 45 °C.

Wash mycetoma grains several times with sterile saline to remove most contaminants before culturing.

Incubate cultures at 28 °C. Most yeasts grow in 2–3 days, though cryptococci are rather slower. Common moulds, too, will grow up in a few days while cultures for dermatophytes should be kept for two weeks though most are identifiable after a week. It is useful, then, to examine all mould cultures once or twice a week, depending on their number, the time available and enthusiasm for the subject.

Cultures from food, soil, plant material, etc.

Inoculate plates of malt agar with chloramphenicol and Czapek-Dox agar. Zygomycetes, if present, will outgrow everything else but are markedly restrained by Czapek-Dox. To grow moulds from physiologically dry materials such as jam increase the sucrose content of the Czapek-Dox to 20%. Some yeasts can tolerate low pH and can be isolated on malt agar to which 1% lactic acid has been added after melting and cooling to 50 °C. Incubate cultures at room temperature and at 30 °C.

Sampling air for mould spores

A cheap, simple device for this purpose is the Porton impinger in which particles from a measured volume of air are trapped in a liquid medium on which viable counts are done. Slit or SAS samplers (p. 245) are also useful. More information is gained by using the Andersen sampler, which sizes airborne particles using the principle that the higher the wind speed the smaller the particle which will escape the air stream and be impacted on an agar surface. Plastic petri dishes give lower counts than glass ones but they make the whole machine much easier to handle.

The medium and time of incubation will depend on the organism sought. For general use one run with Czapek-Dox and one with chloramphenicol malt agar will serve; for thermophilic actinomycetes use glucose peptone agar without antibiotics; incubate one set at 40 °C, the other at 50 °C.

Mould growths on and from foods, etc.

Examine with a lens or low-power microscope *in situ*—to see the arrangement of spores *etc.* before disturbing the growth. Remove a small piece of mycelium with a wire or needle, the end of which has been bent at right angles to give a short, sharp hook. Do not dig in the middle of a colony, or you will see nothing but spores. Use the point of the needle to cut out a piece of the growth near the edge where sporulation is just beginning. Transfer it to a drop of PVA mountant on a slide, apply a cover-slip then examine after 30 min when the stain has penetrated the hyphae.

PVA mountant

Polyvinyl alcohol, MW 125 000	8 g
Water	70 ml
Glycerol	20 ml
Tween 80 10% aqueous	10 ml

Mix, heat to 80 °C to dissolve then let stand overnight.

For staining, add:

Methyl blue CI 42780	20 mg
Lactic acid	1 ml

The rather low refractive index, 1.37, gives a good contrast with fungal cells and the mount hardens if left at 37 °C overnight or in a couple of days at room temperature.

Slide culture

Cut small blocks of medium (0.5 by 0.5 cm) from a poured plate and transfer them to the centres of sterile slides. Inoculate the edges of each block and apply a cover-slip. Make a moist chamber by cutting two lengths from a plastic drinking straw and glueing them, 2 in apart, to the bottom of a 100-mm square plastic petri dish with a drop of chloroform. Drop a rectangle of filter-paper soaked in 10% glycerol between the supports. Up to three prepared slide cultures may be incubated in one dish. Watch the progress of growth using the low power of the microscope when visible growth has developed. When typical structures are seen, prepare a slide and a cover-slip each with a drop of mountant. Lift the cover-slip off the slide culture and mount on the prepared slide, then lift off and discard the agar block: the slide is now mounted on the prepared cover-slip. Examine both preparations after 30 min.

To study aerobic growth and spore formation, cut a strip about 1.5 cm wide from a roll of 2.5 cm Sellotape, then attach by a narrow end to a mounted needle, forming a flag. Blot the surface of the colony with this near the growing edge, then mount the tape, sticky side down, in a drop of mountant and examine at once.

Yeasts

Culture in duplicate on chloramphenicol malt with and without cycloheximide. *Candida* species will grow in the presence of cycloheximide while other yeasts will not, which is a useful differential feature. Incubate cultures at 30°C. To study yeast morphology use corn meal agar: streak the organism on a plate of the medium, cover with a sterile cover-glass and incubate at 30 °C. Examine daily through the bottom of the petri dish, using a low-power microscope. Look for cell shape, mycelium and spore formation.

To isolate yeasts from highly acidic environments add 1% lactic acid to melted and cooled malt agar before pouring plates, or use buffered yeast agar plus lactic acid (p. 71).

To encourage ascospore formation inoculate sodium acetate agar made by adjusting the pH of 0.5% sodium acetate to 6.5 with acetic acid, then solidifying it with agar. Stain the growth for spores after a week.

Fermentation tests

These differ from those used in bacteriology in that the sugar concentration is 3%. Durham tubes are always used and a 10-ml volume of medium is preferable to encourage fermentation rather than oxidation.

Auxanographic tests

Fermentation tests alone are insufficient to identify species of yeasts. Make a thick suspension by adding about 5 ml of sterile water to a malt agar slope: it is not necessary to wash the organisms. Add about 0.25 ml of this suspension to each bottle of auxanogram (carbon assimilation) medium (p. 69), then pour a plate. When set, place discs of carbon or nitrogen sources on the surface of the agar. It is convenient to arrange the galactose, lactose, maltose and sucrose discs in a square pattern on the surface of the carbon source plate with the glucose in the centre as a control of growth. On the nitrogen source plate put a disc of sodium nitrate, with an asparagine disc as the positive control. These controls guard against fastidious organisms which decline to grow and against inadvertent sterilization by medium poured too hot. Sterilize coloured paper discs in petri dishes, then add one drop of 5% sugar solution, 5% sodium nitrate or 2% asparagine and dry the discs at 37 °C. Incubate the plates at the optimum temperature for the yeast concerned.

Some yeast strains, particularly of *Trichosporon* spp. may fail to grow. In this case repeat the test using a medium fortified with a few drops of 5% yeast extract per 20-ml bottle.

For more detailed information on methods *see* Benecke and Rogers[1], Haley and Calloway[2] and Emmons *et al.*[3].

Serological methods

Methods of antigen production

Growing the organism

For production in quantity, the method of choice is some form of chemostat. A dilution rate of 0.05–0.08 is satisfactory. Once growth is established, the oxygen tension should be maintained at a high level; at least 80% saturation. *Aspergillus* and similar organisms grow well at 30 °C, *Candida* species at 35 °C and the thermophilic actinomycetes at 42 °C. A large inoculum is essential for success. A suitable starter culture of a filamentous mould such as an aspergillus can be produced in a stirred batch culture. For this, 300 ml of glucose peptone broth are autoclaved in a 500-ml Erlenmeyer flask containing a 5-cm polypropylene covered magnetic stirring bar. The cotton wool plug should not be too tight. The culture, after inoculation with a few spores or a fragment of mycelium, is stirred, slowly at first, then when growth is established sufficiently fast for the vortex of air to touch the magnetic follower and disperse air bubbles through the medium, thus ensuring adequate aeration. A dense growth once obtained, it is used as the inoculum for the chemostat or for a larger flask. The inoculum of thermophilic actinomycetes or *Candida* species is most easily prepared from plate cultures on solid glucose peptone agar or from a series of slopes of this medium. In this case also, the inoculum should be heavy, and the inoculated chemostat culture left to grow up overnight before the medium flow is re-started.

Aspergillus and other robust organisms may be harvested by filtering the culture through a circle of nylon, cut from an old shirt, on a Buchner funnel of convenient size. The mat of mycelium should be sucked as free of medium as possible, then washed once with distilled water. On removal from the funnel the nylon can be peeled off the cake of mycelium like a wet postage stamp, then the cake transferred to a tared container, weighed and kept at −20 °C.

Thermophilic actinomycetes and yeasts are best recovered by centrifuging in a continuous rotor if it be available. These too are weighed and kept at −20 °C. Alternative methods are as follows.

For aspergilli and similar organisms 3 litres of glucose peptone broth are autoclaved in a 5-litre Erlenmeyer flask containing a suitable magnetic follower. A dense growth in 300 ml of medium, as described above, is used as a starter. Adequate growth is obtained in 2–3 days stirring vigorously from the outset. The incubation period should not be too long, or the mycelium will soften, begin to autolyse and the yield of antigen will be less.

Thermophilic actinomycetes can be grown in petri dishes or a layer of glucose peptone agar with added Marmite and glycerophosphate at 43 °C. Lots of 70 ml of medium should be poured in 6-in petri dishes. As soon as they are set, they should be inoculated with 2–3 ml of actinomycete suspension in peptone water, which is run over the surface, providing a film of moisture in which the organisms will spread to cover the surface, and reducing drying of the agar during 48-h incubation at 42 °C. The growth from the plates is then scraped off with the end of a 3 × 1-in slide into a tared beaker.

Candida species can also be grown on plates. The inoculum should be less in volume, about 0.5 ml, and the plates should be incubated for two nights and a day: e.g. inoculated on Monday evening; then the growth harvested after incubation at 37 °C till Wednesday morning. The growth should then be washed off with saline and packed down in the centrifuge.

In every case the organisms are kept at −20 °C, for at least one night, so as to make them more easily disrupted. The time can be extended to several weeks if need be, providing a convenient point at which to interrupt the preparation.

The next step is to break up the organisms and so to release their soluble constituents. This may be done in several ways. The simplest is to make a thick suspension of the organisms in volatile buffer, then to mix it with an equal bulk of sand in a large mortar. The whole is frozen at −20 °C, beaten with a pestle till liquid, then the process repeated. The liquid, clarified by centrifuging, constitutes the antigen. Two other methods of disintegration are available; ultrasonic rupture and shaking with ballotini. The former method is unsuitable for yeasts; filamentous organisms should be suspended at a concentration of 20% w/v in volatile buffer by allowing the cake to thaw then macerating it with the buffer in an MSE, Ato-Mix or other suitable blender for a couple of minutes. This suspension may then be exposed to ultrasound. The MSE 100-W machine will process batches of 30–40 ml; the treatment time is 20 min and the tube containing the suspension must be surrounded by an ice-bath or the temperature will rise almost to boiling point. Alternatively, the suspension may be run through a Dyno-Mill, an apparatus in which the container remains still, but the contained ballotini and suspension to be ground are stirred by blades which rotate at 2000 rev/min. Continuous operation is achieved by a pair of separator plates with a narrow gap, too small to allow the ballotini to pass. The macerated suspension of fungus may be pumped through this machine at a rate of 2 litres/h. Equally satisfactory results are obtained if a 20% suspension of yeast cells is used. The other methods for disrupting yeasts also use ballotini. They are the Mickle shaker, when overnight treatment in a cold room is necessary; and the Braun MSK homogenizer, in which the bottle containing yeasts and ballotini is made to describe a circle at 4000 rev/min, while it is cooled by liquid CO_2. Neither of these methods is as easy, reliable or reproducible as disintegration in the Dyno-Mill.

Once the organisms have been disintegrated the treatment is in every case the same. The suspension is frozen overnight to aggregate insoluble material. This is then removed from the thawed suspension by centrifuging, and the supernatant liquor is concentrated with concomitant removal of low molecular weight impurities by dialysis at 4 °C against daily changes of 10% polyethylene glycol 6000 in volatile buffer. When the volume of extract has been reduced to the desired level, it is frozen overnight again, centrifuged for 1 h at 15 000 rev/min, then the supernatant fluid freeze dried in pre-weighed ampoules. As a rough guide, 1 g of *Candida* or actinomycete taken should yield 5 ml of antigenic extract, and 1 g of fungal thallus 0.5–1 ml of extract. An extract so prepared should be effective in double diffusion tests at 6 mg/ml and at 24 mg/ml in immunoelectrophoresis.

Complement fixation tests

These tests are useful in histoplasmosis and coccidiomycosis. Both yeast and mycelial phase antigens of histoplasma are used. Those who have facilities and expertise for growing and handling dangerous pathogens can prepare histoplasma mycelial antigen as described above. The cultures must be inoculated in an exhaust protective cabinet and killed by exposure for three days to 1:5000 thiomersal before processing. A yeast phase produced on chocolate cystine agar can be used to inoculate glucose peptone broth fortified with 0.1% of cystine. The culture is then stirred or shaken at 37 °C for 3–4 days and killed with thiomersal. The washed yeasts are suspended in saline with 1:5000 thiomersal and this whole yeast cell suspension constitutes the antigen.

The preparation of *Coccidioides immitis* antigen is best left to the experts: laboratory infections are almost always fatal.

Tests for Candida antibodies

Agglutination of whole yeast cells

The organisms are grown on glucose peptone agar plates for 24 h at 37 °C. Prolonged incubation should be avoided as it encourages mycelial formation. The growth is washed off in saline, centrifuged, washed three times with saline and the packed cells are suspended to about 20% by eye. A Wintrobe haematocrit tube is filled with this suspension and the packed cell volume estimated. The stock solution is then distributed in bottles sufficient to make a 2% v/v suspension by the addition of 10 ml of phosphate-buffered saline with azide. These bottles of stock suspension are kept at −20 °C and one is diluted for use as required.

Sera need no inactivation, except for cryptococcus latex tests. Volumes of 0.4 ml of phosphate-buffered saline are dispensed into a WHO tray and serial twofold dilutions of serum made as usual. One drop (0.02 ml) of yeast suspension is added to each well and the tray gently shaken to mix and then left at room temperature for 24 h. The deposit is then re-suspended by gentle shaking and the end-point read at + agglutination. Trace agglutinations are ignored.

Tests for cryptococcal antigen and antibody

Antigen

Polystyrene latex, coated with rabbit anti-cryptococcus globulin, is agglutinated by cryptococcal capsular polysaccharide. Because of extensive cross-reaction, type A globulin acts as a 'polyvalent' reagent.

Preparation of globulin

Rabbits are immunized with a small capsule strain of cryptococcus until their serum has an agglutination titre of 1 in 500 or better. The animals are then bled and the globulin fraction of their serum isolated using ammonium sulphate or caprylic acid. After dialysis, it is freeze-dried and reconstituted at a concentration of 4% for use.

Sensitization of latex

The stock latex solution is made from Dow or Difco latex 0.81. A 1:20 dilution is normally satisfactory. When this is further diluted 1:100 in a round 12-mm tube, it should have an absorbance of 0.29–0.31 at 650 nm. To this stock an equal volume of the maximally reactive dilution of globulin in glycine-saline is added. This dilution is determined by a chess-board titration using latex sensitized with 1:100, 1:200, 1:400, 1:600 and 1:800 globulin against a known positive human serum or CSF. Alternatively, diluted washings from cryptococcal cells can be used.

The routine reagent can usually be made by mixing equal volumes of 1:250 globulin and 1:10 latex and leaving the mixture at 4 °C overnight.

Screening test

Eight drops of diluent are placed in each of two wells in a WHO plastic tray for each specimen. Three areas are marked off on a microscope slide with a felt pen. Two drops of serum or CSF are added to the left-hand area of the slide and two drops to the first well. After mixing, two drops of this 1:5 dilution are added to the centre area of the slide and two to the second well, making this 1:25. Two drops of the 1:25 dilution from the second well are added to the right-hand area of the slide. One drop of sensitized latex is added to each area on the slide and the slide rocked for 3 min. A slight granularity always appears: this should be disregarded. Definite agglutination is considered positive.

When the screening test is positive or if cryptococci have been isolated, doubling dilutions are tested to obtain an end-point.

Tests for circulating antibody

A suspension of serotype A is made and the test carried out as for candida agglutination (above) using, however, a medium that discourages florid capsule formation. The following has been found suitable:

Glucose	10 g
Tryptone	10 g
$NaH_2PO_4.2H_2O$	5 g
Agar	15 g
Water	1000 ml

Dissolve by heating, distribute and sterilize at 115 °C for 10 min.

Diluent for cryptococcal latex agglutination

Glycine	7.31 g
Sodium chloride	10 g
1 N sodium hydroxide solution	3.5 ml
Water	to 1000 ml

Leave the solution overnight to equilibrate then check the pH, which should be 8.2.

Use this buffer as it is to dilute the latex; for diluting test and control sera, add 0.1% of bovine albumin. (This buffer is an excellent culture medium; keep the stock frozen at −20 °C.)

Routine double diffusion tests

Agar base

This is made by dissolving 2 g of agar (Oxoid No. 1) in 100 ml of water by autoclaving. A 100-ml volume of buffer is heated to 50 °C and mixed with the agar. The medium is kept in a water-bath at 50 °C and discarded if not used within 48 h of preparation.

Buffer

Boric acid (H_3BO_3)	10 g
Powdered borax ($Na_2B_4O_7.10H_2O$)	20 g
EDTA, disodium salt	10 g
Water	to 1000 ml

The pH should be 8.2. This buffer is, of course, double strength; it is diluted with agar for plates and slides or with an equal volume of water for use in electrophoresis tanks.

Plates

A 3-ml volume of agar is measured into a plastic petri dish and a Perspex jig with metal pegs is fitted in place of the lid to produce the pattern of wells shown in *Figure 8.1.* The large wells are 6 mm in diameter and the smaller wells 2 mm. The distance between the central and peripheral wells is also 6 mm. The test serum occupies the central well, antigens the pairs of large and small wells (shown shaded) and appropriate control sera the top and bottom wells. Two antigens can thus be tested and a single solution suffices as the relative volumes (60 and 6 μl in the two holes) give high and low concentration gradients in the agar. This arrangement also allows reactions of identity to be obtained between test and control sera, thus eliminating some anomalous reactions, particularly among the aspergilli.

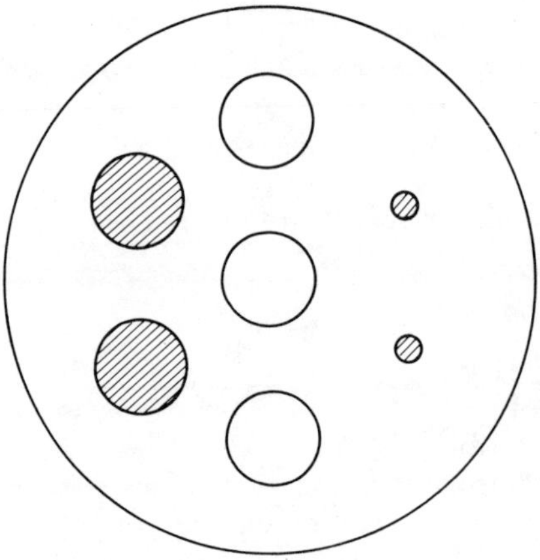

Figure 8.1 Double diffusion tests

Electrophoresis

Slides for electrophoresis

A clean slide is thinly coated with buffer agar, using a clean finger, and allowed to dry. A 2.5-ml volume of agar is run on and allowed to set with the slide on a level surface. The pattern of slots and holes shown in *Figure 8.2* is

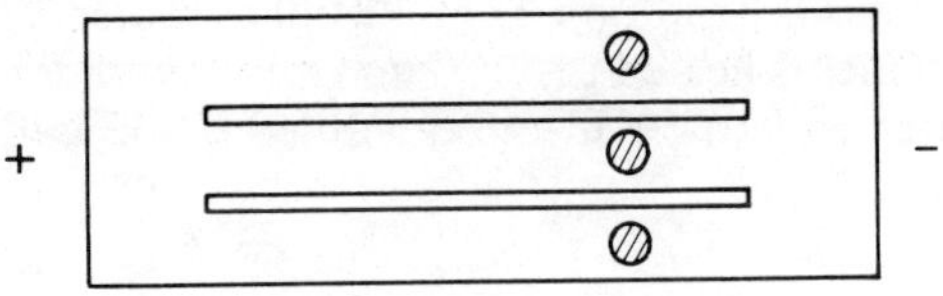

Figure 8.2 Immunoelectrophoresis slide

used for immunoelectrophoresis. The holes are 2 mm in diameter and the slots are 1 mm wide. The holes are filled with antigen and then run in an electrophoresis tank for 1¼ h with a voltage gradient of 30 V from end to end of the slide. It is best to measure the voltage directly with a voltmeter; this eliminates variations due to the buffer level and concentration in the tank, length of connecting wicks and ambient temperature. After a run, the slots are filled with serum and the whole is incubated in a moist chamber.

Antigens

These are used at a concentration of 6 mg/ml for double diffusion, and 24 mg/ml for electrophoretic tests, where the volume of antigen is relatively small.

Double diffusion and immunoelectrophoresis reactions are allowed to proceed for three days at 28 °C (room temperature is satisfactory: at 37 °C the plates dry out). All unreacted serum must then be washed out. The plates are therefore stood on edge in the washing fluid for two days, allowing at least 200 ml of fluid per plate. A convenient solution is:

Borax ($Na_2B_4O_7.10H_2O$)	4 g
Sodium chloride	4 g
Water	to 1000 ml

The plates are then soaked for about 2 min in distilled water, the agar discs are transferred to 3 × 1-in microscope slides and the overhanging agar is cut off. They are then dried in an oven at 60 °C. When dry they are stained with one of the usual protein stains. The following is suitable.

Diluent

Methanol	500 ml
Glacial acetic acid	100 ml
Water	400 ml

Stain

0.1% Coomassie Brilliant Blue R in the above solution.

They are stained for 10 min and then rinsed in methylated spirit and differentiated in two changes of the diluent, 10 min in each. The dried slides can be stored in a histological slide file.

Electrophoresis slides are treated in the same way except that a narrow rubber band is needed to hold them on the slide while they are being washed, and the slots must be filled with agar before the slide is dried or they will expand and mask the arcs of the reaction.

Counterimmunoelectrophoresis

This is a more sensitive method than immunodiffusion and better for handling large numbers of sera. It requires much practice and a close attention to detail and is not suitable for occasional use. For details of this and other serological methods *see* Mackenzie *et al.*[4].

Antifungal drugs

Methods of sensitivity testing and assay

Medium

5-Fluorocytosine is sulphonamide-like, in that it requires a medium free of antagonists. This excludes the use of peptone and meat extract. Some organisms, particularly cryptococci, are reluctant to grow on a medium whose only source of nitrogen is inorganic. Growth is much improved by using casamino acids, a nitrogen source free of nucleic acids and their components, supplemented by sufficient yeast extract to give good growth without antagonizing 5-fluorocytosine. Difco or Oxoid yeast extract is suitable for this; Marmite is not. A small amount of glycerophosphate is added as a buffer. This does not precipitate on autoclaving, and the sterilized medium remains clear. The other antifungal drugs have no special requirements but it is obviously simpler to use one medium for all. For preparation of the medium *see* p. 68.

Drug solutions

5-Fluorocytosine is freely soluble in water. A stock solution containing 1000 μg/ml will keep indefinitely if frozen at −20 °C. When this solution is thawed for use, a few crystals of 5-fluorocytosine may remain undissolved. Complete solution is assured by placing it in the 56 °C water-bath for 10 min before use. Amphotericin B is available as Fungizone, a complex with sodium deoxycholate in 50-mg quantities. Dissolve contents of one bottle in 10 ml of sterile water (avoid saline at all stages, it precipitates the amphotericin). This stock solution will keep for six months if frozen at −20 °C. For a working solution make two tenfold dilutions in 0.4% sodium deoxycholate solution, to give a concentration of 50 μg/ml. This keeps for several months if frozen at −20 °C. Pimaricin must be dissolved in dimethyl formamide; Nystatin and the imidazole compounds in equal volumes of acetone and water. Intermediate dilutions of these drugs are opalescent, and should be prepared just before use; if kept for more than about 30 min there is an appreciable loss through precipitation on the glass of the bottle. Solutions of the imidazole compounds—20 mg of solid weighed into a sterile bottle then taken up in 10 ml of water plus 10 ml of acetone—have been found to keep for four months in the refrigerator. There is no point in keeping them at −20 °C since the acetone prevents them from freezing.

Working solutions for sensitivity testing

5-Fluorocytosine: 100 μg and 10 μg/ml
Miconazole, ketoconazole, pimaricin and nystatin (the last in units): 50 μg and 5 μg/ml
Amphotericin B: 10 μg and 1 μg/ml.

Sensitivity testing

Melt the measured volumes of agar, then cool them to 48 °C. Use 10-ml lots for slopes, adding drug solutions as follows:

Miconazole, ketoconazole, pimaricin and nystatin

50 μg/ml				0.1	0.2	0.4	0.8 ml
5 μg/ml	0.1	0.2	0.4				ml
Concentration	0.05	0.1	0.2	0.5	1.0	2.0	4.0 μg/ml

Amphotericin B

10 μg/ml				0.1	0.2	0.4	0.8 ml
1.0 μg/ml	0.1	0.2	0.5				ml
Concentration	0.01	0.02	0.05	0.1	0.2	0.4	0.8 ml

5-Fluorocytosine

(The drug may be added to measured volumes of agar before autoclaving.)

100 μg/ml				0.1	0.2	0.4	0.8 ml
10 μg/ml	0.1	0.2	0.5				ml
Concentration	0.1	0.2	0.5	1.0	2.0	4.0	8.0 μg/ml

The volume of antibiotic added has been neglected. The maximum error introduced is 8%, too small a difference to affect the result. Divide each 10-ml lot into two slopes, and include a pair of control slopes without drugs. For plates, 20-ml lots of medium are used and the above concentrations are of course doubled.

The inoculum

Grow the organism on the assay medium at 30 °C. Results are less satisfactory if this step is omitted. Yeasts will grow well enough after 24 h, and the growth should be suspended in sterile water, then this suspension added, drop by drop, to 10 ml of sterile water till a faint but distinct opalescence results. One drop of this diluted suspension is run down each slope, and incubation continued at 30 °C until growth on the control slopes is adequate. *Candida* species reach this stage in 24 h: cryptococci need a further 24-h incubation. Filamentous organisms; aspergilli, histoplasmata and the like, are grown on the same medium again at 30 °C. The young growth is stripped off the surface of the medium after 2–3 days before spore formation begins and macerated in a 'bijou' macerator with 3 ml of sterile water, using 15-s bursts until enough of the growth is disintegrated to give an opalescent suspension. The macerator bottle is allowed to stand for 10 min, so that aerosols within may settle. The macerating dog is then removed, replaced by the cap of a sterile bijou bottle and then used to cap that bottle. The contents of the bottle are now mixed gently, large lumps are allowed to settle, then the supernatant suspension is taken off and treated exactly like the yeast suspension, i.e. it is diluted till just opalescent. One drop is used as inoculum and the tubes are incubated as before until the growth of the controls is adequate.

If the sensitivities of several yeast strains are required at the same time, up to four organisms can be accommodated by using duplicate drops of suspension on plates marked out in sectors as in viable counting. As the drops are concentrated on a small area, a 1/25 dilution of the suspension described above is necessary to obtain comparable results. In all cases the aim should be to produce discrete colonies, not a confluent growth.

When several strains are to be treated it is convenient to prepare the slopes or plates containing 5-fluorocytosine, miconazole or ketoconazole in the afternoon, then keep them in the refrigerator overnight. If plates are used, they can be dried next morning while the amphotericin series is being prepared. 5-Fluorocytosine or miconazole incorporated in agar slopes will keep for three weeks in the refrigerator at 4 °C with no detectable loss of potency. Where lots of tests are done, quantity production is feasible—and easier.

Short method for yeast sensitivity tests

In practice, yeast sensitivities generally fall within the four lowest drug concentrations given above. Using a square petri dish with 25 compartments, four drug concentrations and a control can be put in the five columns across one plate using volumes of 1.2 ml. If the centre row is left empty, two organisms can be tested in duplicate. Plates containing 5-fluorocytosine or ketoconazole will keep for a week in the refrigerator.

Amphotericin B, miconazole and ketoconazole assay

Standard solutions

Amphotericin B, 2.0 μg/ml; miconazole, 4.0 μg/ml or ketoconazole, 4.0 μg/ml is diluted 1/10 in horse serum. The appropriate standard is diluted in duplicate and in parallel with duplicate dilutions of the patient's serum in one of the following ways.

(1) Four rows of eleven 75 × 12 mm test-tubes with aluminium caps (Oxoid) are sterilized in a rack, 0.5 ml of infected broth is added to each; 0.5 ml of standard to the first tube of rows 1 and 2 and 0.1 ml of the test serum to rows 3 and 4. Each one is then serially diluted to tube 10, leaving tube 11 as a growth control.
(2) A sterile Microtitre plate (M24AR) with lid, (M42R) used with 0.1-ml volumes of diluent and sera. The whole operation is made much easier by using a four-place multichannel automatic pipette with autoclaved tips for dispensing and diluting. The infected broth may be dispensed from a 5-cm plastic petri dish or a sterile 50-ml beaker.

In either case, the results of amphotericin B assay can be read after overnight incubation. Assays of imidazoles should be left for 24 h, then the end point taken where growth is sharply reduced by comparison with the controls, neglecting traces of growth.

Methods of assay

5-Fluorocytosine assay

5-Fluorocytosine diffuses readily in agar while amphotericin B, miconazole and ketoconazole, the other drugs commonly encountered, diffuse little if at all. Fortunately, the expected levels of these latter drugs are low and it is possible to estimate serum levels by dilution in broth in parallel with a standard prepared in horse serum.

Media

The liquid medium for dilution tests is given on p. 68.

Organisms

The test organism is *Saccharomyces cerevisiae* NCPF 3178. It is the least granular when grown on glucose peptone agar and safer to use than a potential pathogen. A distinctly cloudy suspension is made in 10 ml of sterile distilled water: 0.1 ml is used in broth for the liquid assay and the rest added to 110 ml of 5-fluorocytosine assay medium melted and cooled to 45 °C.

Pour the inoculated medium into a bioassay plate 250-mm square (e.g. Nunc 1015).

Standard and test sera

Prepare standards containing 10, 15, 25, 40 and 60 μg/ml of 5-fluorocytosine.

Keep frozen (will last for six months). Dilute patient's serum as follows (any amphotericin B which is present will be neutralized within the hour):

Horse serum	0.4 ml
Patient's serum	0.5 ml
Ergosterol 1% in acetone	0.1 ml

If the patient is being treated with an imidazole drug, use 0.1 ml of 10% Tween 80 in place of ergosterol.

The test

Lay the dish on a sheet of paper (template) having a pattern of 25 positions to which test or standard numbers are assigned at random. Punch holes in the medium with a 4-mm cork borer and remove the agar. Add 15 μl volumes of test and standard solutions in duplicate in the pattern shown on the template. Incubate at 30 °C overnight.

Measure the diameters of the zones of inhibition. Plot the standards on one decade semilogarithmic paper and obtain the concentration of drug in the test serum by interpolation.

References

1. BENECKE, E.S. and ROGERS, A.L. (1971) *Medical Mycology Manual.* 3rd edn. Minneapolis: Burgess
2. HALEY, L.D. and CALLOWAY, C.S. (1978) *Laboratory Methods in Medical Mycology.* 4th edn. Washington: US Department of Health, Education and Welfare
3. EMMONS, C.W., BINFORD, C.H., UTZ, J.P. and KWON-CHUNG, K.J. (1970) *Medical Mycology.* 3rd edn. Philadelphia: Lea and Febiger
4. MACKENZIE, D.W.R., PHILPOT, C.M. and PROCTOR, A.G. (1980) *Basic Serodiagnostic Methods for Diseases Caused by Fungi and Actinomycetes.* Public Health Laboratory Service Monograph No. 12. London: HMSO

Chapter 9

Counting micro-organisms

It is often necessary to report on the size of the bacterial population in a sample. Unfortunately, industries and health authorities have been allowed to attach more importance to these 'bacterial counts' than is permitted by their technical or statistical accuracy.

If a *total count* is required, many of the organisms counted may be dead or indistinguishable from other particulate matter. A *viable count* assumes that a visible colony will develop from each organism. Bacteria are, however, rarely separated entirely from their fellows and are often clumped together in large numbers particularly if they are actively reproducing. A single colony may therefore develop from one organism or from hundreds or even thousands of organisms. Each colony develops from one *viable unit*. Because any agitation, as in the preparation of dilutions, will break up or induce the formation of clumps, it is obviously difficult to obtain reproducible results. Bacteria are seldom distributed evenly throughout a sample and as only small samples are usually examined very large errors can be introduced.

Many of the bacteria present in a sample may not grow on the medium used at the pH or incubation temperatures employed or in the time allowed.

Accuracy is often demanded where it is not needed. If it is decided that a certain product should contain less than, say, ten viable bacteria/g, this suggests that, on average, of ten tubes each inoculated with 0.1 g, seven would show growth and three would not, and out of ten tubes each inoculated with 0.01 g only one or two would show growth. If all the 0.1-g tubes or five of the 0.01-g tubes showed growth, there would be more than ten organisms/g. It does not matter whether there are 20 or 10 000: there are too many. There is no need to employ elaborate counting techniques.

In viable count methods, it is recognized that large errors are inevitable, even if numbers of replicate plates are used. Some of these errors are as indicated above, inherent in the material, others in the technique. Errors of $\pm 90\%$ in counts of the order of 10 000 to 100 000/ml are not unusual even with the best possible technique. It is, therefore, necessary to combine the maximum of care in technique with a liberal interpretation of results. The figures obtained from a single test are valueless. They can be interpreted only if the product is regularly tested and the normal range is known.

Physical methods are used to estimate total populations, i.e. dead and living organisms. They include direct counting and measurements of turbidity. Biological methods are used for estimating the numbers of viable units. These include the plate count, roll-tube count, drop count, surface colony count, dip slide count, contact plate, membrane filter count, most probable number estimations, ATP luminescence and the measurement of electrical impedance due to bacterial metabolism.

Counting chamber method

The Helber counting chamber is a slide 2–3 mm thick with an area in the centre called the platform and surrounded by a ditch, which is 0.02 mm lower than the remainder of the slide (*Figure 9.1*). The top of the slide is ground so that when an optically plane cover-glass is placed over the centre depression, the depth is uniform. On the platform an area of 1 mm^2 is ruled so that there are 400 small squares each 0.0025 mm^2 in area. The volume over each small square is 0.02×0.0025 mm^3, i.e. 0.000 05 ml.

Add a few drops of formalin to the well mixed suspension to be counted. Dilute the suspension so that when the counting chamber is filled there will be about five or ten organisms per small square. This requires initial trial and error. The best diluent is 0.1% peptone water containing 0.1% lauryl sulphate

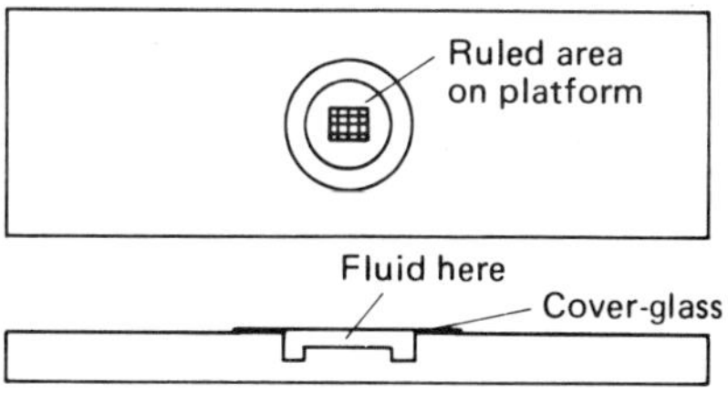

Figure 9.1 Bacterial counting chamber (depth is exaggerated)

and (unless phase contrast or dark field is used for counting) 0.1% methylene blue. Always filter before use.

Place a loopful of suspension on the ruled area and apply the cover-glass, which must be clean and polished. The amount of suspension must be such that the space between the platform and the cover-glass is just filled and no fluid runs into the ditch; again, this requires practice. If the cover-glass is applied properly, Newton's rings will be seen. Allow 5 min for the bacteria to settle.

Examine with a 4-mm lens with reduced light, dark field or phase contrast if available. Count the bacteria in 50–100 squares selected at random so that the total count is about 500. Divide the count by number of squares counted. Multiply by 20 000 000 and by the original dilution factor to obtain the total number of bacteria/ml. Repeat twice more and take the average of the three counts. Clumps of bacteria, streptococci, etc., can be counted as units or each cell counted as one organism.

With experience, reasonably accurate counts can be obtained but the chamber and cover-glass must be scrupulously cleaned and examined microscopically to make sure bacteria are not left adhering to either.

The Breed count

Named after its originator, this is a rough but useful technique. Breed slides have areas of 1 cm^2 marked on them. Into a square, place 0.01 ml of the fluid to be counted, for example milk, with a commercially available microsyringe or loop (Astell). Allow to dry and stain with methylene blue. Examine with an oil immersion lens giving a known field diameter, which is usually about 0.16 mm. The area seen is thus πr^2 or 3.14×0.8^2, i.e. 2 mm^2, and as the total area is 100 mm^2, one field represents 1/5000 part of the whole area, or 1/5000 part of the original 0.01 ml. One organism per field equals 5000/0.01 ml or 500 000/ml. Count the number of organisms in several fields in different parts of the ruled area and use the equation

$$\text{Count/ml} = \frac{N \times 4 \times 10^4}{\pi d^2}$$

where N is the number per field and d is the diameter of the field.

Opacity tube method

International Reference Opacity Tubes, containing glass powder standards are available from Wellcome. These are numbered narrow glass tubes of increasing opacity and a table is provided equating the opacity of each tube with the number of organisms/ml. The unknown suspension is matched against the standards in a glass tube of the same bore. It may be necessary to dilute the unknown. These physical methods should not be used with Risk Group III organisms.

Viable counts

In these techniques, the material containing the bacteria is serially diluted and aliquots of each dilution are placed in or on suitable culture media. Each colony developing is assumed to have grown from one viable unit, which, as indicated earlier, may be one organism or a group of many.

Diluents

Some diluents, e.g. saline or distilled water, may be lethal for some organisms. Diluents must not be used direct from a refrigerator as cold-shock may prevent organisms from reproducing. Peptone water, 0.1%, is the best diluent.

Pipettes

The 10-ml and 1-ml straight-sided blow-out pipettes are commonly used and are specified for some statutory tests. Disposable pipettes are labour saving. Mouth pipetting must be expressly forbidden, regardless of the nature of the material under test. A method of controlling teats is given on p. 29. Automatic pipettors or pipette pumps, described on p. 29, should be used. Some of these can be pre-calibrated and used with disposable polypropylene pipette tips.

Pipettes used in making dilutions must be very clean, otherwise bacteria will adhere to their inner surfaces and may be washed out into another dilution. Siliconed pipettes may be used. Fast-running pipettes and vigorous blowing-out should be avoided; they generate aerosols. As much as 0.1 ml may remain in pipettes if they are improperly used.

Preparation of dilutions

Pipette 9-ml amounts of diluent solution into sterile test-tubes with suitable caps. These are the *dilution blanks*. Do not tube and then sterilize these unless screw-capped bottles are used; autoclaving tubes with aluminium or polypropylene caps may alter the volume.

When diluting liquids, for example milk for bacterial counts, proceed as follows.

Mix the sample by shaking. With a straight-sided pipette dipped in half an inch only, remove 1 ml. Deliver into the first dilution blank, about half an inch above the level of the liquid. Wait 3 s, then blow out. Discard the pipette. With a fresh pipette, dip half an inch into the liquid, suck up and down ten times to mix, but do not blow bubbles. Raise the pipette and blow out. Remove 1 ml and transfer to the next dilution blank. Discard the pipettes. Continue for the required number of dilutions, and remember to discard the pipette after delivering its contents, otherwise the liquid on the outside will contribute to a cumulative error. The dilutions will be:

Tube no.	*1*	*2*	*3*	*4*	*5*
Dilution	1:10	1:100	1:1000	1:10 000	1:100 000
Vol. of original fluid/ml	0.1 (or 10^{-1})	0.01 (10^{-2})	0.001 (10^{-3})	0.0001 (10^{-4})	0.00001 (10^{-5})

When counting bacteria in solid or semisolid material, weigh 10 g and place in a Stomacher Lab-Blender (Seward), or other blender. Add 90 ml of diluent and homogenize.

Alternatively, cut into small pieces with a sterile scalpel, mix 10 g with 90 ml of the diluent and shake well. Allow to settle. Assume that the bacteria are now evenly distributed between the solid and liquid.

Both of these represent dilutions of 1: 10, i.e. 1 ml contains or represents 0.1 g, and further dilutions are prepared as above. The dilutions will be:

Tube no.	*1*	*2*	*3*	*4*
Dilution	1:100	1:1000	1:10 000	1:100 000
Weight of original material	0.01 (or 10^{-2})	0.001 (10^{-3})	0.0001 (10^{-4})	0.00001 g (10^{-5} g)

There are mechanical aids for preparing dilutions, e.g. the Droplette, described on p. 133 (Seward) and the Compupet (Warner; General Diagnostics). Trotman[8] reviews several instruments.

Plate count

Melt nutrient agar or other suitable media tubed in 10-ml amounts. Cool to 45 °C in a water-bath.

Set out petri dishes, two or more per dilution to be tested and label with the dilution number. Pipette 1 ml of each dilution into the centre of the appropriate dishes, using a fresh pipette for each dilution. Do not leave the dish uncovered for longer than is absolutely necessary. Add the contents of one agar tube to each dish in turn and mix as follows. Move the dish gently six times in a clockwise circle of diameter about 150 mm. Repeat counter-clockwise. Move the dish back-and-forth six times with an excursion of about 150 mm. Repeat with to-and-fro movement. Allow the medium to set, invert and incubate for 24–48 h.

Economy in pipettes

In practice, the pipette used to transfer 1 ml of a dilution to the next tube may be used to pipette 1 ml of that dilution into the petri dish.

Counting colonies

To count on a simple colony counter select plates with between 30 and 300 colonies. Place the open dish, glass side up, over the illuminated screen. Count the colonies using a 75-mm magnifier and a hand-held counter. Mark the glass above each colony with a felt tip pen. Calculate the colony or viable count/ml by multiplying the average number of colonies per countable plate by the reciprocal of the dilution. Report as 'colony forming units/ml' (cfu) or as 'viable count/ml' not as 'bacteria/g or /ml'.

If all plates contain more than 300 colonies rule sectors, e.g. of one-quarter or one-eighth of the plate, count the colonies in these and include the sector value in the calculations.

For large work loads semi- or fully automatic counters are essential. In the former the pen used to mark the glass above the colonies is connected to an electronic counter which displays the numbers counted on a small screen (Astell, Anderman, Gallenkamp, Horwell, Technilab). In the fully automatic models a TV camera or laser beam scans the plate and the results are displayed or recorded on a screen or read-out device (New Brunswick, Don Whitley, Exotech).

Roll-tube count

Instead of using plates, tubes or bottles containing media are inoculated with diluted material and rotated horizontally until the medium sets. After incubation, colonies are counted.

Tube the medium in 2–4-ml amounts in 25-ml screw-capped or Astell bottles. The medium should contain 0.5–1.0% *more* agar than is usual. Melt and cool to 45 °C in a water-bath. Add 0.1 ml of each dilution and rotate horizontally in cold water until the agar is set in a uniform film around the walls of the bottle. This requires some practice. An alternative method employs a slab of ice taken from the ice tray of a refrigerator. Turn it upside down on a cloth and make a groove in it by rotating horizontally on it a bottle similar to that used for the counts but containing warm water. The count tubes are then rolled in this groove. If the roll-tube method is to be used often, the Astell Roll Tube apparatus, which includes a water-bath, saves much time and labour.

Incubate roll-tube cultures inverted so that condensation water collects in the neck and does not smear colonies growing on the agar surface. To count, draw a line parallel to the long axis of the bottle and rotate the bottle, counting colonies under a low-power magnifier.

The roll-tube method is popular in the dairy and food industries. Machines for speeding up and taking the tedium out of this method are available from Astell and Colab (Inolex). The Colworth Diluter Dispenser (Seward) may be used in order to add the inoculum to the roll-tubes, which are then placed on these machines for rolling and setting. Astell also make counting boxes for roll-tubes.

A roll-tube method for counting organisms in bottles (container sanitation) is given on p. 243.

Drop count method

In this method, introduced by Miles and Misra[1] and usually referred to by their names, small drops of the material are placed on agar plates. Colonies are counted in the inoculated areas after incubation.

Prepare 50-dropper pipettes. These are pasteur pipettes passed through a Starratt Standard Wire Gauge hole No. 59 (0.95 mm) and cut off accurately above the hole. When tested they should deliver 0.02 ml, i.e. 50 drops/ml. A tolerance of ± 2 drops is permissible. They are available commercially (Bilbate). Alternatively use hypodermic needles (19 gauge) purchased before the bevels and points are ground. Clean well with chloroform to remove grease before testing, as it affects accuracy. Attach to a short piece of glass tubing by a silicone rubber or PVC sleeve and use with a rubber teat. When tested, about 70–80% of these deliver between 48 and 52 drops/ml. Clean 50-drop pipettes of either kind in hot detergent, rinse in chloroform and sterilize in glass tubes.

Dry plates of suitable medium very well before use. Drop at least five drops from each dilution of sample from a height of not more than 2 cm (to avoid splashing) on each plate. Replace the lid but do not invert until the drops have dried. After incubation, select plates showing discrete colonies in drop areas, preferably one which gives less than 40 colonies per drop (10–20 is ideal). Count the colonies in each drop, using a hand magnifier. Divide the total count by the number of drops counted, multiply by 50 to convert to 1 ml and by the dilution used.

This method lends itself to arbitrary standards; for example, if there are less than ten colonies between five drops of sample, this represents less than 100 colonies of that material. If these drops are uncountable (more than 40 colonies), then the colony count/ml is greater than 2000.

The 'droplette' method

This accurate and rapid method was introduced by Sharpe and co-workers[2,3]. Serial, replicate dilutions of the sample are made mechanically in agar medium in 0.1-ml amounts and 0.1-ml drops are placed in petri dishes automatically. The apparatus, known as the 'Colworth Droplette', is manufactured by Seward and offers great savings in time and labour.

It consists of a diluter-dispenser, which accurately and automatically meters ungraduated disposable polypropylene or pasteur pipettes in 0.1–1.7-ml amounts, a viewer, which enables the agar drops to be seen on a grid screen at about 9-cm diameter, and an electromechanical counter, which records, with a felt-tipped pen, when the colony images are touched.

Surface count method

This is not very accurate but is useful for rough estimates of bacterial numbers, e.g. in urine examinations or 'in-line' checks in food processing plant.

Place 0.1 ml or another suitable amount of the sample measured with a standard loop (Astell, Nunc) or a micropipette in the centre of a well-dried plate of suitable medium and spread it with a loop or spreader all over the surface (p. 30). Incubate and count colonies.

When counting colonies, difficulties arise with spreading organisms and large 'smears' of small colonies caused when a large viable unit is broken up but not dispersed during manipulation. Treat these as single units. Usually, other colonies can be seen and counted through spreaders.

The spiral plate method

The apparatus is described on p. 90. The results are said to compare favourably with those obtained by the surface spread plate method[4] and the Droplette method.[5]

Membrane filter counts

Liquid containing bacteria is passed through a filter that will retain the organisms. The filter is then allowed to absorb the culture medium and incubated, when the colonies which develop can be counted.

The filter-carrying apparatus is made of metal, glass or plastic and consists of a lower funnel, which carries a fritted glass platform surrounded by a silicone rubber ring. The filter disc rests on the platform and is clamped by its periphery between the rubber ring and the flange of the upper funnel. The upper and lower funnels are held together by a clamp.

The filters are thin, porous cellulose ester discs, varying in diameter and about 120 μm thick. The pores in the upper layers are 0.5–1.0 μm diameter enlarging to 3–5 μm diameter at the bottom. Bacteria are thus held back on top, but culture medium can easily rise to them by capillary action. Filters are stored interleaved between absorbent pads in metal containers. A grid to facilitate counting is ruled on the upper surface of each filter.

Membrane filter apparatus is supplied by Gallenkamp, Millipore, Thomas and several other companies. Disposable apparatus is made by Millipore and Falcon (Becton Dickinson). Several very useful booklets on membrane filtration are published by Millipore. References to some of these are made elsewhere in this book. Leaflets are also available from Falcon (Becton Dickinson), Oxoid and Gelman. Membrane filters are supplied by Oxoid, Gelman, Millipore, Fisons and others.

Shallow metal incubating boxes with close-fitting lids, containing absorbent pads to hold the culture media, are also required.

The most convenient size of apparatus for counting colonies takes filters 47-mm diameter, but various sizes larger or smaller are available.

Sterilization

Assemble the filter carrier with a membrane filter resting on the fritted glass platform, which in turn should be flush on top of the rubber gasket. Screw up the clamping ring, but not to its full extent. Wrap in kraft paper and autoclave for 15 min at 121 °C. Alternatively, loosely assemble the filter carrier, wrap and autoclave; sterilize the filters separately interleaved between absorbent pads in their metal container also by autoclaving. This is the most convenient method when a number of consecutive samples are to be tested; in this case, sterilize also in the same way the appropriate number of spare upper funnels.

Sterilize the incubation boxes, each containing its absorbent pad by autoclaving.

Re-use of membrane filters

Membranes are expensive and can often be re-used several times provided that they have not been used for pathogens and are handled with care (entomological forceps are the best tool for this). Wash membranes well with running water. Steep in laboratory detergent solution for several hours and boil in 3% hydrochloric acid. Wash again with running water, blot and store interleaved in absorbent pads as described above.

Culture media

Ordinary or standard culture media do not give optimum results with membrane filters. It has been found necessary to vary the proportions of some of the ingredients. Membrane filter versions of standard media are noted on p. 62. A 'Resuscitation' medium to revive attenuated bacteria is also desirable for some purposes.

Method

Erect the filter carrier over a filter flask connected by a non-return valve to a pump giving a suction of 25–50 mmHg. Check that the filter is in place and tighten the clamping ring. Pour a known volume, neat or diluted, of the fluid to be examined into the upper funnel and apply suction.

Pipette about 2.5–3 ml of medium on a Whatman 5 cm, No. 17, absorbent pad in an incubation container or petri dish. This should wet the pad to the edges but not overflow into the container.

When filtration is complete, carefully restore the pressure. Unscrew the clamping ring and remove the filter with sterile forceps. Apply it to the surface of the wet pad in the incubation box so that no air bubbles are trapped. Put the lid on the container and incubate. A fresh filter and upper funnel can be placed on the filter apparatus for the next sample.

For total aerobic counts, use tryptone soya membrane medium. For counting coliform bacilli, either 'presumptive' or *E. coli*, *see* Water examination, p. 253. For counting *C. perfringens* in water, *see* p. 258.

For anaerobic counts, use the same medium incubated anaerobically, or roll the filter and submerge in a 25-mm diameter tube of melted thioglycollate agar at 45 °C and allow to set. To count anaerobes causing sulphide spoilage, roll the filter and submerge in a tube of melted iron sulphite medium at 42 °C and allow to set. For yeast and mould counts, use the appropriate Sabouraud-type or Czapek-type media.

Counting

Count in oblique light under low-power magnification. If it is necessary to stain colonies to see them, remove the filter from the pad and dip in a 0.01% aqueous solution of methylene blue for half a minute and then apply to the pad saturated with water. Colonies are stained deeper than the filter. The grid facilitates counting. Report as membrane colony count per standard volume (100 ml, 1 ml, etc.).

Count anaerobic colonies in thioglycollate tubes by rotating the tube under strong illumination. Count black colonies in iron sulphite medium.

Bacteria in air

Membrane filters may be used if the filter carrier is fitted with an impinger. Methods are given on p. 245.

Dip slides

This useful method, commonly used in clinical laboratories for the examination of urine, now has a more general application, especially in food and drink industries for viable coliform and yeast counts. The slides are made of plastic and are attached to the caps of screw-capped bottles. There are two kinds: one is a single- or double-sided tray containing agar culture media; the other consists of a membrane filter bonded to an absorbent pad containing dehydrated culture media. Both have ruled grids, either on the plastic or the filter, which facilitate counting. They are sold by Gibco, Medical Wire, Millipore, Oxoid and others. The Millipore leaflet on dip slides is very useful.

The slides are dipped into the samples, drained, replaced in their containers and incubated. Colonies are then counted and the bacterial load estimated.

Most probable number (MPN) estimates

These are based on the assumption that bacteria are *normally* distributed in liquid media, that is, repeated samples of the same size from one source are expected to contain the same number of organisms *on average*: some samples will obviously contain a few more, some a few less. The average number is the *most probable number*. If the number of organisms is large, the differences between samples will be small; all the individual results will be nearer to the average. If the number is small the differences will be relatively larger.

If a liquid contains 100 organisms/100 ml, then 10-ml samples will contain, on average, ten organisms each. Some will contain more, perhaps one or two samples will contain as many as 20; some will contain less, but a sample containing none is most unlikely. If a number of such samples are inoculated into suitable medium, every sample would be expected to show growth.

Similarly, 1-ml samples will contain, on average, one organism each. Some may contain two or three and others will contain none. A number of tubes of culture media inoculated with 1-ml samples would therefore yield a proportion showing no growth.

Samples of 0.1 ml, however, could be expected to contain only one organism per ten samples and most tubes inoculated would be negative.

It is possible to calculate the most probable number of organisms/100 ml for any combination of results from such sample series. Tables have been prepared (*Tables 9.1–9.4*) for samples of 10 ml, 1 ml and 0.1 ml using five tubes or three tubes of each sample size and, for water testing, using one 50-ml, five 10-ml and five 1-ml samples.

This technique is used mainly for estimating coliform bacilli, but it can be used for almost any organisms in liquid samples if growth can be easily observed, e.g. by turbidity, acid production. Examples are yeasts and moulds in fruit juices and beverages, clostridia in food emulsions and rope-spores in flour suspensions.

'Black tube' MPN counts for anaerobes are described on p. 370.

Mix the sample by shaking and inverting vigorously. Pipette 10-ml amounts into each of five tubes (or three) of double-strength medium, 1-ml amounts into each of five (or three) tubes of single-strength medium and 0.1-ml amounts (or 1 ml of a 1:10 dilution) into each of five or three tubes of single-strength medium. For testing water, also add 50 ml of water to 50 ml of double-strength broth.

TABLE 9.1[a]. MPN/100 ml, using one tube of 50 ml, and five tubes of 10 ml

50-ml tubes positive	*10-ml tubes positive*	*MPN/100 ml*
0	0	0
0	1	1
0	2	2
0	3	4
0	4	5
0	5	7
1	0	2
1	1	3
1	2	6
1	3	9
1	4	16
1	5	18+

[a] *Tables 9.1* to *9.3* indicate the estimated number of bacteria of the coliform group present in 100 ml of water, corresponding to various combinations of positive and negative results in the amounts used for the tests. The tables are basically those originally computed by McCrady[6] with certain amendments due to more precise calculations by Swaroop[7]; a few values have also been added to the tables from other sources, corresponding to further combinations of positive and negative results which are likely to occur in practice. Swaroop[7] has tabulated limits within which the real density of coliform organisms is likely to fall, and his paper should be consulted by those who need to know the precision of these estimates.

These tables and the accompanying information are reproduced from The Bacteriological Examination of Water Supplies, Reports on Public Health and Medical Subjects, No. 71 (1957), by permission of the Controller of Her Majesty's Stationery Office, London.

Double-strength broth is used for the larger volumes because the medium would otherwise be too dilute.

Incubate for 24–48 h and observe growth, or acid and gas, etc. Tabulate the numbers of positive tubes in each set of five (or three) and consult the appropriate table.

TABLE 9.2. MPN/100 ml, using one tube of 50 ml, five tubes of 10 ml and five tubes of 1 ml

50-ml tubes positive	*10-ml tubes positive*	*1-ml tubes positive*	*MPN/100 ml*
0	0	0	0
0	0	1	1
0	0	2	2
0	1	0	1
0	1	1	2
0	1	2	3
0	2	0	2
0	2	1	3
0	2	2	4
0	3	0	3
0	3	1	5
0	4	0	5
1	0	0	1
1	0	1	3
1	0	2	4
1	0	3	6
1	1	0	3
1	1	1	5
1	1	2	7
1	1	3	9
1	2	0	5
1	2	1	7
1	2	2	10
1	2	3	12
1	3	0	8
1	3	1	11
1	3	2	14
1	3	3	18
1	3	4	20
1	4	0	13
1	4	1	17
1	4	2	20
1	4	3	30
1	4	4	35
1	4	5	40
1	5	0	25
1	5	1	35
1	5	2	50
1	5	3	90
1	5	4	160
1	5	5	180+

TABLE 9.3. MPN/100 ml, using five tubes of 10 ml, five tubes of 1 ml and five tubes of 0.1 ml

10-ml tubes positive	*1-ml tubes positive*	*0.1-ml tubes positive*	*MPN/100 ml*
0	0	0	0
0	0	1	2
0	0	2	4
0	1	0	2
0	1	1	4
0	1	2	6
0	2	0	4
0	2	1	6
0	3	0	6
1	0	0	2
1	0	1	4
1	0	2	6
1	0	3	8
1	1	0	4
1	1	1	6
1	1	2	8
1	2	0	6
1	2	1	8
1	2	2	10
1	3	0	8
1	3	1	10
1	4	0	11
2	0	0	5
2	0	1	7
2	0	2	9
2	0	3	12
2	1	0	7
2	1	1	9
2	1	2	12
2	2	0	9
2	2	1	12
2	2	2	14
2	3	0	12
2	3	1	14
2	4	0	15
3	0	0	8
3	0	1	11
3	0	2	13
3	1	0	11
3	1	1	14
3	1	2	17
3	1	3	20
3	2	0	14
3	2	1	17
3	2	2	20
3	3	0	17
3	3	1	20
3	4	0	20
3	4	1	25
3	5	0	25
4	0	0	13
4	0	1	17

TABLE 9.3 (*continued*)

10-ml tubes positive	*1-ml tubes positive*	*0.1-ml tubes positive*	*MPN/100 ml*
4	0	2	20
4	0	3	25
4	1	0	17
4	1	1	20
4	1	2	25
4	2	0	20
4	2	1	25
4	2	2	30
4	3	0	25
4	3	1	35
4	3	2	40
4	4	0	35
4	4	1	40
4	4	2	45
4	5	0	40
4	5	1	50
4	5	2	55
5	0	0	25
5	0	1	30
5	0	2	45
5	0	3	60
5	0	4	75
5	1	0	35
5	1	1	45
5	1	2	65
5	1	3	85
5	1	4	115
5	2	0	50
5	2	1	70
5	2	2	95
5	2	3	120
5	2	4	150
5	2	5	175
5	3	0	80
5	3	1	110
5	3	2	140
5	3	3	175
5	3	4	200
5	3	5	250
5	4	0	130
5	4	1	170
5	4	2	225
5	4	3	275
5	4	4	350
5	4	5	425
5	5	0	250
5	5	1	350
5	5	2	550
5	5	3	900
5	5	4	1600
5	5	5	1800+

TABLE 9.4. MPN/100 ml, using three tubes each inoculated with 10, 1.0 and 0.1 ml of sample

Tubes positive			*MPN*	*Tubes positive*			*MPN*	*Tubes positive*			*MPN*
10 ml	*1.0 ml*	*0.1 ml*		*10 ml*	*1.0 ml*	*0.1 ml*		*10 ml*	*1.0 ml*	*0.1 ml*	
0	0	1	3	1	2	0	11	2	3	3	53
0	0	2	6	1	2	1	15	3	0	0	23
0	0	3	9	1	2	2	20	3	0	1	39
0	1	0	3	1	2	3	24	3	0	2	64
0	1	1	6	1	3	0	16	3	0	3	95
0	1	2	9	1	3	1	20	3	1	0	43
0	1	3	12	1	3	2	24	3	1	1	75
0	2	0	6	1	3	3	29	3	1	2	120
0	2	1	9	2	0	0	9	3	1	3	160
0	2	2	12	2	0	1	14	3	2	0	93
0	2	3	16	2	0	2	20	3	2	1	150
0	3	0	9	2	0	3	26	3	2	2	210
0	3	1	13	2	1	0	15	3	2	3	290
0	3	2	16	2	1	1	20	3	3	0	240
0	3	3	19	2	1	2	27	3	3	1	460
1	0	0	4	2	1	3	34	3	3	2	1100
1	0	1	7	2	2	0	21	3	3	3	1100+
1	0	2	11	2	2	1	28				
1	0	3	15	2	2	2	35				
1	1	0	7	2	2	3	42				
1	1	1	11	2	3	0	29				
1	1	2	15	2	3	1	36				
1	1	3	19	2	3	2	44				

From Jacobs and Gerstein's *Handbook of Microbiology*, D. Van Nostrand Company, Inc., Princeton, N.J. (1960)

Automated methods

These depend either on electrical impedance measurements, e.g. the Coulter Counter, the Bactometer (Bactomatic), Bactobridge (Lewis Williams), or ATP luminescence (Lumac). They are good investments in laboratories (e.g. food and drink processing) for the rapid monitoring of large numbers of samples.[8, 9, 10]

References

1. MILES, A.A. and MISRA, S.S. (1938) *Journal of Hygiene, Cambridge*, **38,** 732
2. SHARPE, A.N. and KILSBY, D.C. (1971) *Journal of Applied Bacteriology*, **34,** 435
3. SHARPE, A.N., DYETT, E.J., JACKSON, A.K. and KILSBY, D.D. (1972) *Applied Microbiology*, **24,** 4
4. JARVIS, B., LACH, V.H. and WOOD, J.M. (1977) *Journal of Applied Bacteriology*, **43,** 149
5. HEDGES, A.J., SHANNON, R. and HOBBS, R.P. (1978) *Journal of Applied Bacteriology*, **45,** 57
6. McCRADY, M.H. (1918) *Public Health Journal, Toronto*, **9,** 201
7. SWAROOP, S. (1951) *Indian Journal of Medical Research*, **39,** 107
8. CODY, P. *et al.* (1978) *Journal of Food Protection*, **41,** 277
9. MARTINS, S.B. and SELBY, M.J. (1980) *Applied and Environmental Microbiology*, **39,** 518
10. STANLEY, P. (1982) *Laboratory Equipment Digest*. February

Chapter 10

Agglutination tests

These tests are performed in small test-tubes (75 × 9 mm). Dilutions are made in physiological saline with graduated pipettes controlled by a rubber teat or (for single-row tests) with pasteur pipettes marked with a grease pencil at approximately 0.5 ml. Automatic pipettes or pipettors are useful for doing large numbers of tests.

Standard suspensions and agglutinating sera can be obtained from BBL, Difco, Pasteur Institute and Wellcome. Standard suspensions are used mainly in the serological diagnosis of enteric fever, which may be caused by several related organisms, and of brucellosis.

Standard agglutinating sera are used to identify unknown organisms.

Testing unknown sera against standard H and O suspensions

To test a single suspension

Prepare a 1:10 dilution of serum by adding 0.2 ml to 1.8 ml of saline. Set up a row of seven small tubes. Add 0.5 ml of saline to tubes 2–7 and 0.5 ml of 1:10 serum to tubes 1 and 2. Rinse the pipette by sucking in and blowing out saline several times. Mix the contents of tube 2 and transfer 0.5 ml to tube 3. Rinse the pipette. Continue with doubling dilutions but discard 0.5 ml from tube 6 instead of adding it to tube 7. Rinse the pipette between each dilution. The dilutions are now:

Tube no.	1	2	3	4	5	6	7
Dilution	1:10	1:20	1:40	1:80	1:160	1:320	0

Add 0.5 ml of Standard suspension to each tube. The last tube, containing no serum, tests the stability of the suspension. The dilutions are now:

Tube no.	1	2	3	4	5	6	7
Dilution	1:20	1:40	1:80	1:160	1:320	1:640	0

To test with several suspensions

In enteric fever investigations up to nine suspensions may be required. Set up six large tubes (152 × 16 mm). To tube 1 add 9 ml of saline, to tubes 2–6 add 5 ml of saline and to tube 1 add 1 ml of serum. Mix and transfer 5 ml from tube 1 to tube 2, and continue in this way, rinsing the pipette between each dilution.

Set up a row of seven small tubes (75 × 9 mm) for each suspension to be tested. Transfer 0.5 ml from each large tube to the corresponding small tube. Work from right to left, i.e. weakest to strongest dilution to avoid unnecessary rinsing of the pipette.

Add 0.5 ml of appropriate suspension to each row of small tubes. These will be the *H* and *O* suspensions of *S. typhi*, *S. paratyphi* A, B and C and Non-specific Salmonella *H* suspension. The dilutions are now 1:20 to 1:640.

Incubation temperature and times

Incubate tests with *O* suspensions in a water-bath at 37 °C for 4 h, then allow to stand overnight in refrigerator.

Incubate *H* tests for 2 h in a water-bath at 50–52 °C (*not* 55–56 °C, as the antibody may be partially destroyed). The level of the water in the bath should be adjusted so that only about half of the liquid in the tubes is below the surface of the water. This encourages convection currents in the tubes, mixing the contents.

Brucella agglutinations

To avoid false negative results due to the prozone phenomenon, double the dilutions for at least three more tubes, i.e. use 1:20 to 1:5120.

Reading agglutinations

Examine each tube separately. Wipe dry and use a hand-lens. If the tubes are scratched, dip them in xylene. The titre of the serum is that dilution in which agglutination is easily visible with a low-power magnifier.

Vi agglutinations

These are used in detecting typhoid carriers.

Test serum

Dilute test serum 1:5 by adding 0.5 ml to 2 ml of saline. Set up eight test-tubes (75 × 12 mm). To tubes 2 to 8 add 1 ml of saline and to tubes 1 and 2 add 1 ml of 1:5 serum. Make doubling dilutions by mixing the contents of tube 2, transferring 1 ml to tube 3, and so on. Discard 1 ml from tube 7. Tube 8 contains no serum. The dilutions are:

Tube no.	1	2	3	4	5	6	7	8
Dilution	1:5	1:10	1:20	1:40	1:80	1:160	1:320	0

Control positive serum

Dilute Standard *S. typhi Vi* serum 1:100 by adding 0.1 ml to 9.9 of saline. Add 1 ml of this to each of seven large tubes. Add saline too give dilutions as follows:

Tube no.	1	2	3	4	5	6	7
Saline (ml)	5	7	9	11	13	15	17
Dilution	1:600	1:800	1:1000	1:1200	1:1400	1:1600	1:1800

The titre given on the bottle should be in the middle of this range. If it is not, adjust the range.

Transfer 1 ml of each dilution to a 75 × 12 mm tube.

Control negative serum

Make dilutions of a known negative serum 1:5, 1:10 and 1:20 in 1-ml amounts in 75 × 12 mm tubes as for test serum. To each tube of each of the three series add 0.05 ml (1 drop) of Standard *Vi* suspension. Mix well and incubate at 37 °C for 2 h. Place in a refrigerator overnight.

Reading

Hold each tube over a concave mirror and observe the deposit:

(1) clear button or organisms: no agglutination;
(2) slight scattering around less well defined button: trace agglutination;
(3) clumps of agglutinated bacteria with turbid supernatant: partial agglutination;
(4) no button, clumped organisms scattered over bottom of tube and supernatant clear: complete agglutination.

The *Vi* titre of the test serum is the dilution that gives agglutination equal to that given by the Provisional *Vi* serum at its given titre (usually 1:1400).

If there is any agglutination in the saline control or negative serum control the test is invalid: the suspension is too sensitive and should be replaced. Normally, test titres of less than 1:5 are obtained. Higher titres may indicate typhoid infection.

Testing unknown organisms against known sera

Preparation of O suspensions

The organism must be in the smooth phase. Grow on agar slopes for 24 h. Wash off in phenol-saline and allow lumps to settle. Remove the suspension and dilute it so that there are approximately 1000×10^6 bacteria/ml by the opacity tube method (p. 130). Heat at 60 °C for 1 h. If this antigen is to be stored, add 0.25% chloroform.

If the bacteria are known to contain *Vi* antigen, grow on agar containing 1:500 phenol. This discourages *Vi* antigen formation.

If a K antigen is suspected, heat the suspension at 100 °C in a water-bath for 1 h (but the B antigen is thermostable).

Preparation of H suspensions

Check that the organism is motile and grow it in nutrient broth for 18 h, or in glucose broth for 4–6 h. Do not use glucose broth for the overnight culture as the bacteria grow too rapidly. Add formalin to give a final concentration of 1.0%. Heat at 50–55 °C for 30 min (this step may be omitted if the suspension is to be used at once). Dilute to approximately 1000×10^6 organisms/ml by opacity tube method.

Agglutination tests

Use the same technique as that described under 'Testing unknown sera ...' but as most sera are issued with a titre of at least 1:250 (and labelled accordingly) it is not necessary to dilute beyond 1:640. In practice, as Standard sera are highly specific and an organism must be tested against several sera, it is usually convenient to screen by adding 1 drop of serum to about 1 ml of suspension in a 75 × 9 mm tube. Only those sera which give agglutination are then taken to titre.

Slide agglutination

This is the normal procedure for screening with *O* sera.

Place a loopful of saline on a slide and next to it a loopful of serum. With a straight wire, pick a colony and emulsify in the saline. If it is sticky or granular, or autoagglutinates, the test cannot be done. If the suspension is smooth, mix the serum in with the wire. If agglutination occurs it will be rapid and obvious. Dubious slide agglutination should be discounted.

Only *O* sera should be used. Growth on solid medium is used for slide agglutination and this is not optimum for the formation of flagella. False-negative results may be obtained wiht *H* sera unless there is fluid on the slope.

Suspensions of organisms in the R phase are agglutinated by 1:500 aqueous acriflavine solution.

Chapter 11

Clinical material

J. D. Jarvis

The laboratory investigation of clinical material for the presence of microbiological pathogens is subject to a variety of external hazards. In addition, there are fundamental errors and omissions that may occur in any laboratory. Vigorous performance control testing programmes and the routine use of controls will reduce the incidence of error but are unlikely to eliminate it entirely.

Most samples for microbiological examination are not collected by the laboratory staff, and it must be stressed that however carefully a specimen is processed in the laboratory, the result can be only as reliable as the sample will allow. The laboratory should provide specimen containers that are suitable, i.e. leak-proof, stable, easily opened, readily identifiable, aesthetically satisfactory, suitable for easy processing and preferably capable of being incinerated. It will be apparent that these qualities are mutually exclusive and the best compromise to suit particular circumstances must be made. Containers are discussed on p. 27. The specimen itself may be unsatisfactory because of faulty collection procedures, and the laboratory must be ready to explain its requirements and give a reasoned explanation if cooperation is to be obtained from ward staff and doctors.

Specimens should be:

(1) collected without extraneous contamination;
(2) collected before antibiotic therapy;
(3) representative; pus rather than a swab with a minute blob on the end; faeces rather than a rectal swab. If this is not possible, then a clear note to this effect should be made on the request form, which must accompany each sample;
(4) ideally collected and sent to the laboratory in a clear plastic bag to avoid leakage during transit. The request form accompanying each specimen should not be in the same bag but clearly displayed to minimize sorting errors;
(5) delivered to the laboratory without delay. If this is not possible, appropriate specimens must be delivered in a transport medium (*see* p. 60). For other specimens, storage at 4 °C will hinder bacterial overgrowth. If urine samples cannot be delivered promptly and refrigeration is not possible,

boric acid preservative may be considered as a last resort. Dip slide culture is to be preferred. Sputum and urine specimens probably account for the greater part of the laboratory work load and the problems associated with their collection are particularly difficult to overcome because the trachea, mouth and urethra have a normal flora.

It must be stressed that in any microbiological examination, only that which is sought will be found and that it is essential to view the sample microbiologically. Every specimen should be evaluated to consider bacteria, fungi, parasites and viruses. The examination of pathological material for viruses and parasites is not within the scope of this book, but the possibility of infection by these agents should be considered and material referred to the appropriate laboratory

The laboratory has an obligation to ensure that its rules are well publicized so that the periods of acceptance of specimens are known to all. Any tests that are processed in batches should be arranged in advance with the laboratory. This may be important for microbiological assay, etc., and is essential for the monitoring of antibiotic assays because time will be needed for the laboratory to subculture the correct strains of organisms required.

Blood cultures

Blood must be collected with scrupulous care to avoid extravenous contamination. In patients with a recurrent fever, blood culture is an important diagnostic procedure and the highest success rate is associated with the collection of cultures just as the patient's temperature begins to increase rather than at the peak of the rigor. In patients with septicaemia, the timing is less important and a rapid answer and a sensitivity result is essential.

Castenada blood culture bottles are biphasic, i.e. they have a slope of solid medium and a broth which may also contain Liquoid (sodium polyethanol sulphonate: Roche) which neutralizes antibacterial factors including complement. Large amounts of blood (up to 50% of the total volume of the medium) may be examined in Liquoid cultures; otherwise, it is necessary to dilute the patient's blood at least 1:20 with broth. The advantage of the Castenada system is that frequent subculture is avoided; the bottles are tipped so that the blood-broth mixture washes over the agar slope every 72 h. Colonies will appear first at the interphase and then all over the slant. The bottles should be retained for at least four and preferably six weeks. The Castenada system avoids repeated subculture and lessens the risk of laboratory contamination. This is important because patients receiving immunosuppressive drugs are likely to be infected with organisms not usually regarded as pathogens. No organism can therefore be automatically excluded as a contaminant.

Vacuum blood collection systems, e.g. Vacutainers (Becton Dickinson) also help to reduce contamination. Some of these ensure a 10% carbon dioxide atmosphere in the container. This is important when organisms such as *Brucella* spp. are sought. If other media are used they should be incubated in an atmosphere containing at least 10% carbon dioxide (p. 92).

A variety of enriched media is available for blood culture. Fewer workers now use Robertson's cooked meat medium. This medium has been used for many years for blood cultures but because it is turbid it must be subcultured repeatedly to see if there is any growth.

Blood cultures should be subcultured on to appropriate media when growth is visible, i.e. colonies appear on the solid phase of the Castenada bottle, or turbidity is evident. In any case, subculture at 24, 48 and 72 h and then at weekly intervals. Great care is needed when subcultures are made or contamination will result. It is best to do this in a microbiological safety cabinet.

Rapid systems for the detection of growth in blood cultures are now available. The 'Bactec System'[1] (Laboratory Impex), which uses radiolabelled glucose in the medium, is widely accepted. However, expert and early subculture (4 h) may be as effective in detecting up to 80% of significant isolates. This percentage will vary with differing types of patient (cardiac, post-operative, geriatric, neonatal, etc.).

Likely pathogens

Indifferent viridans streptococci, non-haemolytic streptococci, enterococci, *Brucella*, *Staphylococcus aureus*, *Neisseria meningitidis*, *Clostridium*, *Salmonella*, *Bacteroides*. In ill or immunosuppressed patients, any organisms, especially if recovered more than once, may be pathogens. In patients with septicaemia following major surgery, a mixed growth of enteric organisms may be found.

Commensals

None.

Cerebrospinal fluid (CSF)

It is important that these samples should be examined with the minimum of delay. As with other fluids, a description is important. It is essential to note the colour of the fluid; if it is turbid, then the colour of the supernatant after centrifugation may be of diagnostic significance. A distinctive yellow tinge (xanthochromia) usually confirms the earlier presence of blood (it is necessary, of course, to compare the colour of the supernatant with a water blank in a similar tube).

Before any manipulation, the specimen must be carefully examined for clots, either gross or in the form of a delicate spider's web that is usually associated with a significantly raised protein and traditionally is taken as an indication that one should look particularly for *Mycobacterium tuberculosis*. The presence of a clot will invalidate any attempt to make a reliable cell count.

If the cells and/or protein are raised (> three lymphocytes/mm^3 and/or > 40 mg%), a CSF sugar determination should be compared with the blood sugar level collected at the same time. This and the type of cell may indicate the type of infection:

Raised polymorphs with very low sugar—usually indicate bacterial infection or a cerebral abscess.

Raised lymphocytes with normal sugar—usually indicate viral infection.

Raised lymphocytes with lowered sugar—usually indicate tuberculosis.
Raised polymorphs or mixed cells with lowered sugar—usually indicate early tuberculosis.

Make three films from the centrifuged deposit, spreading as little as possible. Stain one by the Gram method, one with a cytological stain and retain the third for examination if indicated for acid-fast bacilli. A prolonged search may be necessary if tuberculosis is suspected.

Plate on blood agar and incubate aerobically and anaerobically and on a chocolate agar plate for incubation in a 5–10% carbon dioxide atmosphere. Examine after 18–24 h. If organisms are found, set up a direct sensitivity test including penicillin, ampicillin, chloramphenicol and sulphonamides. Culture for *M. tuberculosis* as indicated.

Countercurrent immuno-electrophoresis of supernatant is a valuable and rapid technique especially in patients partially treated with antibiotics where culture may be unsuccessful. Examine urine also by this technique after concentration by dialysis, e.g. MINICON technique (Amicon). This is especially helpful with *Streptococcus pneumoniae*. Commercially prepared latex suspensions to detect antigens from organisms commonly associated with meningitis are now available (Wellcome).

Likely pathogens

Haemophilus influenzae, *N. meningitidis*, *Streptococcus pneumoniae*, *Listeria monocytogenes*, *M. tuberculosis*, and in very young babies coliform bacilli and *Pseudomonas aeruginosa*. *Staphylococcus epidermidis* and some micrococci are often associated with infections in patients with devices (shunts) inserted to relieve excess pressure.

Opportunists

Any organism introduced during surgical manipulation involving the spinal canal may be involved in meningitis.

Commensals

None.

Dental specimens

These may be whole teeth, when there may have been an apical abscess, or scrapings of plaque or carious material. Plate on selective media for staphylococci, streptococci and lactobacilli.

Likely organisms

S. aureus, streptococci, especially *S. mutans*, *S. milleri*, lactobacilli.

Ear discharges

Swabs should be small enough to pass easily through the external meatus. Many commercially available swabs are too plump and may cause pain to patients with inflamed auditory canals.

A direct film stained by the Gram method is helpful, as florid overgrowths of faecal organisms are not uncommon and it may be relatively difficult to recover more delicate pathogens. Plate on blood agar and incubate aerobically and anaerobically overnight at 37 °C. Tellurite medium may be advisable if the patient is of school age. A medium that inhibits spreading organism is frequently of value; CLED, or MacConkey, chloral hydrate or phenethyl alcohol agar are the most useful.

Likely pathogens

S. pyogenes, *S. aureus*, *Haemophilus*, *Corynebacterium diphtheriae*, *P. aeruginosa*, coliform bacilli, anaerobic Gram-negative rods.

Commensals

Micrococci, diphtheroids, *S. epidermidis*.

Eye discharges

It is preferable to plate material from the eye directly on culture media rather than to collect it on a swab. If this is not practicable then a swab in transport medium is essential.

Examine a direct Gram-stained film before the patient is allowed to leave. Look particularly for intracellular Gram-negative diplococci and issue a tentative report if they are seen so that treatment can be started without delay; this is of particular importance in neonates. Infections occurring a few days after birth may be caused by *Chlamydia trachomatis*, which should be suspected if films have excess numbers of monocytes or the condition does not resolve rapidly. Direct staining with Giemsa stain or the immunofluorescence technique are indicated.

Plate on blood agar plates and incubate aerobically and anaerobically and on a chocolate agar or 10% horse blood Columbia agar plate for incubation in a 5% carbon dioxide atmosphere. It is not usually possible to do 'direct' sensitivities as the primary growth is frequently sparse; if, however, the film shows many organisms, direct sensitivity tests including penicillin and chloramphenicol should be attempted.

Likely pathogens

S. aureus, *S. pneumoniae*, viridans streptococci, *N. gonorrhoeae*, *Haemophilus*, *Chlamydia*, *Moraxella*, rarely coliform organisms, *C. diphtheriae*, *P. aeruginosa*.

Commensals

S. epidermidis, micrococci, diphtheroids.

Faeces and rectal swabs

Faeces are always more reliable than rectal swabs for bacteriological examination. Swabs that are not even stained with faeces are useless. If swabs must be used, they should be moistened with sterile saline or water before use and care must be taken to avoid contamination with perianal flora. It may be useful to collect material from babies by dipping the swabs into a recently soiled napkin (diaper). Swabs should be sent to the laboratory in transport medium unless they can be delivered within 1 h.

Because of the random distribution of pathogens in faeces, it is customary to examine three sequential specimens.

Plate on DCA, MacConkey and bismuth sulphite agar and inoculate selenite or tetrathionate broth. Incubate for 18–24 h and subculture selenite and tetrathionate broths on to DCA. It is essential that the MacConkey agar will support the growth of *S. aureus*, which may be causative in staphylococcal enterocolitis.

Plate stools from children, young adults and diarrhoea cases on Campylobacter Medium plus antibiotic supplements. Incubate for 48 h at 42 °C and at 37 °C in 7–10% CO_2 in hydrogen or nitrogen.

If the patient is under three years old, plate on MacConkey agar and blood agar and look for enteropathogenic *Escherichia coli.*

Plate on TCBS and inoculate alkaline peptone water if cholera is suspected, if the patient has recently returned by air from an area where cholera is endemic or if food poisoning due to *Vibrio parahaemolyticus* is suspected.

Yersinia enterocolytica is now recognized as an enteric pathogen. Plate on Yersinia Selective Agar and incubate at 32 °C for 18–24 h. *See also* Chapter 29.

In cases of food poisoning due to *S. aureus*, plate stools on blood agar and on one of the selective staphylococcal media (p. 59) and inoculate salt meat broth. Incubate all cultures overnight at 37 °C and subculture the salt meat broth to blood agar and selective staphylococcal medium. It should be noted that here the symptoms relate to a toxin so the organisms may not be recovered.

In food poisoning where *Clostridium perfringens* may be involved, plate the stools on blood agar and incubate anaerobically. Include aerobically incubated plates as controls. Inoculate two tubes of Robertson's cooked meat broth. Heat one tube at 80 °C for 30 min. Incubate both tubes overnight and subculture to blood agar. Incubate aerobically and anaerobically. *See* Chapter 14.

Likely pathogens

Salmonella, *Shigella*, *Vibrio*, enteropathogenic *E. coli*, *C. perfringens* (both heat resistant and non-heat resistant), *S. aureus*, *Bacillus cereus*, *B. subtilis*, *Campylobacter*, *Yersinia*, *Aeromonas*, *P. aeruginosa*, *C. botulinum* and *C. difficile*.

Commensals

Coliform bacilli, *Proteus*, *Clostridium*, *Bacteroides*, *Pseudomonas*.

Nasal swabs (*see also* Throat swabs)

The area sampled will influence the recovery of particular organisms, i.e. the anterior part of the nose must be sampled if the highest carriage rate of staphylococci is to be found. For *N. meningitidis*, it may be preferable to sample further in and for the recovery of *Bordetella pertussis* in cases of whooping cough a perinasal swab is essential.

Inoculate as soon as possible onto charcoal agar (*O*) containing 10% horse blood and 40 mg/l cephalexin and incubate at 35 °C in a humid atmosphere.

Direct films are of no value. If *M. leprae* is sought, smears from a scraping of the nasal septum may be examined after staining with ZN stain or a fluorescence stain.

Plate on blood agar and tellurite media and incubate as for throat swabs.

Likely pathogens

S. aureus, *S. pyogenes*, *N. meningitidis*, *B. pertussis*, *Haemophilus*, *C. diphtheriae*.

Commensals

Diphtheroids, *S. epidermidis*, *S. aureus*, *Neisseria*, aerobic spore bearers and small numbers of Gram-negative rods (*Proteus* and coliform bacilli).

Pus

As a general rule, pus rather than swabs of pus should be sent to the laboratory. Swabs are variably lethal to bacteria within a few hours because of a combination of drying and toxic components of the cotton wool released by various sterilization methods. Consequently, speed in processing is important; a delay of several hours may allow robust organisms to be recovered while the more delicate pathogens may not survive. Swabs should be moistened with broth before use on dry areas.

Examine Gram-stained films and films either direct or of concentrated material (p. 374), for acid-fast rods. Because the distribution of organisms in pus is random, wash all pus samples and examine them for the 'sulphur granules' of *Actinomyces israelii*. Shake the pus with two or three changes of sterile physiological saline (it is important to avoid both contamination and aerosols in this procedure). This will produce a relatively clear supernatant with the granules depositing very quickly. Plate on blood agar and incubate aerobically and anaerobically. Include a chloral hydrate plate in case *Proteus* is present. Prolonged anaerobic incubation in a 90% hydrogen–10% carbon dioxide atmosphere of blood agar cultures, possibly with neomycin or nalidixic acid may allow the recovery of strictly anaerobic *Bacteroides* spp. These organisms may be recovered frequently if care is taken and the number of sterile collections of pus consequently reduced. Culture in thioglycollate broth may be helpful. The dilution effect may overcome specific and non-specific inhibitory agents.

The recovery of anaerobes will be enhanced by the addition of menadione and vitamin K to the medium. Wilkins and Chalgren's medium (*O*) is also of value.

Likely pathogens

S. aureus, *S. pyogenes*, peptostreptococci, peptococci. *Mycobacterium*, *Actinomyces*, *Pasteurella*, *Yersinia*, *Clostridium*, *Neisseria*, *Bacteroides*, *B. anthracis*, *Listeria*, *Proteus*, *Pseudomonas*, *Nocardia*, *Fungi*, other organisms in pure culture.

Commensals

None.

Serous fluids

These fluids include pleural, synovial, pericardial, hydrocele, ascitic and bursa fluids.

Record the colour, volume and viscosity of the fluid. Centrifuge at 3000 rev/min for at least 15 min preferably in a sealed bucket and note the colour of the supernatant. Transfer the supernatant to another bottle. Make Gram-stained films of the deposit and also stain films for cytological evaluation. Plate on blood agar and incubate aerobically and anaerobically at 37 °C for 18–24 h. Culture on blood agar anaerobically plus 10% carbon dioxide atmosphere and incubate for 7–10 days for *Bacteroides*. Use additional material as outlined above. Prolonged incubation may be necessary for these strictly anaerobic organisms.

Gonococcal arthritis must not be overlooked; inoculate Thayer Martin or a similar medium and incubate in a 10% carbon dioxide atmosphere for 48 h at 37 °C. Culture, if appropriate, for mycobacteria.

Likely pathogens

S. pyogenes, *S. pneumoniae*, peptostreptococci, peptococci, *S. aureus*, *N. gonorrhoeae*, *M. tuberculosis*, *Bacteroides*, *Coxiella burnetii*.

Commensals

None.

Sputum

Sputum is a difficult specimen. Ideally, it should represent the discharge of the bronchial tree expectorated quickly with the minimum of contamination from the fauces and the mouth, and delivered without delay to the laboratory.

Because of the irregular distribution of bacteria in sputum it may be advisable to wash portions of purulent material to free them from contaminating mouth organisms before examining films and making cultures.

Homogenization of the specimen with Sputolysin, *N*-acetylcysteine or pancreatin, for example, followed by dilution enables the significant flora to be assessed. In sputum cultures not so treated initially small numbers of contaminating organisms may overgrow pathogens.

The following method is recommended.

Add about 2 ml of sputum to an equal volume of Sputolysin (Calbiochem) in a screw-capped bottle and allow to digest for 20–30 min with occasional shaking, e.g. on a Vortex Mixer. Inoculate a blood agar plate with 0.01 ml of the homogenate, using a standard loop (e.g. a Nunc plastic loop) and make a Gram-stained film. Add 0.01 ml with a similar loop to 10 ml of peptone broth. Mix this thoroughly preferably using a Vortex Mixer. Inoculate blood agar and MacConkey agar with 0.01 ml of this dilution. Incubate the plates overnight at 35 °C; incubate the blood agar plates in a 10% carbon dioxide atmosphere. Chocolate agar may be used but should not be necessary for growing *H. influenzae* if a good blood agar base is used.

Five to 50 colonies of a particular organism on the dilution plate are equivalent to 10^6–10^7 of these organisms/ml of sputum and this level is significant except for viridans streptococci and *Neisseria*; one to five colonies are equivalent to 10^5–10^6 organisms/ml and this level is equivocal.

Set up direct sensitivity tests if the Gram film shows large numbers of any particular organisms.

Methods of examining sputum for mycobacteria are described in Chapter 36.

For the isolation of *Legionella* and similar organisms from lung biopsies or bronchial secretions use blood agar or one of the commercial legionella media plus supplement.

Likely pathogens

S. aureus, *S. pneumoniae*, *H. influenzae*, coliform bacilli, *Klebsiella pneumoniae*, *Pasteurella*, *Mycobacterium*, *Candida* spp. *Branhamella catarrhalis*, *Mycoplasma*, *Pasteurella/Yersinia* spp., *Y. pestis*, *Legionella pneumophila*, *Aspergillus*, *Histoplasma*, *Cryptococcus*, *Blastomyces*.

Commensals

S. epidermidis, micrococci, *Neisseria*, coliforms, *Candida* in small numbers, *S. viridans*.

Throat swabs

Collect throat swabs carefully with the patient in a clear light. It is customary to attempt to avoid contamination with mouth organisms. Evidence has been advanced that the recovery of *Streptococcus pyogenes* from the saliva may be higher than from a 'throat' or fauces swab. Make a direct film, stain with dilute carbol fuchsin and examine for Vincent's organisms, yeasts and mycelium.

Plate on blood agar and incubate aerobically and anaerobically for a minimum of 18 preferably 48 h in 7–10% CO_2 at 35 °C. Plate on tellurite medium and incubate for 48 h in 7–10% CO_2 at 35 °C.

Likely pathogens

S. pyogenes, *Corynebacterium diphtheriae*, *C. ulcerans*, *S. aureus*, *Candida albicans*, *Neisseria meningitidis*, *Borrelia vincenti*.

Commensals

Neisseria, indifferent viridans streptococci, *S. epidermidis*, diphtheroids, *S. pneumoniae*, probably *Haemophilus influenzae* (*H, influenzae* Pittmans type b in the epiglottis is certainly a pathogen).

NB. For details of the logistics of mass swabbing, e.g. in a diphtheria outbreak, *see* the paper by Collins and Dulake[2].

Tissues, biopsy specimens, post-mortem material

As a general rule, retain all material submitted from post-mortem until the forensic pathologists agree to their disposal. If it is essential to grind up tissue, obtain approval.

Tissue specimens may be ground up after suitable selection procedures in a blender or, if the specimen is very small, in a Griffith's tube or with glass beads on a Vortex mixer. The Stomacher Lab-Blender is best for larger specimens as it saves time and eliminates the hazard of aerosol formation. Homogenizers, etc., are described on p. 15. Examine the homogenate as described under 'Pus'. It may be necessary to resort to animal inoculation to recover organisms from these materials.

Urethral discharges

Swabs of discharge should be collected into a transport medium or plated directly at the bedside or in the clinic. Examine direct Gram-stained films. Look for intracellular Gram-negative diplococci. In the sexually active male, the commonest cause of urethral discharge is gonorrhoea. In the female, the recovery of organisms is more difficult and the interpretation of films may occasionally present problems because over-decolorized or dying Gram-positive cocci may resemble Gram-negative diplococci.

Plate on blood agar plates for aerobic and anaerobic incubation and chocolate or Columbia agar with and without antibiotics (vancomycin, colistin and nystatin) or on Thayer Martin medium and incubate in a 5% carbon dioxide atmosphere.

Likely pathogens

N. gonorrhoea, *Mycoplasma*, foreign bodies, occasionally *S. pyogenes*, anaerobic cocci and *Chlamydia trachomatis*.

Commensals

S. epidermidis, micrococci, diphtheroids, small numbers of coliform organisms.

Urine

The healthy urinary tract is free from organisms over the greater part of its length; there may, however, be a few transient organisms present, especially at the lower end of the female urethra.

Collection

Catheters may contribute to urinary infections and are no longer considered necessary for the collection of satisfactory samples from females.

A 'clean catch' mid-stream urine (MSU) is the most satisfactory specimen for most purposes if it is delivered promptly to the laboratory. Failing prompt delivery, the specimen can be kept in a refrigerator at 4 °C for a few hours or, exceptionally, a few crystals of boric acid may be added.

Dip slides (Oxoid, Gibco, Medical Wire) are now widely used for assessing urinary tract infections. They do, however, preclude examination for cells and other elements.

Examination

This includes counting leucocytes, erythrocytes and casts, estimating the number of bacteria/ml and identifying them. Normally, there will be none or less than 20/mm^3 of leucocytes or other cells and a colony count of less than 10^4 organisms/ml.

Mix the urine by rotation, never by inversion. Count the cells using a Fuchs Rosenthal or similar counting chamber and inoculate a blood agar plate and a MacConkey, CLED or EMB plate with 0.001 ml using a standard loop (Nunc disposable loops or micropipettes with disposable tips are satisfactory alternatives). Also 0.1 ml of a 1 in 100 dilution of urine may be used. Spread with a glass spreader, incubate overnight at 37 °C and count the colonies:

<10 colonies	$\equiv 10^4$ organisms/ml	—not significant
10–100 colonies	$\equiv 10^4$–10^5 organisms/ml	—doubtful significance
>100 colonies	$\equiv >10^5$ organisms/ml	—significant bacteraemia

If dip slides are submitted, incubate them overnight and follow the manufacturer's directions for counting.

Note that the accuracy and usefulness of bacterial counts on urines is limited by the age of the specimen. Only fresh specimens are worth examining.

Other limitations are the volume of fluid recently drunk and the amount of urine in the bladder.

For methods of examining urine for tubercle bacilli, *see* Chapter 36.

Likely pathogens

E. coli, other enterobacteria, *Proteus*, staphylococci including *S. epidermidis*, enterococci, *Salmonella* (rarely), leptospira, mycobacteria.

Commensals

Small numbers ($<10^4$/ml) of almost any organisms. It is unwise to be too dogmatic about small numbers of organisms isolated from a single specimen: better to repeat the sample.

Vaginal discharges

Puerperal infections

A high vaginal swab is usually taken. Swabs that cannot be processed very rapidly after collection should be put into transport medium. Examine direct Gram-stained films. If non-sporing square-ended Gram-positive rods, not in pairs, and which appear to be capsulated are seen they should be reported without delay, as they may be *C. perfringens*, which may be an extremely invasive organism.

Plate on blood agar and incubate aerobically and anaerobically overnight at 37 °C and also plate MacConkey or other suitable selective medium. Plate any swabs suspected or harbouring *C. perfringens* on a neomycin half-antitoxin Nagler plate for anaerobic incubation.

To isolate *Gardnerella vaginalis* (*Haemophilus*/*Corynebacterium vaginalis*) add 10% human blood to the medium. Incubate for up to 48 h at 35 °C in humid conditions. Characteristic colonies have a hazy, diffuse β haemolysis. 'Clue cells' (epithelial cells with many small rods which may be Gram-variable or Gram-negative) should correlate with the isolation of this organism.

Likely pathogens

S. pyogenes, *L. monocytogenes*, *C. perfringens*, *Bacteroides*, anaerobic cocci, *Candida* spp., excess numbers of enteric organisms, *G. vaginalis*.

Commensals

Lactobacilli, diphtheroids, micrococci, *S. epidermidis*, small numbers of coliform bacilli, yeasts.

Non-puerperal infections

Examine a Gram-stained smear and also a 'hanging drop' preparation for the presence of *Trichomonas vaginalis*. If the examination is to be immediate, then a saline suspension of vaginal discharge will be satisfactory, otherwise a swab into transport medium is essential. A wet preparation may still be examined for actively motile flagellates but culture in Trichomonas medium usually yields more reliable results, especially in cases of light infection.

If *N. gonorrhoeae* infection is suspected in the female, then swabs from the urethra, cervix and posterior fornix should be collected and placed into transport medium. Many workers do not make films from these swabs because the difficulties of interpretation are compounded by the presence of charcoal from the buffered swabs; they rely upon cultivation using VCN (vancomycin colistin, nystatin) or VCNT (VCN + trimethoprim) medium as a selective inhibitor of organisms other than *N. gonorrhoeae*.

Plate on blood agar and incubate, aerobically and anaerobically, Sabouraud agar (when looking for *Candida*) and chocolate agar or another suitable medium (e.g. Thayer Martin) for incubation in a 5% carbon dioxide atmosphere to encourage isolation of *N. gonorrhoeae*.

If *L. monocytogenes* infection is suspected plate on serum agar with 0.4 g/l of nalidixic acid added.

Likely pathogens

N. gonorrhoeae, *C. albicans* and related yeasts in moderate numbers, *S. pyogenes*, *L. monocytogenes*, *Haemophilus*, *Bacteroides*. anaerobic cocci, possibly *Mycoplasma*, *T. vaginalis*.

Commensals

Lactobacilli, diphtheroids, enterococci, small numbers of coliform organisms.

Wounds (superficial)

Examine Gram-stained films and culture on blood agar and proceed as for pus, etc. Culture also on Lowenstein–Jensen medium and incubate at 30 °C because some mycobacteria that cause superficial infections do not grow on primary isolation at 35–37 °C.

Likely pathogens

S. aureus, streptococci, *M. marinum*, *M. ulcerans*, *M. chelonei*.

Commensals

Staphylococci, pseudomonas and a wide variety of bacteria and fungi.

Wounds (deep) and burns

The collection of these specimens and their consequent rapid processing are two vital factors in the recovery of the significant organisms. Burns cases frequently become infected with staphylococci and streptococci as well as with various Gram-negative rods, especially *Pseudomonas* spp. It may be difficult to recover Gram-positive cocci from the overgrowing Gram-negative enteric organisms.

Plate on blood agar and incubate aerobically and anaerobically. Plate also on MacConkey agar and a medium that will inhibit the spreading of *Proteus* spp., such as chloral hydrate, Dispersol LN and phenethyl alcohol blood agar. Robertson's cooked meat medium may be useful.

Likely pathogens

Clostridium, *S. pyogenes*, *S. aureus*, peptococci, peptostreptococci, *Bacteroides*, Gram-negative rods.

Commensals

Small numbers of a wide variety of organisms.

References

1. CARLSON, L.G. and PLORDE, J.J. (1982) *Journal of Clinical Pathology*, **16,** 590
2. COLLINS, C.H. and DULAKE, C. (1983) *Journal of Infection*, **6,** 227

Chapter 12

Fluorescent antibody techniques

A. Cremer

Fluorescent antibody techniques were introduced into the field of immunology by Coons *et al.*[1]. The principle is that proteins, including serum antibodies, may be labelled with fluorescent dyes by chemical combination without alteration or interference with the biological or immunological properties of the proteins. These proteins may then be seen in microscopical preparations by fluorescence microscopy.

In the apparatus used, the preparation is illuminated by ultraviolet or ultraviolet blue light. Any fluorescence emitted by the specimen passes through a barrier filter above the object. This filter is designed so as to transmit only the visible fluorescence emission. Microscopes suitable for this purpose are now readily available, and it is a simple matter to convert standard microscopes for fluorescence work. Fluorescent dyes are chosen for this work in preference to ordinary dyes as they are detectable in much smaller concentrations. They are available in a form which simplifies the conjugation procedure considerably. Many fluorochromes have been tried but the only ones in frequent current use are fluorescein isothiocyanate (FITC) and Lissamine rhodamine B (RB200). Of these, FITC is the most commonly used. It gives an apple-green fluorescence.

In microbiology, the fluorescent antibody method is usually employed by using immune serum globulin conjugated with fluorochrome to locate its corresponding antigen. It has been used in bacteriology, virology, parasitology, mycology and auto-immunology, often as an adjunct to traditional serological procedures, but there are certain instances where a fluorescent antibody technique is now the method of first choice.

The fluorescence microscope

The main object of the fluorescence microscope is to transmit as much as possible of the fluorescence emitted by the specimen to the eye. As this fluorescence is seldom of high intensity, great care must be taken that as little of it as possible is lost. For this reason, monocular microscopes are preferred as the presence of unnecessary prisms in an optical system may considerably diminish the brightness of the fluorescence. Similarly, the exicting light must reach the specimen with as much intensity as possible. For this reason, the sub-stage light path utilizes ultraviolet-transmitting glass, a high-intensity light source and surface-reflecting mirrors. Mercury vapour discharge lamps were once used almost exclusively to provide the exciting illumination but recently the iodine quartz lamp has proved to be an excellent substitute. It has the advantage of being very much cooler in use, and can be switched on and off at will without the necessity for a warming-up or cooling-off period. These lamps are very much cheaper and, in general, have a much longer operative life than the mercury vapour lamps.

A dark-ground condenser directs the primary illumination away from the objectives so that no direct light reaches the ocular, thus giving a much darker background and more contrast with the fluorescing object. Ideally, the condenser should be of the immersion type fitted with a toric lens. This provides intense uniform illumination of the field and not just a hollow cone, as it utilizes most of the primary illumination. If such a condenser is not available, a cardioid-type dark ground condenser gives good results for most purposes.

The objectives used need to have the highest possible numerical aperture to ensure the greatest possible illumination. They must be constructed so that they are themselves non-fluorescent. Achromatic lenses are to be preferred to apochromats as they very rarely show any fluorescence and will transmit all the usual colour ranges encountered in fluorescence microscopy. When a dark-ground condenser is used, it is an advantage to have iris diaphragms on all objectives of higher magnification than about ×40.

If the apparatus is to be used for inspecting slides stained with a fluorescent stain, such as auramine, when searching specimens for acid-fast bacilli, the optical system used should be slightly different from that used for the fluorescent antibody technique. Unlike the prepared antigen slides, a very small number of organisms may be fluorescing and it is important that these should be found. For this reason, and as fluorescence makes organisms appear larger than normal, a lower-power objective may be used, which facilitates the inspection of as much of the preparation as possible in a short time. A bright-field condenser may also be employed as the dark-ground condenser is more inclined to show brightly shining artefacts and also the auramine staining usually gives a more intense fluorescence than is obtained with FITC.

If it is thought necessary to use immersion objectives, it is best to avoid the usual immersion oils. These become slightly fluorescent after exposure to air and this process is hastened by intense illumination. AnalaR-grade glycerol is used as the immersion medium but even this slightly impairs the performance of high-power lens systems as it does not have the ideal refractive index, although lenses made for glycerol immersion can be purchased. For this reason, it is better to use dry lenses of the highest available magnification. For magnifications higher than ×40, it is also necessary to mount the specimen under a cover-glass. The mounting medium used for this purpose is buffered glycerol.

Two filter systems are necessary in fluorescence microscopy. The primary or 'exciter' filter is placed between the light source and the specimen so as to exclude all but the light of wavelength necessary to stimulate fluorescence emission in the fluorochrome being used. This stimulating light is usually ultraviolet or ultraviolet blue and so the exciter filters are usually blue. The secondary or 'barrier' filter is usually placed in the eye-piece, although it may be anywhere between the specimen and the eye. This ensures that only the light from the fluorescing specimen reaches the eye. These filters must be chosen with care and will differ with the fluorochrome, the type of light source and the optical system used.

The equipment described here will give optimum results but is expensive. The essentials are a good monocular microscope, preferably one that does not utilize a prism in the body tube, a high-intensity light source, and the necessary filters. The filters alone are not costly. If possible, however, it is advisable to set aside one microscope for fluorescence tests as this minimizes the need for optical adjustments when the system is in use.

The fluorescent antibody test

There are many factors in the preparation of the specimen that may affect fluorescence one way or another and of these the commonest is probably variation in pH. For this reason, all dilutions and washing procedures are carried out in buffer solution. This is available as a lyophilized powder from Difco and Mercia. Difco, BBL, the Pasteur Institute and Wellcome, supply a variety of fluorescent antibody reagents. The catalogues and handbooks of these companies contain much useful information.[2–4]

In microbiology, two fluorescence methods may be applied, a direct method or the indirect method, which is also known as the sandwich technique. The direct test can be used to identify microbes while the indirect test can be used both for identification and for the examination of patients' serum for antibodies.

The direct test depends upon the attachment of a fluorochrome-conjugated known specific antiserum to a microbe that may be in a pathological exudate, in a culture or in tissue. The fluorescent antibody combines specifically with the microbe, which can then be viewed with the optical system previously described.

The indirect test is performed in two stages. In the first stage, the patient's serum is applied to a known antigen. When antibody is present, it will combine with the antigen but cannot be seen as it is non-fluorescent. In the second stage, fluorescent anti-human globulin is applied, which will attach to any human globulin left in the washed preparation, and antibody being 'sandwiched' between it and the antigen. When the indirect technique is used for identification, ordinary specific antisera are used and, after washing, a fluorescent antiserum is applied which contains antibodies to the globulins of the animal species in which the specific antiserum was prepared.

The direct test

A typical example of the direct fluorescent antibody test is that used to identify the pathogenic *Neisseria*.

Method

(1) Take a colony of the organism to be tested and emulsify on a clean slide.
(2) Dry in air without heating and fix by immersion in 3% formalin–saline buffer solution.

(3) Wash in saline buffer for 10 min with two changes of solution. The washing may be carried out by simply immersing the slides in a Coplin jar. Blot gently to dry.
(4) Apply a drop of the conjugated antiserum, which is kept deep-frozen, and spread it over the specimen.
(5) Incubate in a moist petri dish for 30 min, preferably being rotated at 100 rev/min on a mechanical rotator.
(6) Wash in saline buffer for 10 min with three changes of solution. Blot dry, mount in buffered glycerol and apply a clean cover-slip.
(7) Inspect for fluorescence under the fluorescent microscope. If fluorescing cocci are seen, this means that the specific antibody has reacted with the unknown antigen. A positively reacting organism may be either *N. gonorrhoeae* or *N. meningitidis*, as these two species have antigens in common.

It is essential with all fluorescent antibody techniques, as with most serological tests, to test known positive and negative controls with each batch of investigations.

The indirect test

The most commonly practised form of this test is the absorbed fluorescent treponemal antibody test.[5]

(1) Reconstitute a lyophilized culture of *Treponema pallidum* and make smears by taking a fairly large loopful of the suspension and spreading it in a small circle on a thin slide, between 0.8 and 1.0 mm in thickness. The PTFE-coated slides made by Hendley are particularly suitable for this purpose. Dry in air and fix in 10% methanol for 5 min. Fixed slides can be stored for several months in a deep-freeze.
(2) Dilute the previously inactivated sera for testing 1/5 in FTA sorbent and allow to adsorb for 30 min.
(3) Place one or two drops of each adsorbed serum on a circle of a slide containing the Treponema, and incubate at 37 °C for 30 min in a moist chamber.
(4) Wash the slide in at least three changes of fluorescent antibody test buffer solution of pH. 7.2 (Mercia) and gently blot dry.
(5) Spread over each inoculum one or two drops of fluorescein-conjugated anti-human globulin and incubate as before.
(6) Wash in buffer as before. Blot gently and mount in buffered glycerol with a clean cover-slip. Include a positive, weakly positive and negative control with each batch of tests.

Inspect with a fluorescence microscope using a dark-ground condenser. If anitibody is present in the tested serum, it will react with the organisms and they will become coated with human globulin. The fluorescent anti-human globulin will then react with the globulin and it will fluoresce. A positive test indicates the presence of treponemal antibodies in the patient's serum. In a completely negative test there is always the possibility that no treponema are present on the slide. This can be easily checked by removing the exciter filter. As a dark-ground condenser is used, the presence of the spirochaetes is easily seen.

The advantage of the indirect test is that a single conjugated species-specific antiserum can be used for demonstrating the presence of many antigens, provided that the specific antisera used are prepared in the same species.

The fluorescent antibody technique can be applied in microbiology to any organism against which it is possible to prepare specific high-titre antisera. Most of the medically important bacteria now have conjugated antisera available commercially (*see below*). The technique is capable, in certain circumstances, of greatly shortening the time taken to identify an organism. Organisms recovered at autopsy from animals into which they have been inoculated may sometimes be identified in a few hours by this method rather than the several days that cultural identification may require. In certain instances, microcolonies of organisms that have been growing for only 2.5 h may be identified by cover-slip impression techniques with an homologous conjugated antiserum.

It is essential, however, to realize that although the fluorescent antibody technique has proved its worth in the field of microbiology, and is the only technique that combines the specificity of an antigen–antibody reaction with the speed and precision of microscopy, it has its limitations. As Nairn[6] so wisely pointed out, it is easily forgotten that the technique is employed to detect antigens and not organisms. Pathogenic organisms have antigens not only in common with each other but also in common with harmless commensals of the same or related genera. False-positive reactions are therefore inevitable unless the antisera employed are very highly specific. However, reagents and techniques are continually being improved, and there can be little doubt that the use of such techniques will increase as specificity improves, and will eventually encroach even upon the field of quantitative antibody determinations.

More information on these techniques is given by Heimer[7], Chadwick and Slade[8] and Wick *et al.*[9].

Incident light fluorescence

Apart from the ever increasing number of applications of fluorescent antibody techniques in microbiology, perhaps the most interesting development in recent years is incident light microscopy.

The greatest problem that arises with both dark-ground and bright-field transmitted light is the relatively low brightness of the fluorescence that results from absorption of both the exciting light from the illuminating system and the longer wavelength light emitted by the fluorescing specimen. This is mainly due to the necessary passage of the light through the various optical components of the microscope system and also, particularly in dark-ground equipment, to the necessity for accurate alignment of the lenses.

For incident light microscopy, the light source is above the specimen and both the exciting light and the light transmitted from the fluorescing specimen are passed through the objective. This eliminates any loss of illumination that would have occurred due to passage through a condenser, and also means that the system is always optimally aligned. Such a system is the Zetopan equipment, which is manufactured by Reichert (*Figure 12.1*).

Incident light microscopy is clearly a system of bright-ground microscopy. Theoretically, any such system may be used for fluorescence microscopy,

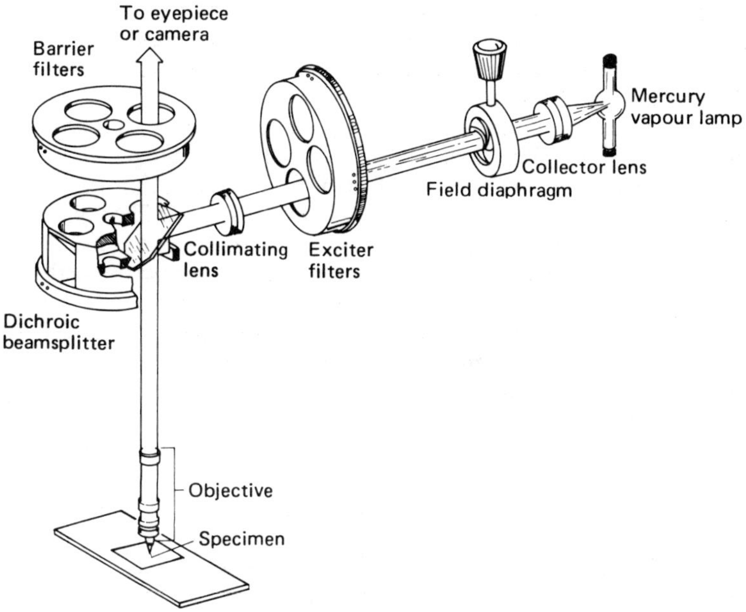

Figure 12.1 Incident light fluorescence: Mercury vapour lamp provides rich desirable ultraviolet radiation sources. Collector lens directs light to the exciter filter, which transmits selected wavelengths. Dichroic beamsplitter reflects virtually all of the desirable short wavelength, exciting light down through the objective to the specimen. Fluorochromes in specimen react to excitation wavelengths and emit longer wavelengths of visible fluorescence up through the system to the eye-piece. The barrier filter passes the emitted light from the specimen and blocks out unwanted background illumination. (Reproduced by permission of the British American Optical Co. Ltd)

provided that the light source is suitable. In practice, however, the fluorescence normally obtained is very poor, and particularly if ultraviolet light is used as the stimulating source. The cause of this poor fluorescence is the beam-splitting mirror used in the opaque illuminator. At the very best, the system reflects 50% of incident illumination with visible light wavelengths, and even less with ultraviolet light. As the light coming from the specimen will suffer a further 50% loss as it passes back through the illuminator, the low brightness is easily explained.

The improvement of this system for fluorescence microscopy originates from work by Brumberg[10] and Ploem[11]. They concentrated on altering the reflecting characteristics of the mirror to achieve improved light recovery with the incident light system, particularly in use for fluorescence microscopy. They replaced the original beam-splitting mirror with a multi-layer interference filter that produces up to 90% reflection of the exciting illumination and transmits a similar amount of light from the specimen. With an ideal beam-splitting mirror, only certain portions of the spectrum are completely reflected, while the remainder are transmitted without loss. Compared with the original methods of obtaining incident light, this system gives an appreciable increase in brightness.

Hypothetically, this system eliminates the need for both the exciter and the barrier filters, because in theory only the shortwave exciting light reaches the specimen, and only the fluorescing light of longer wavelength reaches the eye. In practice, however, such perfection with the beam-splitting dichroic mirror is impossible to achieve as the spectral spacing of the two wavelengths of light is not sufficiently wide. Consequently, it is still necessary to use an exciter filter and a barrier filter; these are built into the system.

The exciter filter carries out its usual function of transmitting only that wavelength of light which will stimulate the fluorochrome used. The difference in this system is that the light source and filters are placed above the specimen, and the exciting light reaches the specimen through the objective and not through a condenser. Light of the optimum wavelength is then passed through the dichroic beam splitter, which reflects it with great efficiency towards the specimen. The resulting fluorescence, together with any residual exciting light, again passes through the beam splitter, where most of the exciting light is reflected towards the lamp, while a small proportion passes through towards the eye-piece. The fluorescent light of longer wavelength passes unaltered through the beam-splitter and straight on towards the eye-piece. The barrier filter below the eye-piece removes any remaining exciting light, thus protecting the eye against any possibility of contact with ultraviolet light, and also allowing only the required fluorescent wavelength to be observed.

By altering the dichroic lenses and permutating the exciter and barrier filters, this system can be used for a variety of fluorochromes.

References

1. COONS, A.H., CREECH, H.J. and JONES, R.N. (1941) *Proceedings of the Society for Experimental Biology and Medicine* **47,** 200
2. *BBL Manual of Products and Laboratory Procedures* (1968) 5th edn. Cockeysville, MD: BBL Division of Beckton Dickinson Ltd
3. *Difco Supplementary Literature* (1968) Detroit: Difco Laboratories
4. *Laboratory Diagnostic Reagents* Beckenham: Wellcome Reagents Ltd
5. WILKINSON, A.E. and RAYNER, C.F.A. (1966) *British Journal of Venereal Diseases*, **42,** 8
6. NAIRN, R.C. (Editor) (1975) *Fluorescent Protein Tracing*. Edinburgh: Churchill-Livingstone
7. HEIMER, G.V. (1967) Fluorescent antibody techniques. In *Progress in Microbiological Techniques*. Edited by C.H. Collins. London: Butterworths
8. WICK, G., TRAILL, K.N. and SCHAUENSTEIN, K. (Editors) (1982) *Immunofluorescence Technology*. Amsterdam: Elsevier
9. BRUMBER, E.M. (1959) *Biophysics*, **4,** 97
10. PLOEM, J. (1967) *Zeitschrift für Wisserschaftlicht Mikroscopie*, **68,** 129

Chapter 13

Antibiotic sensitivity and assay tests

A. Cremer

Since the last edition of this book many new methods have been developed for determining the sensitivity of micro-organisms to antibacterial agents, and of assaying those agents in body fluids. Many of these methods involve mechanization or automation, and some of them are described below. This is a constantly expanding field of microbiology and it is largely in the light of the more sophisticated and objective techniques that the limitations of the traditional diffusion methods have become apparent. In particular, it is now clear that it is illogical to use a biological method as a standard for comparing new methods that utilize modern machinery: to do so implies a degree of precision and accuracy which the method does not warrant.

Antibiotic sensitivity tests

The familiar diffusion techniques for testing the antibiotic sensitivity of organisms isolated from clinical specimens have suffered severely from the difficulties of standardization. It is doubtful if tests on the same organism, done in different places, would result in consistent recommendations for treatment.

Recently, however, these techniques have been improved and as a result most workers in the UK have used the Stokes method[1]. In the USA, however, the modified Kirby–Bauer[2] method is preferred and in Sweden the diffusion techniques of the International Collaboration Study[3] are used.

The classical diffusion technique requires that a source of antimicrobial agent is applied to the surface of a solid medium. The diffusion of the agent through the medium inhibits the growth of a sensitive organism growing in it or on it to a degree partially dictated by the susceptibility of the organism. It is well known, however, that a number of other factors will influence the size of the zones of inhibition, and these factors will require control.

Factors affecting zone sizes

Ingredients of culture media

These are the most important. Apart from viscosity and depth in the petri dishes many substances present in culture media may affect the zones of inhibition. These include peptone, tryptone, yeast extract and agar. They may vary in their mineral content, and it is now well known that calcium, magnesium and iron affect the sizes of zones produced by tetracyclines and gentamicin. Sodium chloride reduces the activity of aminoglycosides and enhances the effect of fusidic acid. Carbohydrates may enhance the effect of nitrofurantoin or ampicillin.

Perhaps the most notable ingredients of culture media that influence sensitivity tests are those related to the drugs which act by inhibiting folate metabolism, e.g. sulphonamides and trimethoprim. The action of drugs on even the most susceptible organisms will be affected if a medium containing more than a very small amount of the end products of bacterial folic acid synthesis is used.

The two most common examples of such substances are *p*-aminobenzoic acid and thymidine. The presence of these chemicals in culture media will often not merely decrease the size of the zone of inhibition but, depending on the amounts available, may permit the organism to grow right up to a disc, although the colonies may be smaller than usual. The effect of small amounts of thymidine in a medium, however, can usually be overcome by the addition of haemolysed horse blood. For testing the sensitivity of fastidious organisms or to perform direct sensitivity tests upon clinical specimens and at the same time to observe haemolysis, equal parts of whole blood and haemolysed blood may be added. Chocolate agar may also be used, provided that the basic medium is free from inhibitor, but all media that contain blood or blood products may produce smaller zones with antimicrobials which are particularly protein-bound.

Choice of medium

Consistent and reproducible results are obtained when media especially formulated for sensitivity testing are used, e.g. those marketed by the companies named on p. 56. Ideally plates should be poured flat with an even depth of medium throughout. Very thin or thick plates are unsatisfactory.

Effect of pH

pH has an effect on the size of inhibition zones produced by some antibiotics.

The activity of aminoglycosides is enhanced in alkaline medium and reduced in acid media. The reverse effect is true for tetracyclines. Carbon dioxide in the incubator atmosphere and the presence of fermentable carbohydrates in culture media will also induce acidic conditions.

Size of inoculum

Although many antimicrobials are not markedly affected by the presence of a large number of organisms, heavy inocula reduce inhibition zones to some extent. The ideal inoculum is one which gives an even, dense growth without being confluent. The actual density of a confluent growth is impossible to assess.

Broth cultures of organisms and suitable suspensions made from solid media may be diluted accurately to give optimal inocula for sensitivity testing but in practice satisfactory results can usually be achieved by taking a loopful of a well grown culture, or a suitably made suspension of organisms, and spreading it with a dry sterile swab[4].

Direct sensitivity tests

Sensitivity tests may be set up on primary cultures instead of indirectly on pure suspensions prepared by selection of colonies from such cultures. These direct tests enable a report to be given 24 h earlier.

Colonies which all look alike on the usual primary culture media may show distinctly different populations on the direct sensitivity plate. This is particularly true of staphylococci. When, for instance, a single colony is picked from an aerobic blood agar plate it is a matter of chance which population is being tested and the colony chosen may not necessarily be representative of the majority. Arguments are often put forward that a large proportion of primary sensitivity tests have to be repeated because the inoculum is unsuitable, because there is antibiotic present in the original specimen, or because there is a mixed bacterial population. In practice, however, when primary sensitivity tests are carried out regularly it is surprising how often an answer can be given despite these drawbacks. Those who perform them regularly, moreover, argue with some justification that even when such eventualities render a test unreadable nothing is lost as far as the patient is concerned, and the only wastage is that of a small amount of laboratory time and culture medium.

Certain sources of error peculiar to primary sensitivity tests do, however, need to be noted when these tests are carried out. Firstly, there is the possibility of a penicillinase-producing organism present in a mixed culture 'protecting' an otherwise susceptible organism against the action of a penicillinase-susceptible antibiotic. Penicillinase-producing staphylococci are usually quite easily recognized because the edges of the zones of inhibition which they exhibit show no tendency to produce smaller colonies nearer to the disc and are heaped up and sharp. Any significant organism that is present in company with a penicillinase producer should be subcultured and retested in pure culture.

Bacteria which satellitize other organisms in mixed cultures (e.g. *Haemophilus* spp.) may show a false sensitivity to any antimicrobial substance that prevents the growth of the colonies to which they show satellitism. When this happens the sensitivity test should be repeated with a pure culture.

Although some laboratories do not do many direct sensitivity tests it would seem sensible to include them in investigations on body fluids which are normally expected to be sterile, or are only likely to be infected with a single species.

Other factors

Difficulties in reading sensitivity tests may arise from the presence of swarming *Proteus* species. When there is a clear edge to the zone given by the parent colonies, however, measurements may safely be taken from this and swarming nearer the disc disregarded.

A modification of the method is necessary for testing the methicillin resistance of staphylococci. Most cultures of such organisms contain relatively few resistant cells when incubated at 37 °C and these are not obvious unless a heavy inoculum is used and incubation continued for 48 h. Methicillin resistance is much easier to see if 5% sodium chloride is added to the sensitivity test media[5], or if the cultures are incubated at 30 °C[6]. It is also important that only methicillin disc or strips are used for this test because cloxacillin and flucloxacillin may give false sensitive results. Coagulase-positive staphylococci which prove to be resistant to methicillin are also always resistant to cephalosporins.

The performance of diffusion techniques

Strength of antibiotics

Until very recently there has been little or no general agreement about the strengths of antibiotic discs for use in *in vitro* sensitivity tests. Tables are now available which give recommended disc strengths, which vary according to the method of interpretation being used (*Table 13.1*).

TABLE 13.1. Suitable disc contents (μg)

Antibiotic	*Organisms from urine*	*Organisms from other infections*
Ampicillin	25	10
Carbenicillin	100	100
Cephaloridine	30	5
Chloramphenicol	30	10
Clindamycin	—	2
Erythromycin	—	15
Fucidin	—	5
Gentamicin	10	10
Kanamycin	30	30
Methicillin	—	10 (strips 25)
Nalidixic acid	30	—
Nitrofurantoin	200	—
Penicillin	—	2[a]
Polymyxin B	300[a]	300[a]
Streptomycin	25	25
Sulphafurazole	100	100
Tetracycline	30	10
Trimethoprim	1.25	1.25

Reproduced from p. 522 of *Antibiotics and Chemotherapy*, L.P. Garrod, H.P. Lambert and F. O'Grady. 4th edn. 1973. London, Churchill-Livingstone by permission of authors and publisher.
[a] Units.

Storage and application of discs

Appropriate storage conditions are very important if reproducible results are to be achieved. Discs should always be kept cool and dry, and should be applied firmly to the medium to ensure proper contact and thus even diffusion. Fine-pointed forceps and dissecting needles are convenient tools.

Incubation time

Ideally the incubation time should be the minimum required for the growth of the organism and consistent with the daily routine of the laboratory. Prolonged incubation of a culture may result in decay of the antibiotic and subsequent growth of organisms which have been prevented from reproducing but which are not killed. Thus, it is clear that such methods are suitable only for organisms which reproduce rapidly, and that any inhibitor in the medium or the lack of any optimal physical condition that reduces growth rate will produce artificially large inhibition zones.

Controls

For the best interpretation of results and recognition of any source of error in disc diffusion sensitivity methods the correct use of controls is essential. For the Stokes method which is described in more detail below, control organisms are present on every plate.

Almost the entire range of antibiotic sensitivities performed in the diagnostic laboratory may be adequately controlled by using one of three organisms: *Staphylococcus aureus* (NCTC 6571: the 'Oxford Staphylococcus'), *Pseudomonas aeruginosa* (NCTC 10662), and *Escherichia coli* (NCTC 10418). These organisms may be kept on agar slopes at room temperature and subcultured about once a month. For routine daily use the organisms are most conveniently kept at 4 °C on sterile throat swabs. A jar full of such swabs can be impregnated at one time, and they keep well for at least a week. The control organisms are applied to the plates directly from the swabs. The choice of control organism will depend upon a number of factors. Firstly it must be sensitive to normal doses of the antibiotic being tested. This will also be affected by the site of the infection and the concentration of the antibiotic usually attainable at that site. More resistant organisms for instance, will be likely to respond to treatment with antibiotics excreted by the kidneys in urinary tract infections, because much higher concentrations of such substances occur in the urine. Hence, in these circumstances, a higher content disc may be used and controlled with a more resistant organism such as *E. coli*. This does not hold good, however, when common urinary tract pathogens such as enterococci and coliform organisms are isolated from other parts of the body. Although these organisms may respond well to normal doses of antibiotics excreted in the urine, they are usually only moderately sensitive and thus when isolated from other sites the fully sensitive control must be used to make clear in the report the higher doses are required. The Oxford Staphylococcus is used for all antibiotics except the polymyxins. The standard *E. coli* should be used for these antibiotics and with all organisms from the urine. The standard *P. aeruginosa* is used in tests against pure cultures of known pseudomonads that are tested against only a limited range of antibiotics.

Technical methods

The Stokes method

This method does not require the stringent standardization demanded by those described below. Test and control organisms are compared against the same discs, on the same medium and under the same physical conditions.

For the individual disc technique inoculate the middle third of a suitable culture plate with the specimen or a suitable suspension of organisms in broth using a cotton wool (throat) swab. Inoculate the areas on either side with the control organism, leaving a small gap between the inoculated areas. Place the disc in this gap (*Figure 13.1*). On 90-mm round plates four discs are convenient, while on larger square dishes six discs can be accommodated. After incubation zone sizes are compared by measurement with calipers, a transparent rule, or one of the more sophisticated zone-readers.

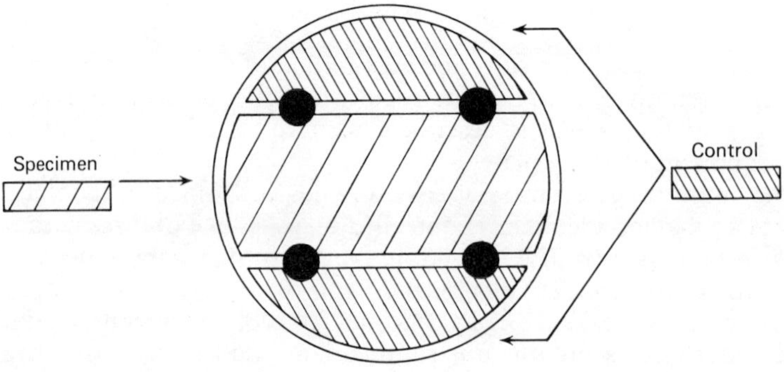

Figure 13.1 Controlled single disc sensitivity method. (Reproduced by permission of Dr E. J. Stokes)

Modifications of the Stokes method

Most of the resistance to the use of the Stokes method has resulted from the introduction of multiple discs such as the Multodisc (Oxoid) and the Mastring (Mast). Laboratory workers quite rightly claimed that the method was unsuitable for these annular disc arrangements particularly with the original Multodisc which had a 'solid centre'. This method is perfectly suited however for rings of discs with a 'hollow centre', and particularly when the method is adapted for use with one of the rotary plating devices (p. 90).

In the rotary plating technique a suitable inoculum of the test organism or specimen is placed in the centre of the plate and as it is rotated by the machine it is spread with a sterile swab to cover an area indicated by the template inscribed on the carrier of the machine, or by the right-angled metal guide which can be attached. In this way the gap for the discs is left between the central inoculum of the test organism and the control organism which is streaked in the same way on the periphery of the plate. The result is read as before.

One of the disadvantages of multiple disc sets, however, is that the user is unable to change the combination of antibiotics on any given set to meet the extra demands that are not unusual in the diagnostic laboratory. This means that a variety of disc sets must be kept, and from time to time unnecessary antibiotics are tested in order to obtain the result needed from an antibiotic that happens to be in the same set. Apart from any other consideration this is clearly not very cost-effective. For this reason greater flexibility is obtainable when the rotary plating technique is carried out using single discs of which many varieties are available. The manual application of the discs is an irksome chore, however, when large numbers of tests are necessary. The operation may be mechanized. Useful equipment for this purpose includes the Manual Disc Dispenser Mark 2 and the fully mechanized Discamat machine, both manufactured by Oxoid. Both use cartridges of individual discs which can be changed easily to give a variety of combinations.

When a manual dispenser is used, the central knob is depressed and the discs are released to lie above holes in the base which are just too small to allow their complete passage. When placed over a suitable plate of medium small metal rods descend upon the discs and press them into the surface of the medium, thus ensuring firm contact with the medium so that they will not fall into the lid when the plate is inverted. The Discamat works on very much the same principle, except that the machine is worked electrically and the discs are placed in position pneumatically.

The Stokes method can also be employed with paper strips instead of discs when a number of strains is to be tested against the same antibiotic. The test organisms are streaked across the plate and the strip is placed at right angles to them. The control organism is streaked across the plate in the same way. The antibiotic may also be applied as a suitable solution of antibiotic in agar filled into cups or ditches cut into the medium.

Interpretation of results

Zones of inhibition produced around antibiotic discs or strips with the test organisms must be compared with those produced by the appropriate control organisms. When zone sizes are clearly not the same to the naked eye it is necessary to measure them with calipers or a transparent rule, or with a commercial zone reader (Luckham, Mast, Oxoid). Bearing in mind the relative densities of inocula, sensitivities may be reported with reasonable limits, as follows: Measurements are taken from the edge of the disc to the edge of the zone. A zone 'radius' that is the same size as or is larger than the control, or is not smaller by more than 3 mm, is reported as *sensitive*. A zone that is more than 3 mm in radius but is smaller than the control by more than 3 mm in radius is reported as *moderately sensitive*. A zone of radius of 2 mm or less is reported as *resistant*.

In this context a report that says that an organism is sensitive to a particular antimicrobial agent implies that the infection should respond to treatment with normal doses under normal circumstances. A report that says that an organism is moderately sensitive or relatively resistant implies that it might respond to treatment with high dosage. An organism which is reported resistant is unlikely to show any clinical response at all.

International collaborative study method

This method relies on the linear relationship that exists for most antibiotics between the diameter of the zone of inhibition and the logarithm of the Minimum Inhibitory Concentration (MIC).

The MIC for at least 100 organisms with widely differing sensitivities to antibiotics are calculated. At the same time, and under the same strictly controlled conditions, the diameter of the inhibition zone produced by a high-content disc of the same antibiotic with each strain is measured. A regression line on semilogarithmic graph paper of the MICs on the ordinate and against the diameters of the inhibition zones on the abscissa is then plotted. The diameter of the zone around an unknown organism under the same conditions is measured and the MIC is read from the graph. Instead of reporting the MIC, the organism is placed in one of four classes of susceptibility calculated locally with reference to the original paper.[3]

Kirby–Bauer technique

Mueller–Hinton agar in plates 5–6 mm deep are used with high-content discs. Carefully standardized inocula are applied with cotton wool (throat) swabs to give a confluent growth. The zone diameters are read and the tables that give the break-point for each antibiotic are consulted. When the exact conditions of the test are not reproduced, tables may be worked out locally. The sensitivity of the organism is reported as *sensitive*, *intermediate*, or *resistant*.[2]

Why does the laboratory sometimes 'get it wrong?'

Despite all the very best efforts of the investigating laboratory using the most accurate methods possible, there are still occasions when an organism which is seen to be sensitive to a given antibiotic in the laboratory causes an infection which fails to respond to treatment in spite of the patient receiving the appropriate dosage by the correct route. This is particularly difficult for a clinician to understand. There are, however, several reasons why it may happen.

The most obvious of these, at least to the laboratory worker, is that for many reasons the antibiotic may not have been absorbed properly and is thus not achieving a therapeutic level in the patient's tissues. It may also be that the antibiotic is not able to reach the site of infection, particularly in the cases of osteomyelitis or lung abscesses. In extensive and contaminated wounds such as those caused by road traffic accidents or war, the presence of detritus in wounds may provide a shield behind which bacteria may survive. It may also be that the laboratory has chosen an irrelevant organism for the sensitivity test. This may well happen when a specimen (e.g. sputum) has been badly taken, or has been delayed in warm conditions before reaching the laboratory. When this happens commensal organisms may outnumber the pathogen, and may be tested as the predominant flora. On occasion the original organism may be eliminated, only to be replaced by another before the clinician is able to detect any clinical change. Despite all efforts by the laboratory there are examples of the most careful culture techniques failing to demonstrate the presence of a pathogenic organism at all.

Occasionally the opposite situation occurs. An organism which has been assessed as being resistant is apparently eliminated by an antimicrobial agent to which it should, in theory, not respond. It is possible that in this case the patient was not infected at all but merely 'colonized'. It is also true to say that organisms do not always behave in the human body in exactly the same way as they do in the laboratory. A classical example of this is shown by typhoid bacilli which are present in blood cultures from patients who have been given tetracycline, but fail to grow, although therapeutically tetracycline may not be used successfully in the treatment of the disease. It is, of course, also true that many infections will resolve spontaneously without antibiotics.

Further information

Further detailed information about antibiotic sensitivity testing may be found in the book by Garrod *et al.*[7], and in the Association of Clinical Pathologists Broadsheet No. 55[8].

Supplies of antibiotic discs

The principal suppliers of discs and strips in Great Britain are probably BBL, Difco, Mast and Oxoid.

Assay of antibiotics in body fluids

It is necessary to ascertain the concentration of antibiotics in the body fluids to ensure that the levels are adequate to deal with the infecting organism but not so high that the patient will suffer ill effects, e.g. with aminoglycosides that there is no damage to the eighth nerve in patients with renal problems.

Traditionally antibiotic assays have been performed using either tube dilution or diffusion methods. As with sensitivity tests, however, many other methods of performing assays have been demonstrated since publication of the last edition of this book, and the most useful of these, in our judgement, will be considered below.

Collection of specimens

If proper assessment of the results is to be made it is essential that the laboratory knows how much time has elapsed between the administration of the last dose of antibiotic and the collection of the specimen. The laboratory also needs to know if any other antibiotic has been given. Blood from a patient with renal insufficiency may contain antibiotic many days after the last dose was given.

Choice of organisms

A list of organisms, with their NCTC numbers is given in *Table 13.2.*

TABLE 13.2. Appropriate organisms for assaying antibacterial drugs

Drug	*Organism*	*NCTC number*	*Optimum pH*
Penicillins } Cephaloridine }	*Staph. aureus*	6571	6.8 (Penicillins, Cephaloridine, Carbenicillin)
	S. lutea	8340	
	B. subtilis	8236	
Carbenicillin[a]	*Ps. aeruginosa*	10490	
Streptomycin	*Staph. aureus*	6571	7.8 (Streptomycin, Kanamycin, Gentamicin)
	Kl. pneumoniae	7242	
	B. subtilis	8236	
Kanamycin } Gentamicin }	*Staph. aureus*	6571	
	B. subtilis	8236	
Tetracycline	*B. cereus*	10320	6.6
	Staph. aureus	6571	
Chloramphenicol	*Esch. coli*	10418	
	S. lutea	8340	
Erythromycin	*Staph. aureus*	6571	7.8
Lincomycin	*B. subtilis*	8236	7.8
Fucidin	*C. xerosis*	9755	6.6
Vancomycin	*Staph. aureus*	6571	7.8
	B. subtilis	8236	
Polymyxins	*Bord. bronchiseptica*	8344	7.3
	Esch. coli	10418	

[a] Usually contains trace amounts of benzylpenicillin and must therefore be assayed against an organism resistant to this.

TABLE 13.2. (*continued*)

Drug	*Organism*	*NCTC Number*	*Optimum pH*
Antifungal	*Sacch. cerevisiae* *C. albicans*	10716	7.3
Trimethoprim	*B. pumilis*	8241	7.3

Reproduced from p. 519 of *Antibiotics and Chemotherapy*, L.P. Garrod, H.P. Lambert and F. O'Grady. 4th edn. 1973. London, Churchill-Livingstone by permission of the authors and publisher.

Stock solutions and standard dilutions of antibiotics

These may be prepared in water, phosphate buffer at a suitable pH or in dimethyl sulphoxide, although one or two antibiotics require special solvents. Standard solutions for use in serum assays should be made in pooled human serum containing no antibiotic, but when this is not available horse serum may be used. The pH at which standard solutions of antibiotics are made is very important. As mentioned above the performance of many antimicrobial substances is greatly affected by the pH of the solution. This applies to tetracyclines and the aminoglycosides. The optimum pH for various antibiotics is given in *Table 13.2*.

pH of specimen

When antibiotics are being assayed in urine the pH of the specimen must be brought to the optimum for the antibiotic concerned. Any dilutions should be prepared in the buffer used for the standard solutions.

Diffusion methods

With most antibiotics a linear relationship exists between the diameter of the zone of inhibition and the logarithm of the concentration of the antibiotic producing the zone. Consequently it is usual to plot the diameters of zones of inhibition given by standard solutions of antibiotics on semilogarithmic graph paper and then to use this graph to read the concentration of antibiotic in the test specimen by finding the zone of inhibition it produces with the assay organisms under exactly similar conditions.

The test fluid and the standard solutions are applied either to preseeded or to surface-flooded plates or to columns of seeded agar in glass tubing. When very high levels of antibiotic are present the specimen should be suitably diluted, as the zones of inhibition given by the specimen should ideally lie within the range given by the standard antibiotic solutions.

In normal circumstances it is unnecessary to prepare a medium especially for antibiotic assay. Any medium that is suitable for antibiotic sensitivity tests may be used, but it is essential that all plates used in any one assay should be from the same batch of medium.

The depth of the medium will affect zone sizes in any test where the antibiotic is applied to the surface of the medium. The thinner the medium the more sensitive the reading is likely to be. The sensitivity of any test may be increased by pre-diffusion of the antibiotic and also by incubation at 30 °C. The density of the inoculum is important as a heavy inoculum will decrease the sensitivity of the test.

The antibiotic may be applied to the culture in a number of ways. Originally porcelain and later stainless steel cylinders were used. Both of these methods have considerable disadvantages.

Fish spine electrical insulating beads are much more convenient, and may be used on very thin plates. Blotting paper discs have also been used. They require very little serum and can be used with a thin layer of agar. They have the distinct disadvantage, however, that a number of types of paper can bind antibiotic resulting in much smaller zones, and when the line is plotted it may not be straight.

The biological method that is currently most widely used involves punching holes in the medium and placing in the holes the fluid to be tested and the necessary standards (*Table 13.3*). The agar does not need to be very thick so that little serum is needed.

TABLE 13.3. Suitable standard solutions, in serum, for serum assays

Drug	*Test organism*	*μg/ml*
Streptomycin[a] }	*B. subtilis* *Staph. aureus* *Klebsiella*	75, 15, 3
Kanamycin }		50, 10, 2
Gentamicin }		10, 2, 0.4
Tobramycin }		10, 2, 0.4
Vancomycin	*B. subtilis* *Staph. aureus*	50, 10, 2
Penicillin	*Staph. aureus*	1, 0.5, 0.25, 0.12
	B. subtilis	10, 2, 0.4
Carbenicillin	*Ps. aeruginosa* NCTC 10490	50, 25, 12.5, 6.2

Reproduced from p. 522 of *Antibiotics and Chemotherapy*, L.P. Garrod, H.P. Lambert and F. O'Grady. 4th edn. 1973. London, Churchill-Livingstone by permission of the authors and publisher.

[a]pH of medium must be raised to 7.8.

Vertical diffusion method

This method, devised by Mitcheson and Spicer[9] is suitable in its original form for assaying all the aminoglycosides and vancomycin.

Assay medium

Use nutrient agar or other suitable medium, but add 1% peptone in water and adjust pH to 7.8. Bottle in 19-ml amounts.

Inoculum

Melt 19 ml of medium and cool to 48 °C. Dilute an overnight broth culture of the Oxford Staphylococcus or other suitable organism 1:100 in broth and add 1 ml to the melted medium. Mix well without causing bubbles.

Test

With a pasteur pipette, introduce the agar into 3-mm internal diameter glass tubing standing upright in Plasticine to form a 2–3-cm column. Allow to solidify and then pipette the standard solutions and the test serum (diluted if necessary) on the tops of the agar columns, using at least three tubes for each. The actual amount at the top of the column is immaterial provided that it is at least 1 mm. Incubate all tubes overnight at 37 °C.

Reading

The edge of the zone of inhibition is marked by the appearance of colonies which are much larger than those distributed through the remainder of the medium. These provide a sharply defined marker for measuring. To determine the exact size of the zone of inhibition remove each tube from the Plasticine, leaving a plug at the bottom. Lay it across two small pieces of Plasticine on a microscope slide and focus the meniscus formed at the top of the agar column with the low power objective and an eye-piece with a hair-line.

Move the stage so that the hair-line is across the beginning of the large colonies which mark the lower edge of the zone of inhibition. Read the vernier scale. Subtract the lesser from the greater reading to give the zone of inhibition in mm.

Average the readings for each standard and unknown. Square the average for each standard and plot against the concentration of the antibiotic on semilogarithmic graph paper. Square the reading of the test and read the concentration from the graph.

Modified method for antibiotics which are active against *Streptococcus pyogenes*

The vertical diffusion method is only really accurate for assaying aminoglycosides because these substances are alone in giving clear-cut zones. Any antibiotic which is active against *S. pyogenes* may be assayed by using a modification of this technique. In the modified technique blood agar and *S. pyogenes* are used instead of plain agar and the Oxford Staphylococcus[10].

Use 17 ml of agar and 1 ml of a well mixed overnight broth culture of *S. pyogenes* diluted to contain approximately 10^6 organisms/ml. Continue as above but measure the inhibition of the zones of haemolysis.

Plate diffusion method

No special medium or apparatus is needed. Holes are punched in nutrient agar with cork borers. For small amounts of test serum the holes should not be more than 2.5 mm deep but with holes 8–9 mm deep smaller amounts of antibiotic may be detected than is possible with any comparable diffusion.

This method, however, is not technically as simple as it seems. Considerable practice is required, particularly in cutting the holes and filling them with solutions and specimens if consistent, reproducible results are to be obtained.

Test organism

It is now customary to give patients several antibiotics together and this invalidates the results of many assay methods. The organism recommended for this method is a *Klebsiella* sp. (NCTC 10896) which is resistant to nearly all antibiotics except aminoglycosides and is thus seldom affected when another agent is present in the specimen.

Seeding the medium

To pre-seed the medium, melt 10 ml of agar and cool it to 48 °C. Add a suspension of the standard organism to give a concentration of about 10^5 cells/ml. Mix well and pour into a 9-cm petri dish on a flat surface. Avoid bubbles. Dry the plates at room temperature for 1 h.

Alternatively make lawn cultures: pour and dry the plates unseeded and flood with a suspension of the organism containing 10^6 cells/ml. Remove excess suspension with a pasteur pipette.

Test

Make holes in the medium with an appropriate cork borer and lift the plugs out carefully without disturbing the surrounding medium. If the medium is disturbed the solutions may seep beneath the agar to give irregular zones. Fill the holes completely with the standard antibiotic solutions and with the serum to be tested and/or dilutions of it. Carry out all tests in triplicate. Incubate at 37 °C for 4 h.

Reading

Measure the zones of inhibition with calipers. Plot the average diameters of the zones of inhibition of the standard solutions against the logarithm of the concentration of antibiotic on semilogarithmic graph paper.

Measure the zones of inhibition of the test serum and use the average to read the concentration of antibiotic from the graph. This method provides an answer after about 4 h of incubation.

Assays in fluid medium

These have the advantages of requiring no special facilities. They are, however, less accurate and more susceptible to contamination than are solid medium methods. More important, because the dilutions are widely spaced the results fall within wide limits and are often erratic.

The composition of the medium for tube dilution assays is most important particularly when assaying aminoglycosides. Results have been shown to vary widely with various brands of nutrient broth. It is therefore recommended that a brand of peptone is used that is known not to depress the activity of the drugs. The reading is made easier when this is used in a medium containing 0.5% glucose and an indicator. Thus the criterion of growth is not simply the presence of turbidity but a change of colour of the medium. The medium should contain 10% serum, but this should be increased to 50% if the antibiotics concerned are known to be heavily protein bound.

Method

Inactivate the patient's serum at 56 °C for 20 min. Prepare a series of doubling dilutions in the medium in sterile test tubes. Prepare also a series of control dilutions of a standard solution of the antibiotic to be assayed in the same medium. These dilutions should be fairly close together, covering the MIC of the organism being used. Inoculate both sets of tubes with a standard drop of the test organism using a suspension containing 10^5 cells/ml. Incubate at

37 °C overnight although the test can often be read earlier than this, particularly when a 'shaking' water bath is used.

Reading

Read the MIC of the test organism from the control row. Calculate the concentration of antibiotic in the patient's serum by multiplying the MIC by the reciprocal of the highest dilution of the test row that inhibits growth.

When the patient is being treated with more than one antibiotic, the same difficulties arise with this method as with any other that utilize inhibition of a microbe to indicate the end-point of the assay. When feasible this may be overcome by inactivating the unwanted antibiotic, or by using an organism which is sensitive only to the antibiotic being assayed.

Bacterial activity

It is sometimes of interest to estimate the bactericidal activity of patients' sera, particularly during the treatment of cases of bacterial endocarditis, and a tube dilution method may be used for this purpose.

Make a series of twofold dilutions of serum in the assay medium. Inoculate these with the organism isolated from the patient's blood (the recommended inoculum is 10^6 organisms/ml). Incubate the tubes overnight and subculture any that show no growth on a suitable solid medium. Compare any growth that appears on the solid medium with that on a control tube plated before incubation.

The greatest dilution giving no growth, or in some laboratories less than ten colonies, is the end point. Serum that is bactericidal in a dilution of 1 in 4 is considered to be satisfactory.

The newer methods of antibiotic assay

There are now a number of ways of performing antibiotic assays by mechanized or automated methods. These eliminate many of the shortcomings of biological assay methods, and in most cases they greatly reduce the time taken to perform the tests. The most important are described below.

The Macro-Vue card test

This is a rapid method of assaying gentamicin and is based upon the phenomenon of agglutination inhibition. Latex particles are sensitized to gentamicin antiserum. When antiserum is added to the latex particles agglutination takes place. The addition of gentamicin to the system inhibits the agglutination reaction. The gentamicin concentration is calculated by comparing the inhibition caused by diluted test sera containing gentamicin with known gentamicin standards. This concentration is obtained by multiplying the lowest standard concentration totally inhibiting agglutination, by the highest reciprocal dilution of the test specimen totally inhibiting the agglutination. (Beckton–Dickinson).

High pressure liquid chromatography (*HPLC*)

This method is extremely precise and rapid. Although of little use in the clinical laboratory HPLC can even quantify the three major components of gentamicin, C1, C1a, and C2. This is of major importance in pharmacokinetic studies. Briefly the assay of antibiotic by HPLC requires the separation of the drug from the serum, derivatization, separation in the chromatograph column and quantitation based on the serum standards (Spectra-Physics).

Enzyme multiplied immunoassay technique (EMIT)

This technique is among the more common of the mechanized methods in use in clinical laboratories. Enzyme-labelled antibiotic competes with patient antibiotic in a protein binding reaction for antibody binding site. When the enzyme-labelled antibiotic is mixed with specific antibody to the antibiotic its enzymatic activity is blocked. When mixed with patient's serum containing unlabelled antibiotic the unlabelled substance will compete with enzyme labelled ones for antibody binding sites, and in so doing displaces labelled antibiotic into the reaction mixture. This free enzyme labelled antibiotic then reacts with the enzyme substrate present. The degree of resulting enzyme activity may be related to the amount of unlabelled antibiotic in the sample (Syva).

Radio immunoassay (RIA)

The basic principle of RIA is similar to that of EMIT. The significant difference is that as the label used in EMIT is an enzyme the label in RIA is a radioactive isotope. Although highly sensitive and specific this method has the disadvantage of needing expensive instrumentation, the reagents have a short shelf-life, and stringent precautions are required for the safe disposal of the isotope (Diagnostic Products).

Fluorescent immunoassay

A similar method to the above is also in existence. This uses fluorescing compounds as the assay label. The instruments used are much cheaper than those required for RIA and there is no problem in disposing of waste materials (Ames).

References

1. STOKES, E.J. (1968) *Clinical Bacteriology*. 3rd edn, London: Edward Arnold
2. BAUER, A.W., KIRBY, W.M.M., SHERRIS, J.C. and TURCK, M. (1966) *American Journal of Clinical Pathology*, **45,** 493
3. ERICSSON, H.M. and SHERRIS, J.C. (1971) *Acta Pathologia Microbiologia Scandinavica*, Section B, Supplement, 217
4. FELMINGHAM, D. and STOKES, E.J. (1972) *Medical Laboratory Technology*, **29,** 198
5. BARBER, M. (1964) *Journal of General Microbiology*, **35,** 183
6. ANNEAR, D.I. (1968) *Medical Journal of Australia*, **1,** 444
7. GARROD, L.P., LAMBERT, H.P. and O'GRADY, F. (1981) *Antibiotic and Chemotherapy*. 5th edn. London: Churchill-Livingstone
8. STOKES, E.J. and WATERWORTH, P.M. (1972) *Association of Clinical Pathologists Broadsheet*, **No. 55,** Revised
9. MITCHISON, D.A. and SPICER, C.C. (1949) *Journal of General Microbiology*, **3,** 184
10. FUJII, R.R., GROSSMAN, M. and TICKNOR, W. (1961) *Pediatrics*, **28,** 662

Chapter 14

Bacterial food poisoning

Before commencing investigations into outbreaks of suspected food poisoning, obtain the following information:

(1) number of people at risk, i.e. how many ate the suspected food;
(2) number of persons actually ill;
(3) nature of illness: (a) vomiting, (b) diarrhoea, (c) nausea, (d) headache, (e) disturbance of the central nervous system;
(4) exact time food was eaten and exact time of onset of symptoms;
(5) description of *all* food eaten in 24 h prior to onset of symptoms, and as much information as possible about the food consumed during the previous four days.

This information will indicate whether the incident merits laboratory investigation. A very small number of people alleged to have been ill when a very large number is at risk is probably not an outbreak of food poisoning. Similarly, stories of single cases of poisoning due to foreign bodies, canned fruit, fresh vegetables, unusual articles of diet or food which has 'gone off' can usually be discredited. In nurseries or institutions, a wide variation in the times of onset usually suggests Winter Vomiting Disease rather than food poisoning. When symptoms appear 30 min or so after eating, chemical rather than bacterial poisoning is indicated. It may not necessarily be the last meal which causes bacterial illness.

The information will suggest one of the types of food poisoning described below and thus indicate the bacteriological examinations required.

Types of food poisoning

Salmonella food poisoning

This is due to *infection* with a salmonella organism and usually the food must contain several million organisms to initiate infection. Diarrhoea and vomiting occur from 12–36 h after eating infected food. Illness may last from two to seven days. Not all the persons who ate the food will develop symptoms; some may become symptomless excreters. Foods to suspect include meat products and egg products, particularly dried or frozen eggs and creams, etc., made from them. These foods become infected from animal or human intestinal sources, directly or indirectly.

Vibrio parahaemolyticus food poisoning

This is rare in the UK but it is one of the commonest causes of food poisoning in Japan and has been incriminated in outbreaks in the USA and Australia.

It is closely associated with eating raw and processed fish products. The organisms has been frequently isolated from sea food, particularly those harvested from warm coastal waters. Symptoms of infection may appear from 2 to 96 h after ingestion, depending on the size of the infecting dose, type of food and acidity of the stomach. They vary from acute gastroenteritis with severe abdominal pain to mild diarrhoea.

Escherichia coli food poisoning

Late in 1971, *E. coli* was responsible for 107 outbreaks of food poisoning in the USA. Serotype 0124 B17 was isolated from the vehicles of infection (Camembert and Brie cheese) and from the stools of the victims[1].

Since then there have been food-associated outbreaks connected with cream, crabmeat cocktail and shrimp and mushroom salad. A variety of serotypes were isolated[1].

Staphylococcal food poisoning

The food poisoning types of staphylococci produce an exotoxin which will withstand heating at 100 °C for 30 min. The toxin, present in food in which the organisms have grown, causes an *intoxication* type of disease with onset in 4–6 h and symptoms of diarrhoea and vomiting lasting 6–8 h. Foods likely to be infected are those which contain a high proportion of salt, meat such as ham, synthetic creams, sauces (particularly Hollandaise) which are never heated above 110°F (40 °C), custards, trifles and occasionally fresh fish or canned vegetables (due to leaker spoilage). Infection occurs usually in the kitchen from infected cuts and abrasions, boils and other lesions or from nasal carriers.

Clostridium perfringens (welchii) food poisoning

Our knowledge of the role of *C. perfringens* in food poisoning has altered in recent years. Originally, only *C. perfringens* forming spores resistant to boiling for 1 h were thought to cause this type of food poisoning. These were invariably non-haemolytic. Recently, both haemolytic and non-haemolytic heat-sensitive strains have been incriminated in food poisoning outbreaks. The incubation period varies from 8–22 h. Young and longstay patients and hospital staff tend to become colonized with *C. perfringens*. Counts of 10^5/g in the faeces of sufferers are not uncommon.

Because the heat resistance of *C. perfringens* spores varies, the organisms survive some cooking processes. Much depends upon the degree of heat that reaches the cell and the period of exposure. If the spores survive and are given suitable conditions, they will germinate and multiply. It is common practice to cook a large joint of meat, allow it to cool, slice it when cold and re-heat it before serving. This is hazardous unless cooling is rapid and re-heating is thorough and above 80 °C. Growth of heat-sensitive strains in cooked food may be the result of too low a cooking temperature or may be due to recontamination after cooking. The usual vehicles for 'welchii' food poisoning are meat and poultry. Stews, made-up meat dishes and large deep-frozen birds are suspect. *C. perfringens* is commonly found in soil; small numbers are normally present in human faeces and in some foods. The symptoms of food poisoning caused by *C. perfringens* are mild diarrhoea, nausea and headache, rarely vomiting, lasting 12–24 h.

***Bacillus cereus* food poisoning**

Cases of food poisoning associated with eating fried rice have been reported from Scandinavia, the USA and Europe. Large numbers of *B. cereus* were isolated from the rice.

Rice for fried rice dishes is often boiled in bulk and left to cool at room temperature. Refrigeration causes the grains to stick together. Before serving, egg is added to the cooked rice, which may have been standing in a warm kitchen for many hours. The mixture is then lightly cooked in a frying pan and served. During the initial cooking, vegetative organisms are destroyed; during standing, spores germinate in ideal conditions for growth. Subsequent heating does not destroy any toxins that may have formed.

B. cereus food poisoning can be confused with staphylococcal or perfringens food poisoning. There are two types[2], diarrhoeal and emetic.

	Diarrhoeal	*Emetic*
Incubation	8–16 h	1–5 h
Symptoms	diarrhoea, nausea	vomiting (v. common) diarrhoea (common)
Duration of illness	12–24 h	6–24 h
Isolation from faeces	sometimes (low numbers)	sometimes
Isolation from vomit	—	large numbers
Nature of toxin	protein	small peptide

The emetic type is more common. Occasionally *B. megaterium* is responsible for similar symptoms.[2]

Botulism

This form of intoxication, caused by the pre-formed toxin of *Clostridium botulinum*, is rare in the UK, but more prevalent in Europe and the USA where home canning of meat and vegetables is common practice. Between 12 and 36 h after eating food containing the toxin, patients develop symptoms including thirst, vomiting, double vision and pharyngeal paralysis, lasting 2–6 days and usually fatal. Vehicles are usually home-preserved meat or vegetables, contaminated with soil or animal faeces and insufficiently heated. Correct commercial canning processes destroy the organisms. Conditions must be optimum for toxin formation: complete anaerobiosis, neutral pH, absence of competing organisms.

Recently there have been several reports (from outside the UK) of botulism caused by the Type E organism that is usually associated with fish and fish products. This type is frequently present in estuarine mud.

Campylobacter food poisoning

These organisms have been growing steadily in importance since 1977. They are not a new cause of infection but efficient methods for their isolation have only recently been developed. The most common vehicle of infection appears to be unpasteurized or inadequately pasteurized milk. Campylobacters are part of the normal flora of many mammals and birds. The organisms are heat sensitive and, in common with salmonellas, control measures such as adequate cooking and prevention of recontamination are indicated.

In England, Wales and Northern Ireland campylobacters caused 13% of the total cases of foodborne disease in 1980. More campylobacters than salmonellas were isolated in 1981[3].

In the late 1970s in the USA outbreaks of infection with *C. fetus* subsp. *jejuni* followed consumption of raw milk, chicken and pork luncheon meat. In Japan in 1980, 800 of 2500 children suffered from enteritis following a meal which included vinegared pork. This was thought to be the most likely cause of infection.[4]

The incubation period has been estimated to be between 2 and 11 days. Symptoms commonly include abdominal pain, diarrhoea, fever, headache, malaise, musculoskeletal pain, rigors and delirium. Vomiting occurs in less than 30% of cases. Sometimes 'flu like' symptoms are present.[4]

Enterococcal food poisoning

Several food associated outbreaks have occurred in which these organisms were the suspected agents.

Yersinia enterocolytica food poisoning

This is an uncommon cause of food poisoning but its ability to grow at 4 °C makes it unusual and a potential hazard. Since 1976, 744 cases of illness attributable to *Y. enterocolytica* have been reported in the USA, Canada and Japan. The symptoms are diarrhoea, fever, abdominal pain and vomiting.[5]

Arizona hinshawii food poisoning

This organism is found in turkeys and reptiles and can produce a salmonella-like illness. Turkey, chicken, cream-filled pastry, ice cream and egg custards have been implicated.[6]

Dose of infecting organisms

To cause food poisoning, a sufficiently large number of organisms must be placed in the food, which must be physically and chemically suitable for their growth. Enough time must elapse between infection of the food and its ingestion to enable the organisms to form a large enough population to cause disease.

The important factors to consider in investigating infected foods are:

(1) source of organism,
(2) nature of food, e.g. moisture content,
(3) temperature of food,
(4) time between preparation and serving.

Examination of pathological material and food

Do a total viable count on all foods suspected of causing food poisoning. In most cases, this will be high. Relate findings to conditions of storage after serving. If possible, also count the number of pathogens.

Salmonella infections

Plate the stools of patients, suspected carriers and food handlers on DCA and bismuth sulphite medium and inoculate selenite medium. Isolation from food and identification of these bacteria is described in Chapter 28.

V. parahaemolyticus infections

Plate stools on thiosulphate citrate bile salt sucrose agar (TCBS) and inoculate alkaline peptone water (APW). Incubate overnight at 37 °C. Subculture enrichment cultures on TCBS.

Homogenize 10–25 g of food in 0.1% peptone water solution: make direct cultures on TCBS and add homogenate to an equal volume of double-strength APW. Make tenfold dilutions in single-strength concentrations of these media. Incubate and plate out as for faeces. Methods for the identification of *V. parahaemolyticus* are given in Chapter 27.

Staphylococcal disease

There is no need to examine the stools of all the patients; a sample of 10–25% is sufficient. Inoculate Robertson's cooked meat medium containing 7–10% sodium chloride (salt-meat medium) with 2–5 g of faeces or vomit and incubate overnight and plate on blood agar or one of the special staphylococcal media (p. 59). Phenolphthalein phosphate polymyxin agar or milk salt (Russian) agar are recommended by Gilbert *et al.*[7]. Examine all food handlers for lesions on exposed parts of the body and swab these as well as the noses, hands and fingernails of all kitchen staff. Swab chopping boards and utensils which have crevices and cracks likely to harbour staphylococci. Use ordinary pathological laboratory swabs: cotton wool wound on a thin wire or stick, sterilized in a test-tube (these swabs can be obtained from most laboratory suppliers). Plate on the solid medium and break off the cotton wool swab in cooked meat medium. Methods for the identification of staphylococci are given in Chapter 31.

It is important that all strains of *S. aureus* isolated from food handlers, implements, foodstuffs and patients are sent to a staphylococcal reference expert. Not all strains of *S. aureus* can cause food poisoning and some of the other strains are usually encountered in this kind of investigation.

Techniques for the detection of enterotoxin have been developed by Gilbert *et al.*[8]. These methods are too complex for routine use.

Examination for *Clostridium perfringens*

Examine the stools from a proportion of cases (10–25% if the outbreak is large). Heat-sensitive *C. perfringens* are found in almost all stools from normal people but large numbers may be demonstrated if the individual is a victim of *C. perfringens* food poisoning. A quantitative or, more practically, a semiquantitative examination is therefore necessary. Make a thick suspension (1:5 to 1:10) of faeces in broth and inoculate:

(1) a neomycin blood agar plate using a calibrated loop as commonly used in urine culture. Incubate anaerobically at 37 °C for at least 16 h but not more than 24 h;
(2) 1–2 ml of suspension into a tube of cooked meat medium. Heat at 100 °C for 1 h, incubate overnight at 37 °C and subculture on neomycin blood agar;
(3) Heat the remainder of the suspension at 80 °C for 10 min and repeat step (1).

These tests give:

(1) total viable counts of *C. perfringens*,
(2) evidence of infection by heat-resistant strains, and
(3) total spore count.

Outbreaks involving both heat-sensitive and heat-resistant strains have been reported.[10]

Prepare a 10% suspension of food in 0.1% peptone water. Make 10^{-1}, 10^{-2} and 10^{-3} dilutions and inoculate all four on neomycin blood agar plate using the Miles and Misra technique (p. 133). Incubate as for faeces. There is no point in heating cooked foods because spores will have germinated. Cooked meat medium containing 70–100 μg/ml of neomycin sulphate should also be inoculated because *C. perfringens* may be present only in small numbers in cooked food.

All strains isolated from food poisoning outbreaks should be sent to a reference laboratory for typing.

It is pointless to examine stools of food handlers and 'contacts' in outbreaks of *C. perfringens* food poisoning as the disease is not spread by human agency.

Examination for *Bacillus cereus*

Soak 20 g of dried material in 90 ml Tryptone Salt solution (*O*) for 50 min at room temperature. Add a further 90 ml of 0.1% peptone water solution. For other materials add 10 ml or 10 g to 90 ml diluent. Homogenize in a

Stomacher. Prepare further dilutions and plate, 0.1 ml amounts on B. Cereus Selective Agar (*O*). Incubate at 37 °C for 24 h then at 25–30 °C for a further 24 h. This medium shows egg yolk reaction, mannitol fermentation and allows sporulation to occur. (The heat resistance of spores grown on laboratory media is lower than in foods.) Do confirmatory tests if necessary by Holbrook and Anderson's method[3] (also described in the Oxoid Manual).

Laboratory investigation of botulism

This is difficult. Methods are given on p. 362. It is best to consult a reference expert as soon as botulism is suspected.

Examination for campylobacters

Prepare a suspension of faeces or food and inculate blood agar base plus 7% lysed horse blood and campylobacter selective supplement (*O*). Incubate at 37 °C in 10–15% CO_2 in hydrogen or nitrogen for up to 48 h.

Examination for enterococci

Plate suspect food on one of the media described in Chapter 32.

Examination for *Yersinia enterocolytica*

Prepare a suspension of faeces or food in M/15 phosphate buffer (*O*) incubate at 32 °C for 24 h. Refrigerate the remaining suspension at 4 °C for up to 21 days and subculture at intervals on Yersinia Selective agar (*O*) Media used for other enteric pathogens can be reincubated at 28–30 °C for 24 h and examined for *Y. enterocolytica* but there is evidence that some strains of this organism are inhibited by bile salts and deoxycholate.

Examination for *Arizona hinshawii*

Do as for salmonellas. This organism is a late lactose fermenter.

Detailed identification of food poisoning bacteria from food

Methods are described under the appropriate headings in this book.

For further information on bacterial food poisoning *see* Hobbs and Gilbert[11] and Corry *et al.*[12].

References

1. MELHMAN, I.J. and ROMERO, A. (1982) *Food Technology*, March, p. 73
2. HOLBROOK, R. and ANDERSON, J. M. (1980) *Canadian Journal of Microbiology*, **26,** 753
3. SHARD, J.B. (1982) *Environmental Health*, June, p. 142
4. DOYLE, M.P. (1981) *Journal of Food Protection*, **44,** 480
5. STERN, N.J. (1982) *Food Technology*, March, p. 84
6. EDWARDS, P.R., FIFE, M.A. and RAMSEY, C.H. (1959) *Bacteriological Review*, **23,** 155

7. GILBERT, R.J., KENDALL, M. and HOBBS, B.C. (1969) Media for the Isolation of Staphylococci from Foods. In *Isolation Methods for Microbiologists*. Edited by D.A. Shapton and G.W. Gould, pp. 9–14. London: Academic Press
8. GILBERT, R.J., WIENEKE, A.A., LANCER, J. and SIMKOVICOVA, M. (1972) *Journal of Hygiene, Cambridge*, **70,** 755
9. COLLEE, J.G., KNOWLDEN, J.A. and HOBBS, B.C. (1961) *Journal of Applied Bacteriology*, **24,** 326
10. SUTTON, R.G.A. and HOBBS, B.C. (1968) *Journal of Hygiene, Cambridge*, **66,** 135
11. HOBBS, B.C. and GILBERT, R.J. (1978) *Food Poisoning and Food Hygiene*. 4th edn. London: Edward Arnold
12. CORRY, J.E.L., ROBERTS, D. and SKINNER, F.A. (Editors) (1982) *Isolation and Identification Methods for Food Poisoning Organisms*. New York: Academic Press

Chapter 15

Food—general principles

Although the bacteriology of food is under consideration by several EEC committees (Codex Alimentarius Commission) there is still no general agreement on methods for laboratory examination. In the UK there are statutory methods for testing milk and officially recognized methods for the examination of water but as yet no official or even recommended methods for examining foods. The techniques used have evolved as a result of cooperation between microbiologists in the food industry and those in the public health services.

In the USA the American Public Health Association publishes standard methods[1] and the Association of Official Analytical Chemists produces a Bacteriological Analytical Manual[2]. There are also the publications of the International Commission on Microbiological Specifications for Foods[3, 4].

Four kinds of microbiological investigations are usually undertaken in the bacteriological examination of food and drink:

(1) estimation of the viable count, i.e. the numbers of colony-forming units (cfu) of bacteria, yeasts and moulds;
(2) estimation of the numbers of coliform bacilli, *E. coli*, (Enterobacteriaceae). This may indicate the standard of hygiene;
(3) detection of specific organisms known to be associated with spoilage. It is becoming apparent that this may be more important than the total or viable count;
(4) detection of food poisoning organisms.

There is, however, no completely satisfactory single method for the examination of foodstuffs. The methods selected and the examinations performed will depend on local conditions, e.g. availability of staff, space and materials, numbers of samples to be examined and the time allowed to obtain a result.

It may even be desirable in some circumstances to examine a large number of samples by a slightly suboptimal method than a smaller number by an exhaustive method.

Many organisms present in foods may be sublethally damaged, e.g. heat, cold or adverse conditions such as low pH and moisture content, high salt or sugar content. On the other hand these conditions may favour the growth of yeasts and moulds.

Standards

Several national and international committees[3,4] are studying this subject. Criteria for any legislation will have to include protection for the public and a fair deal for the manufacturer and make reasonable use of laboratory staff, time, facilities and money. Tests must be standardized and reproducible. Standards must be set by people with laboratory experience. In establishing any criteria for standards, it is vital to take account of the point of sampling. Products sampled in the factory should have a lower bacterial count than those sampled at the point of sale.

Suggested standards are given where appropriate in this section.

Sampling

The usefulness of laboratory tests is largely dependent upon correct sampling procedures. A statistical sampling plan must be applied to each situation so that samples submitted to the laboratory are fully representative of the 'lot'. The choice of plan, whether it be a class attribute or random number scheme, will depend upon the known facts about the material, the degree of hazard to the consumer and economic feasibility. The subject is fully discussed in a publication of the International Commission on Microbiological Specifications for Foods[4]. In factory sampling, it is more important to take a small number of samples at differing times during the day than to take many samples at one time. Where resources are limited, concentrate on the finished product but be ready to work back through a process taking 'in-line' samples if acceptable levels of micro-organisms begin to increase. Use screw-capped aluminium or polypropylene containers, large mono-containers or polythene bags; never use glass jars where any breakages could contaminate the product or cause injury to personnel. Instruments for removing portions of bulk material vary according to the nature of the substance. For frozen samples, use a wrapped and sterilized brace and bit or a clean chopper washed in methylated spirit and flamed. Take care to avoid cuts by ice splinters. Use individually wrapped and sterilized spoons, spatulas or wooden tongue depressors for softer materials. With pre-packaged material, examine several complete packs as offered for sale.

Total and viable counts

These are a useful monitor in processing and may reflect poor handling or storage at retail level.

Methods of performing these counts are given in Chapter 9. No single medium or incubation temperature will give the whole picture and these must be selected according to known storage conditions of the food, for example for chilled meat and frozen fish, a temperature in the psychrophilic range is used. Many spoilage psychrophiles will not grow in the mesophilic range. For these, surface plating is best.

If the food has been refrigerated, the organisms may be 'cold-shocked' and may grow only on media that are enriched. Lower incubation temperatures, e.g. 15–20 °C may be necessary. In general, 30 °C for 48 h is satisfactory, but the incubation time must be extended if lower incubation temperatures are used. Similarly in heated foods, non-sporing mesophiles may be impaired and they also will need enriched media. For psychrophiles 1 °C for five days is useful.

There is often no need to set up complete plate or other counts. If a standard is set, arbitrarily or according to experience, the technique can be modified to give a 'pass' or 'fail' response.

If, for example, an upper limit of 100 000 colonies/g under given conditions of medium, time and temperature is set, then a plate count method that uses 1 ml of a 1:1000 dilution will suffice. More than 100 colonies—easily observed without laborious counting—will fail the sample. Similarly, more than 100 colonies in roll-tubes containing 0.1 ml of a 1:100 dilution or more than 40 per drop in Miles and Misra counts from a 1:50 dilution will suggest rejection. Such counts must be replicated.

Counts from the surface (p. 242) are useful for carcase meat and fish where the organisms are mostly superficial. Breed smears are good rough guides (*see* p. 130). Anaerobes may be counted by plate or 'black tube' MPN methods (*see* p. 369). Mould hyphae and yeasts may be counted directly in a Neubauer or similar counting chamber which is deeper than the Helber model. Exclude foods which depend upon a microbial population for flavour, e.g. yoghurt.

Indicator organisms

Although tests for the presence of coliform bacilli in general and *E. coli* in particular, are very useful it may be desirable to count all the Enterobacteriaceae present because some strains of *Citrobacter* and *Klebsiella* spp. are more heat resistant than *Escherichia* spp. and their presence is a better indication of inadequate heat processing[5]. The presence of streptococci (enterococci) is also regarded by some bacteriologists as indicating suboptimal processing and storage.

General methods

(1) Weigh 10 g of material in to a sterile container or on sterile grease-proof paper. Homogenize with 90 ml of diluent in a Stomacher Lab-Blender or MSE Atomix. But do not compare results obtained by the two methods of mastication. The Stomacher tends to give higher counts[6]. Do counts and inoculate culture media with tenfold serial dilutions of this homogenate. *Table 15.1* shows the amount of food in grammes contained in diluted aliquots of the material. If blenders are not available, good homogenates are obtained if about 20 g of coarse sterile sand are added to the mixture before shaking.

(2) When examining vegetables or materials where the micro-organisms are on the surface, weigh 50 or 100 g into a sterile jar and add 100 ml of diluent. Shake well, stand for 20 min and decant. The organisms are assumed to be washed off and distributed in the diluent, that is, 100 ml now contains the number of bacteria present on 50 or 100 g of the foodstuff.

Plate the homogenate or washings on the various media indicated in *Table 15.1* and also add 5–10 g of original food, chopped if necessary to 50-ml tubes of glucose tryptone broth.

TABLE 15.1. Aliquots of dilutions of food materials

	Initial	*2nd*	*3rd*	*4th*
Dilution	1:10	1:100	1:1000	1:10000
10 ml contains (g)	1	0.1	0.01	0.001
1 ml contains (g)	0.1	0.01	0.001	0.0001

Tenfold serial dilutions of an initial 10 g material +90 ml of 0.1% peptone solution in water

Coliform bacilli, *Escherichia coli* and Enterobacteriaceae

The choice of method lies between counting the numbers present in 1 g of food or simply the presence or absence of the organisms in 1 or 0.1 g.

Coliform count

Weigh 10 g of the sample on sterile grease-proof paper and homogenize it in a Stomacher in 90 ml of peptone water diluent. Add 1 ml of this 1/10 dilution to 9 ml of diluent to make a 1/100 dilution. Add 1 ml of each dilution to a petri dish and pour on 15 ml of melted violet red bile lactose agar cooled to 50 °C. Allow to set and overlay with 10 ml of the same medium. Incubate at 37 °C for 18–24 h and count the dark red colonies. Calculate the number of colony-forming units/g of the food.

Presence or absence of coliform bacilli

Add 10 ml of the 1/10 food suspension to 10 ml of double strength and 1 ml to 10 ml of single strength MacConkey (or similar) broth. Incubate at 37 °C overnight and plate out on MacConkey agar to confirm that coliforms are or are not present in 1 or 0.1 g.

Escherichia coli counts

Attempts to detect *E. coli* at 44 °C in primary cultures often fail. In richer media, however, they usually do grow, and as very few bacteria can produce indole at this temperature a useful but rough indicator of the presence of these organisms can be used.

Add small samples, e.g. 0.1 and 0.5 g to 15-ml tubes of tryptone or similar nutrient broth. Incubate overnight and test for indole. A positive result is presumptive evidence of *E. coli*. Plate these cultures on MacConkey or similar medium to confirm.

Presence or absence of Enterobacteriaceae

This is a modification of the method of Mossel and Harrewijn[7]. Resuscitate damaged cells by incubating 100 ml of a 10% suspension of food in tryptone soya broth at 25 °C for 2 h. Add 10 ml to 10 ml of double strength EE broth, 1 ml to 10 ml of single strength EE broth and 0.1 ml to another 10 ml of single strength EE broth. Incubate at 30 °C for 18–24 h. With a straight wire, inoculate freshly steamed deep tubes (15 ml) of violet red glucose agar.

Incubate at 30 °C overnight. Enterobacteriaceae usually give a line of growth down the centre of the medium, surrounded by a cylinder of purple precipitate. Do oxidase tests on the growth to exclude *Aeromonas* and identify, if necessary, as described in Chapter 28.

Report Enterobacteriaceae present or absent in 1, 0.1 or 0.01 g.

Counting enterococci

Do surface counts as described on p. 134 using azide blood agar, MacConkey or Slanetz and Bartley medium.

Counting clostridia

As counting these organisms usually incurs identifying them as well the techniques are described under *Clostridium*, in Chapter 35.

Yeast and mould counts

Do plate and surface counts on malt or wort agar. Incubate at 25 °C for up to five days.

Spoilage organisms

These, and the methods for culturing them, are given under the headings of the foods concerned.

Food poisoning bacteria

'Routine' examination of all foods for salmonellas and staphylococci, or other food poisoning organisms, is hardly worth while. Only those foods and ingredients known to be vehicles need be tested. They are given in Chapter 14 and under the appropriate headings in the following pages.

References

1. AMERICAN PUBLIC HEALTH ASSOCIATION (1976) *Compendium of Methods for the Microbiological Examination of Foods*. Washington, DC
2. *Bacteriological Analytical Manual* (1978) 5th edn. Washington DC: Association of Official Analytical Chemists
3. THATCHER, F.S. and CLARK, D.S. (Editors) (1974) *Micro-organisms in Foods, 1*. International Commission on Microbiological Specifications for Foods. Toronto: University Press
4. INGRAM, M. (Chairman) (1974) *Micro-organisms in Foods, Vol 2*. International Commission on Microbiological Specifications for Foods. Toronto: University Press
5. MOSSEL, D.A.A., MENGERINK, W.H. and SCHOLTS, H.H. (1962) *Journal of Bacteriology*, **84**, 381
6. JAY, J.M. and MARGITIC, S. (1979) *Applied Environmental Microbiology*, Nov. p. 879
7. MOSSEL, D.A.A. and HARREWIJN, G.A. (1972) *Alimenta*, **11**, 29

Chapter 16

Meat

Fresh and frozen carcase meat

Carcase sampling

The skin or surface of a carcase is usually heavily contaminated. Gram-negative, motile bacteria show a greater attachment to it than do Gram-positive species.[1] Emsweiler[2] described a coring device for sampling deeper tissues. Although deep muscle is usually sterile the meat may have a higher pH if the animal has suffered stress and it may be contaminated.

The most reliable methods are destructive, involving maceration of weighed portions.

To sample the whole carcase by a non-destructive technique use the method devised by Kitchell *et al.*[3]. Sterilize large cotton wool pads, wrapped in cotton gauze in bulk and transfer them to individual plastic bags. Moisten a swab with a 0.1% peptone water and wipe the carcase with it, using the bag as a glove. Take a dry swab and wipe the carcase again. Place both swabs in the same bag, then seal and label it. Add 250 ml of diluent to the swabs and knead, e.g. in a Stomacher, to extract the organisms. Prepare suitable dilutions from the extract and do total counts at 20 °C. Inoculate other media as desired.

The three areas on the carcase that are most likely to be contaminated after butchering are the rump, brisket and forelegs. Murray[4] recommends swabbing a 16-cm^2 area on each part. With good hygiene, counts of less than 150 000 on the brisket, 50 000 on the rump and 25 000 on the forelegs per 16 cm^2 at 20 °C may be achieved immediately after dressing and after three to four days in chill.

'Hot boning' is now becoming more common and this may give rise to different problems.

Routine processing of boned-out meat in the laboratory

It is important to take equal quantities of both fat and lean tissues. Deep tissue near the bone should be examined if bone taint is suspected. To avoid contamination during examination, sear or paint the surface with an antiseptic dye. If an overall picture is required, take core samples from boned-out joints, macerate, dilute and plate for aerobic and anaerobic culture. Express counts as the number of colonies/g of tissue. Most counts should be of the order of, or less than, 100 000 cfu/g.

Most spoilage organisms will be growing on the surface of the meat under aerobic conditions. Scrape a convenient area, e.g. 10 cm^2 with a sterile scalpel, shake the scrapings in 90 ml of warm diluent. Prepare dilutions, plate, incubate and count.

In a meat factory, where the hygiene is good, more than 70% of samples may be expected to have count of less than 1000 cfu/cm^2.

Microbial content

In moist chill conditions, the predominant spoilage organisms are *P.* (*Alteromonas*) *putrefasciens* and other pseudomonads, *Acinetobacter*, *Enterobacter* and *Brochothrix* (*Microbacterium*) *thermosphactum*. Anaerobes do not appear to be important except where meat is held at temperatures above 25 °C, when *Clostridium* spp., notably *C. perfringens*, predominate. About 10% of pork carcases contain *C. botulinum*, 66% carry *C. perfringens*. Fortunately, *C. botulinum* is a poor competitor.

Salmonellas are often found in raw meats, in meat sold as pet-food, in imported frozen boneless beef and in horseflesh. Methods for isolating them are given on p. 243.

In dry chill conditions, mould spoilage is possible in chilled beef (−1 °C) and frozen mutton (−5 °C or less) but moulds do not grow below −10 °C. 'Green spots' are usually due to *Penicillium* spp., 'white spots' to *Sporotrichum* spp., 'black spots' to *Cladosporium* spp. and 'whiskers' to *Mucor* and *Thamnidium* spp.

Comminuted meat

This is fresh meat, minced or chopped and with no added preservative. Surface organisms are therefore distributed unevenly throughout the mass and further contaminations may occur during the process.

Take several random samples.

Viable counts

Do plate counts on dilutions from 10^{-4} to 10^{-8}. Counts are usually very high, but standards must be set by experience. It may be desirable to estimate the number of coliforms present and to determine what proportion of these are *E. coli*. Pierson *et al.*[5] found that the highest presumptive coliform count was obtained using surface overlay plating on violet red bile agar. Conversely, MPN broth counts gave higher numbers of *E. coli* than methods employing solid media. Later workers[6] found that the membrane filter technique gave the best recovery of *E. coli*.

Microbial content

This is similar to that in fresh meat. Salmonellas are very likely to be present.

Fresh sausages

A fresh sausage contains comminuted meat, cereals, spices and sulphur dioxide up to a limit of 450 ppm. This checks the growth of Gram-negative species.

This product should not be confused with European or American sausages, which are smoked.

Viable counts

Remove the casing, if present, and do total viable counts using dilutions from 10^{-4} to 10^{-8}. Incubate at 30 °C for 48 h and at 22 °C for three days. Counts of several million organisms/g may be expected. Skinless sausages may have lower counts than those with casings as the process involves blanching with hot water to remove the casings in which they are moulded.

Sausage casings, whether natural (stripped small intestine of pig or sheep) or artificial, usually do not present any bacteriological problems.

Microbial content

Spoilage organisms are those found in fresh meat and also lactobacilli, coryneforms, microbacteria, micrococci, staphylococci (including *S. aureus*) and yeasts. *B. thermosphactum* is associated with souring. Gardner[7] described a useful medium for the isolation and enumeration of this organism (p. 75).

Salmonellas may be present. We have not found the use of surfactants particularly advantageous. Sometimes they are inhibitory.

Pre-packed fresh meat

Meat in oxygen-permeable wrappers has a similar flora to that of unwrapped meat. In packs with non-permeable wrappers, carbon dioxide gradually replaces oxygen; pseudomonads are suppressed and enterobacteria, *Hafnia*, *B. thermosphactum* and lactobacilli predominate[3].

Meat pies

Meat pies may be 'hot eating', e.g. steak and chicken pies, or 'cold eating', e.g. pork, veal and ham pies and sausage rolls.

'Hot eating' pies

The meat filling is pre-cooked, added to the pastry casing and the whole is cooked again.

Viable counts

Examine the meat content only. Open the pie with a sterile knife to remove the meat. Do counts on dilutions of 10^{-1} and 10^{-2} and incubate at 30 °C for 48 h. A reasonable level is not more than 1000 colonies/g.

Clostridia

Add 1-ml amounts of the 10^{-1} dilution to 15-ml bottles of reinforced clostridial medium melted and at 50 °C. Cool, seal with Vaseline and incubate at 37 °C for 48 h. Examine tubes for blackening. Subculture any black tubes in litmus milk. Plate out on blood agar, incubated anaerobically to identify any other clostridia.

'Cold eating' pies

These contain cured meat, cereal and spices that are placed in the uncooked pastry cases. The pies are then baked and jelly made from gelatin, spices, flavouring and water is added. Provided that the jelly is heated to and maintained at a sufficiently high temperature until it reaches the pie, there should be no problems.

Bad handling, insufficient heating and re-contamination of cool jelly can cause gross contamination by both spoilage organisms and pathogens.

The pies are cooled in a pie tunnel through which cold air is blown. At this stage, mould contamination is likely if the air filters are not properly cleaned. Air in pie tunnels can be monitored by exposing plates of malt agar or similar medium.

Viable counts

Examine only the filling. Do counts on homogenized material using dilutions of 10^{-1} and 10^{-2}. Incubate at 30 °C for 48 h. A reasonable level is less than 1000 colonies/g.

Coliform counts

Examine meat and jelly separately. Use dilutions 10^{-1} and 10^{-2}.

Poultry

Battery rearing produces birds with very tender flesh but the close confinement necessary in this method of production often results in cross-infection. Inadequate thawing of frozen birds followed by relatively little cooking has, on several occasions, resulted in outbreaks of food poisoning.

Viable counts

For the best results divide the whole carcase into portions. Otherwise, sample the skin from under the wing and around the vent. Do total and coliform counts and examine for salmonellas, staphylococci and *C. perfringens*.

To examine a large number of birds, use swabs wound on 25-cm lengths of wire. Sterilize in large test-tubes. Use one for swabbing the outside of the bird and one for swabbing the body cavity. Rinse the swabs and do total and coliform counts by the Miles and Misra or membrane techniques. Inoculate other media as required.

In an attempt to establish advisory standards in the poultry industry, Murray[4] examined a 16 cm^2 area of breast skin for colony count at 22 °C, coli-aerogenes at 30 °C and faecal streptococci and *S. aureus* at 37 °C. He found that in a well run establishment total counts of less than 250 000/cm^2, coli counts less than 1000, faecal streptococci counts of less than 5000 and *S. aureus* counts of less than 100/16 cm^2 could be expected. Low total counts indicate few spoilage organisms and low coli counts indicate good hygiene. Few faecal streptococci suggest good evisceration technique and good hygiene.

To test for salmonellas examine the whole carcase. Swabbing or rinsing in a plastic bag is suitable. Collect the rinsings and add them to double strength enrichment media. Proceed as for salmonella isolation (p. 291).

Total counts and examinations for pathogens may be made on slush ice in the spin chillers and on fluid in the plastic bags in which the birds are packed.

Cured meat

Cured raw meat

This is pork or beef that has been treated with salt and nitrite. The haemoglobin is altered so that the characteristic pink colour is produced when the meat is cooked. This is sold as bacon, gammon, cured shoulder, salt beef or brisket.

Viable counts

Do counts on dilutions of 10^{-2} and 10^{-3}. Incubate at 30 °C for 48 h. Also do counts using diluent and media containing 4% of sodium chloride to detect halophiles. Incubate these for three days.

Inoculate salt meat broth with 0.1-g amounts, incubate at 37 °C for 18 h and plate on phenolphthalein phosphate polymyxin agar (PPPA) or milk salt agar (MSAR)[8] medium to detect *S. aureus*.

Microbial content

Besides halophilic denitrifying bacteria, lactobacilli, micrococci, staphylococci, *B. thermosphactum* and moulds may be present under chill conditions. *Pseudomonas* spp. are inhibited by salt. Above 25 °C, *C. putrefasciens* may be found. This organism gives a characteristic sweet/sour odour[3]. 'Mild cure', 'sweet cure', 'tender cure', etc., are trade terms which may indicate some degree of heat or the addition of sweetening substances. The balance of the flora may be altered by such processes.

Bone taint

This may occur in carcase leg joints or in rib areas. Organisms usually responsible are: vibrios, micrococci, streptococci, proteus, clostridia, enterobacteria, alkaligenes and arthrobacteria.

Vacuum packaging

Vacuum packaging is used for a variety of products in an effort to present them to the customer in as fresh a condition as possible.

Vacuum-packed bacon

The packs used for bacon are not usually permeable to oxygen because oxidation causes fading of the colour. The initial bacterial load, the salt content and storage temperature have a marked influence on the shelf-life.

Low salt content (5–7%) spoilage of bacon gives a characteristic scented sour odour due to the action of lactobacilli, pediococci, streptococci and leuconostoc. High salt content (12%) spoilage gives a cheesy odour which is associated with micrococci or *B. thermosphactum*. Spoilage by enterobacteria or vibrios gives a sulphurous smell.

Viable counts

Swab the outside of the pack with alcohol and open with sterile scissors. Do counts as for cured raw meat.

Microbial content

The main groups of organisms found immediately after packing are micrococci and coagulase-negative staphylococci. During storage at 20 °C, lactobacilli, Group D streptococci and pediococci become dominant. Yeasts may also be found. If large numbers of Gram-negative rods are present it is usually an indication that the initial salt content was low.

Vacuum-packed cooked meats

These meats are invariably cured products and, as with bacon, the absence of oxygen is important. These products, however, always carry a risk of infection with *C. botulinum*. Spores of this organism may survive cooking and during storage germinate and grow to produce toxin in an otherwise sterile and oxygen-free environment. Type E is known to grow at low temperatures. Fortunately, the salt and nitrite contents combined with low-temperature storage help to reduce the danger. In addition, many producers pasteurize their vacuum-packed cooked meat products in the package. This allows for spore germination after the initial cooking and killing of the vegetative forms on pasteurization. Nevertheless, *C. botulinum* has been reported in vacuum-packed frankfurters and cooked ham.

Laboratory examination

Do surface counts on blood agar and on MacConkey agar for enterococci. These are useful organisms for assessing the efficiency of pasteurization. Total counts should be less than 1000 cfu/g.

Cured, cooked meats

These meats include ham, luncheon meats, brawns, tongues, salamis and continental-type sausages. Except in delicatessen stores, these are usually sold in vacuum packs.

Viable counts

Do plate counts on dilutions of 10^{-1}–10^{-3}, and use medium to which 5% of horse blood has been added immediately prior to pouring. This enables the streptococci that cause 'greening' to be counted in addition to the other organisms. Counts of 1000 or less/g are reasonable.

Inoculate salt meat medium with 0.1-g amounts and incubate at 37 °C for 18 h. Plate on PPPA or MSAR medium to detect *S. aureus*.

Microbial content

Contaminants are usually heat-resistant organisms, including spore-bearers, group D streptococci. Coliforms, micrococci and staphylococci may be introduced after heating. In canned hams, coryneforms, lactobacilli and yeasts may be found. Staphylococci may proliferate just beneath the casing.

Brines

Injection brine

This is prepared freshly for each batch of meat.

Sampling

Collect from storage tank, store at <7 °C before counting. Do total count using nutrient agar containing 4% NaCl, incubate at 22 °C for 72 h.

Advisory standard ($\times$ 10^{-3} cfu)[9]

Good	<0.5
Fair	0.5–1.0
Poor	1.1–5.0
V. poor	>5.0

Cover brine

Sampling

Sample when fresh and between batches of meat.

Direct microscopical count (DMC)

Use a Helber haemocytometer (p. 129) and examine by dark field or phase contrast microscopy.

Total viable count

Add 4% NaCl to the medium.

E. coli count

Dilute the sample 1:10 with 4% NaCl in 0.1% peptone and use the membrane filter technique.

Advisory standard[9]

	DMC ($\times 10^6$ cfu/ml)	*Total count* ($\times 10^3$ cfu/ml)	*E. coli* ($\times 10^3$ cfu/ml)
Good	<50	<50	<1
Fair	50–100	50–100	1–10
Poor	101–150	101–500	11–100
V. poor	>150	>500	>100

Pseudomonas and vibrio counts may also be a useful monitor of cover brine quality.[10]

References

1. BUTLER, J.L., STEWART, J.C., VANDERZANT, C., CARPENTER, Z.L. and SMITH, G.C. (1979) *Journal of Food Protection*, **42,** 401
2. EMSWEILER, B.S., NICHOLS, J.E., KOTULA, A.W. and ROUGH, D.K. (1978) *Journal of Food Protection*, **41,** 546
3. KITCHELL, A.G., INGRAM, G.C. and HUDSON, W.R. (1973) Microbiological Sampling in Abattoirs. In *Sampling—Microbiological Monitoring of Environments*. Edited by R.G. Board and D.W. Lovelock, pp. 43–59. London: Academic Press
4. MURRAY, J.G. (1969) *Journal of Applied Bacteriology*, **32,** 123
5. PIERSON, C.J., EMSWEILER, B.S. and KOTULA, A.W. (1978) *Journal of Food Protection*, **41,** 263
6. RAYMAN, M.K. and ARIS, B. (1981) *Canadian Journal of Microbiology*, **27,** 147
7. GARDNER, G.A. (1966) *Journal of Applied Bacteriology*, **29,** 455
8. GILBERT, R.J., KENDALL, M. and HOBBS, B.C. (1969) Media for Staphylococci from Foods. In *Isolation Methods for Microbiologists*. Edited by D.A. Shapton and G.W. Gould, pp. 9–15. London: Academic Press
9. GARDNER, G.A. (1973) Microbiological Examination of Curing Brines. In *Sampling—Microbiological Monitoring of Environments*. Edited by R.G. Board and D.W. Lovelock, pp. 21–27. London: Academic Press
10. GARDNER, G.A. (1980) *Journal of Applied Bacteriology*, **49,** 69

Chapter 17

Fish

Fresh fish

Counts

Do counts from the whole surface of the fish or fillet either by swabbing a defined area (*see* p. 195) or by washing the sample with sterile 0.1% peptone +1% NaCl in plastic bag. Use the washings to prepare dilutions in salt peptone and do counts on Marine agar (*D*). Incubate at 20 °C for five days. Coliform counts are sometimes useful.

Surface slime is usually heavily infected. The count is increased by careless handling and contact with dirty ice and decreased by salting in barrels, hypochlorite and ice mixtures.

Culture

Inoculate glucose tryptone media, MacConkey media, and media containing 5–10% of sodium chloride for halophilic bacteria.

Inoculate PPPA or MSAR or other media for staphylococci and also plate out the salt broths on these media. *S. aureus* is not uncommon in fish.

Microbial content

Large numbers of pseudomonas, alteromonas, achromobacter, flavobacteria, coryneforms, acinetobacter, aeromonas and cytophaga, often associated with slime, and micrococci may be found. Photobacteria, luminescent in the dark, are often present. The flesh count increases rapidly after filleting but in good plants may be as low as 100 000 cfu/g.

Smoked fish

Since 1976 there has been a marked increase in the incidence of scomboid fish poisoning (non-bacterial). The symptoms resemble those of food poisoning, i.e. diarrhoea and vomiting, headache, giddiness, rash on head and neck, 10 min to a few hours after ingestion of scomboid fish, e.g mackerel, tuna, etc. Histamine compounds and some spoilage of the fish is always involved. The condition can result from eating canned fish.

Shellfish and raw crustaceans

These are estuarine animals and are therefore liable to gross faecal pollution. Oysters, which are eaten raw, are self-cleansing 'filter-feeders' and, if placed in tanks of chlorinated sea-water, will eliminate any enterobacteria from their bodies. Cockles, whelks, winkles, prawns and shrimps are boiled for retail sale, but this cooking is usually not controlled. The presence of *E. coli* is usually accepted as evidence of less than optimum hygienic conditions of cultivation or of insufficient heat treatment.

Roll-tube method

This is applicable to both shellfish and raw crustaceans.

Place oysters or mussels in the freezing compartment of a refrigerator overnight. This makes them easy to open and no fluid will be spilled. Scrub and clean the outside of the shell and rinse in boiled water. Hold with the concave shell down and open with a sterile oyster knife. Cut the frozen flesh into small pieces (to release intestinal contents) with a sterile scalpel and remove it into a sterile 200-ml Pyrex measuring cylinder. Push the flesh down with a sterile glass rod and continue until there is about 100 ml of mush. Similarly obtain about 100 ml of the flesh of winkles, cockles or whelks. Add an equal volume of diluent, stopper and shake well. For prawns and shrimps, place 100 g in a blender with 100 ml of diluent and homogenize.

Allow the gross material to settle for 15–20 min. Melt three roll-tubes each containing 2 ml of MacConkey roll-tube agar and cool to 45–50 °C. To each tube add 1 ml of the supernatant liquor and roll the tubes. When cool, incubate overnight in a water-bath at 44 °C inverted and completely submerged (inversion prevents water of syneresis from washing off colonies developing near the bottom of the tube).

Count the large red coliform colonies. Less than five colonies on each of the 1-ml tubes (i.e. less than 5/g of shellfish) is regarded as satisfactory, from 5–15 colonies suspicious and more than 15 colonies as unsatisfactory.

To detect salmonellas in oysters, add 100 ml of mush to 100 ml of double-strength selenite (*see* Chapter 28).

To detect *S. aureus* which may be present in imported frozen prawns and shrimps (infected during handling), add 10-ml aliquots of the mush to 50-ml tubes of broth containing 10% of sodium chloride. Incubate overnight at 37 °C and plate on blood agar and PPPA or MSAR medium (*see* Chapter 31). For *V. parahaemolyticus* add 10 g emulsified flesh to 100 ml of alkaline peptone water containing 3% salt and proceed as on p. 272.

The 'percentage clean' method for oysters and mussels

Take ten shellfish, scrub and clean the outside; and wash in sterile water. Hold with the concave shell down and open with a sterile oyster knife. Take care not to spill any liquor. Cut the flesh into small pieces with a sterile scalpel and mix with the liquor already present. Add 0.2 ml of liquor from each shellfish to a tube of single-strength MacConkey broth. Add 0.2 ml from three shellfish to a tube of glucose broth and 1 ml from three others to a tube of litmus milk.

Incubate at 44 °C for 24 h and examine the MacConkey tubes for acid and gas, the glucose broth microscopically for streptococci (enterococci) and the litmus milk for *C. perfringens*.

If coliform bacilli (44 °C+) are absent from all ten shellfish, they are '100% clean'. If they are absent from eight out of ten, they are '80% clean', and so on. It is generally regarded that '80–100% clean' is satisfactory, '70% clean' is suspicious and '60% clean' or less is unsatisfactory.

Alternative methods of isolating enterococci are given in Chapter 32.

For more information about the bacteriology of shellfish, *see* Clegg and Sherwood[1], Knott[2], Sherwood and Scott Thomson[3] and the Joint FAO/WHO Food Standards Programme[4,5].

Frozen sea-food

Most contamination is introduced in the factory in cutting, battering, packing, etc. Enterobacteria and staphylococci also may be introduced at this stage. If the food is pre-cooked, the bacterial count is reduced but this treatment is usually not enough to kill all enterobacteria, staphylococci and anaerobes. Counts may be very high. *S. aureus*, coliforms, *V. parahaemolyticus* and salmonellas may be present. The flora is usually mixed and reflects processing rather than the raw materials.

Ready-cooked deep-frozen prawns and shrimps

These are imported from the Far East and are an increasingly common article of diet in the West.

The following method, devised by Mitchell[6], allows a simple quantitative examination to be made on this product. Chisel 20 g of fish from a frozen block into a sterile screw-capped jar. Place in a water-bath at 44 °C for 10–30 min to hasten release of the juice. Prepare 1:50 and 1:500 dilutions and plate on well dried blood agar and MacConkey plates using the Miles and Misra method. Incubate overnight at 35°C. This gives the total viable and coliform counts/ml of extruded juice. Homogenize the remaining tissue preferably in a Seward Stomacher Lab-Blender and inoculate salt meat broth for *S. aureus*, alkaline peptone water for *V. parahaemolyticus* and selenite broth for salmonellas. Incubate overnight at 37°C and plate on suitable solid media.

The bacterial population of these crustaceans will consist of spore-forming bacteria that have survived boiling and organisms introduced after cooking.

Counts on isolated samples may not give significant results. Ideally, 5–10 samples per batch should be obtained at the port of entry. Suggested standards, based on a weighing and macerating technique and using an incubation temperature of 35 °C are: counts up to 100 000 cfu/g, release unconditionally; 100 000 to 1 000 000 cfu/g, release with a warning to use immediately on thawing, and over 1 000 000 cfu/g detain.

References

1. CLEGG, L.F.L. and SHERWOOD, H.P. (1947) *Journal of Hygiene, Cambridge*, **45,** 504.
2. KNOTT, F.A. (1951) *Memorandum on the Principles and Standards Employed by the Worshipful Company of Fishmongers in the Bacteriological Control of Shellfish in the London Markets.* London: Fishmongers Company
3. SHERWOOD, H.P. and THOMPSON, S. (1953) *Monthly Bulletin of the Ministry of Health and Public Health Laboratory Service*, **12,** 103
4. CODEX ALIMENTARIUS COMMISSION (1978) *Recommended International Code of Hygienic Practice for Molluscan Shellfish*, FAO/WHO
5. CODEX ALIMENTARIUS COMMISSION (1978) *Recommended International Code of Hygiene Practice for Shrimps and Prawns*, FAO/WHO
6. MITCHELL, N.J. (1970) *Journal of Applied Bacteriology*, **33,** 523

Chapter 18

Milk and dairy products

Milk

Statutory tests in the UK

The Milk (Special Designations) Regulations have been changed several times over the last 30 years and have been related to the schemes for the eradication of tuberculosis from herds in this country. By the end of 1964, all cows' milk produced reached the standard required of the former Tuberculin Tested Grade and this designation was discontinued. Four grades of milk are now marketed: Untreated, Pasteurized, Ultra-Heat-Treated and Sterilized.

Untreated milk

This is produced from a herd of animals that have been individually tuberculin tested and subjected to periodical veterinary inspections. The premises in which the milk is handled are inspected and licensed by the Local Authority under the direction of the Ministry of Agriculture, Fisheries and Food. Retail milk is required to pass the half-hour methylene blue test, or such other tests as the Ministry may prescribe.

Pasteurized milk

This is treated in one of two ways. The 'holder' method involves raising the temperature of the milk to between 145 °F(63 °C) and 150 °F (66 °C), maintaining it at that temperature for at least 30 min and then cooling it immediately to not more then 50 °F (10 °C). In the 'high temperature short treatment' or 'flash-point' method, the milk is heated at 161 °F (72 °C) for at least 15 s and then immediately cooled to below 50 °F (10 °C). This milk is tested by the methylene blue test and the phosphatase test or such other tests as the Ministry may prescribe.

Ultra-heat-treated (UHT) milk

By a continuous process, milk is heated to 270 °F (133 °C) for 1 s and immediately placed in sterile containers. Most bacteria are destroyed. This milk is required to pass the colony count test, which is a simple sterility test.

Sterilized milk

This is filtered, clarified, homogenized and bottled and then heated at 212 °F (100 °C) for such a period that it passes the turbidity test.

Statutory tests, to be recognized in a court of law, must be carried out exactly as described in the appropriate Statutory Instrument. Descriptions of these tests and all relevant information are contained in the following publications obtainable from HM Stationery Office.

(1) *Statutory Instrument, 1963, Food and Drugs, Milk and Dairies: The Milk (Special Designation) Regulations, 1963.* This contains details of sampling, delivery to the laboratory and the technique of the methylene blue and phosphatase tests.
(2) *Statutory Instrument, 1965, Food and Drugs, Milk and Dairies. The Milk (Special Designation) (Amendment) Regulations, 1965.* This prescribes the method of producing ultra-heat-tested milk and the technique of the colony count test.

Precise specifications are given for pipettes, tubes, media, etc. For other purposes, the simplified tests described below can be used, in which the techniques are basically the same but the ordinary bacteriological equipment is used.

Sampling and delivering samples to the laboratory

Sample retail milk in the containers in which it is sold. For churn samples, use plungers and dippers supplied for this purpose by Astell Laboratories or other dairy equipment firms. This apparatus can be sterilized in the laboratory or dairy. Plunge the churn well and take a (100–150 ml) sample with a dipper into a sterile screw-capped or glass-stoppered bottle.

Deliver samples to the laboratory on the day of sampling as soon as possible after sampling in an insulated container (not in refrigerated containers or CO_2 boxes). Suitable containers are supplied by the firms mentioned above.

Mixing samples and opening containers

Mix by shaking the sample up and down 12 times and inverting it 12 times. Flame the tops of ordinary milk bottles and remove the foil caps with flamed forceps. Swab the tops or corners of cartons with spirit and cut off a corner with flamed scissors. Make a hole in plastic containers with a flamed, hot cork-borer.

Methylene blue test

Both raw (untreated) and pasteurized milks are required to pass the half-hour test (but *see* Reductase test, *over*).

Deliver samples to the laboratory on the day of sampling in an insulated box. On arrival at the laboratory, place about 50 ml of milk in a stoppered bottle. Store at atmospheric shade temperature in a louvred cabinet on a north wall.

From 1st May to 31st October. Leave samples at atmospheric shade temperature until 9.30 a.m. the following day.

If the atmospheric shade temperature during any such period of storage exceeds 70 °F, the statutory test is declared void.

From 1st November to 30th April. Leave the samples at atmospheric shade temperature until 5 p.m., then place in a water-bath maintained at 65 ± 2 °F (18–19 °C) (this is near the ambient : *see* Chapter 2) until 9.30 a.m. the following day.

Between 9.30 and 10.0 a.m., mix samples by inverting and shaking and pour 10 ml of milk into a 152 × 16 mm sterile test-tube marked at 10 ml (reductase tube). Add to each tube 1 ml of a methylene blue solution prepared by dissolving one standard tablet (BDH) in 200 ml of cold, sterile glass-distilled water and making up to 800 ml (store in the dark in a refrigerator; pour out a little each day for current use). Pipette this solution down the side of the tube not wetted by the milk. Stopper the tubes with a boiled rubber bung or Astell seal. Mix by inversion and place in a water-bath at 37 °C for half an hour. The level of the water must be above that of the milk and the water-bath must have a lid to protect the dye from the light.

Compare the colour with that of a tube containing 10 ml of milk and 1 ml of water. If the dye is decolorized wholly or to within 5 mm of the top, the milk fails the test.

There is a correlation between the bacterial content and the reduction of methylene blue. The milk in this test is stored under theoretically domestic conditions and should remain sweet overnight.

Phosphatase test

The Aschaffenberg-Mullen test has replaced the more complex Kay-Graham method.

Test at once or refrigerate the milk overnight. Make up the reagent as follows. Dissolve 3.5 g of anhydrous sodium carbonate (AnalaR) and 1.5 g of sodium bicarbonate (AnalaR) in 1 litre of water. Store in a refrigerator. To 100 ml of this solution add 0.15 g of disodium *p*-nitrophenyl phosphate. Keep in the dark and in a refrigerator and use within seven days. There must be no yellow colour.

To 5 ml of the reagent in a 152 × 16 mm test-tube, add 1 ml of milk. Stopper with a rubber stopper and mix by inversion. Incubate for 2 h at 37 °C in a water-bath. Include controls of boiled milk and also boiled milk containing about 2% of raw milk. Determine the actual amount by titration, i.e. by adding varying proportions of raw milk to pasteurized milk. Do a phosphatase test on these and make the control in the proportion which gives a reading of slightly less than 42 μg. Preserve this mixture with 0.5 ml% of saturated mercuric chloride. Store in a refrigerator. Label the bottle 'Poison'.

Compare the test sample with the boiled sample using the Lovibond Comparator and disc designed for this purpose. This is an all-purposes model supported on a stand on which the test-tubes are placed in a sloping position and viewed by reflected light. Unheated milks and insufficiently treated milks give a yellow colour due to *p*-nitrophenol released from the substrate by the action of phosphatase. In properly pasteurized milk, phosphatase has been destroyed and the reading on the disc will be 10 μg or less of p-nitrophenol/ml of milk.

Clean the glassware used in this test in chromic acid solution.

Colony count test for milk treated by the UHT method

Incubate the sample at 30–37 °C for 24 h. Open the carton as described above and remove about 10 ml into each of two sterile test-tubes or small screw-capped bottles. Refrigerate one of these.

With a flamed standard platinum–iridium loop (BS 19) of 4 mm internal diameter (Astell Laboratories), remove one loopful (about 0.01 ml) from the other bottle into 5 ml of melted yeast extract milk agar at 45–50 °C in the screw-capped bottle. Mix by rotation and allow the agar to set with the bottle on its side. Incubate at 30–37 °C for 48 h. Count the number of colonies. The test is satisfactory if there are less than ten; if there are more than ten colonies, repeat the test with the refrigerated sample.

The turbidity test

Weigh 4.0 g of ammonium sulphate (AnalaR) into a small flask or bottle. Add 20 ml of milk and shake for 1 min. Stand for 5 min and filter through a 12.5-cm Whatman No. 12 folded filter-paper into a test-tube. When 5 ml filtrate have collected, place the tube in a boiling water-bath for 5 min and then cool. A properly sterilized milk gives no turbidity.

Sterilization alters the protein constituents and all coagulable protein is precipitated by the ammonium sulphate. If heating is insufficient, some protein remains unaltered and is not precipitated by the ammonium sulphate. It coagulates, giving turbidity when the filtrate is boiled.

US standard methods for milk examination[1]

Three routine tests are prescribed:

(1) plate count at 32 °C;
(2) direct microscopical count, for raw milks if a high count is expected; and
(3) the coliform test.

Plate count

Use Standard Methods agar to prepare pour-plates. Incubate at 32 ± 1 °C for 48 ± 3 h.

Direct microscopical count

This is recommended only for raw milks with a fairly high count. The technique is similar to the Breed count (*see* p. 130).

Coliform test

This test is used after pasteurization in order to detect re-contamination. There are two methods.

(1) *Solid media method.* Use a pour-plate technique with either violet red bile or deoxycholate lactose agar. Mix 1.0 and 0.1 ml of milk with melted and cooled agar, allow it to set and overlay it with a further 3–4 ml of sterile medium. Cover the surface of the inoculated agar completely so as to prevent surface growth. Incubate at 32 °C for 24 ± 2 h and count all dark red colonies that are 5 mm or more in diameter as coliforms.

(2) *Liquid media method.* Inoculate two sets of five tubes containing brilliant green bile broth, one set with 10-ml volumes and one set with 1-ml volumes. Incubate at 32 °C for 48 ± 3 h. Streak a loopful from each tube showing fermentation on to eosin methylene blue or Endo agar. Inoculate typical colonies grown on this medium to nutrient agar for Gram stain and a lactose broth to demonstrate gas production. Use the Most Probable Number (MPN) tables on p. 139 to report.

Examination for psychrophiles

Do counts on Standard Methods agar as for mesophile counts but incubate at 7 ± 1 °C for ten days.

Examination for thermoduric organisms

Pasteurize 5 ml of milk at 62.8 °C for 30 min. Do pour-plates with Standard Methods agar containing 10% of sterile milk. Incubate at 23 ± 2 °C for 48 h. Express the result as Laboratory Pasteurization Count (LPC)/ml.

Other tests used in the UK (not statutory)

Reductase test

This is a methylene blue reduction test but is called the reductase test here in order to distinguish it from the present statutory test.

To 10 ml of milk in a reductase tube, add 1 ml of the standard methylene blue solution described above. Pipette the dye down the side of the tube opposite to that down which the milk was poured. Include controls of boiled milk plus methylene blue, and boiled milk plus 1 ml of water. Mix the contents of the tubes by inversion.

Incubate at 37 °C in a water-bath for 5 h or more, and inspect and invert every 30 min. The speed at which the dye is reduced is an index of bacterial content and keeping quality.

Under the old Milk Regulations, a consumer milk, examined immediately after sampling, and a producer milk kept overnight at atmospheric shade temperature were required not to decolorize the dye in 5.5 h in the winter and 4.5 h in the summer.

The coliform test

Set up two sterile test-tubes containing 9 ml of 0.1 % peptone water for each sample. Mix the sample by inversion and shaking. Dip a 1-ml pipette not more than 25 mm into the milk and remove 1 ml. Add it to the first dilution tube, delivering above the level of the diluent. Discard this pipette. With a fresh pipette, suck up and down ten times to mix the contents of the first dilution tube and transfer 1 ml to the second tube. Discard this pipette. With a fresh pipette, mix the contents of this tube and deliver 1 ml to each of three tubes of single-strength MacConkey broth. Each tube receives 0.01 ml of milk.

Incubate at 37 °C for 48 h and note acid and gas as evidence of presumptive coliform bacilli.

Formerly, designated raw milks were required to have no coliforms in two out of three tubes but many dairies insist on a higher standard than this.

Bromcresol purple (0.4 g/l) is a better indicator than neutral red in MacConkey broth for this purpose.

Plate counts

The standard plate count (Chapter 9) using Yeastrel milk agar is most useful if done at least in duplicate. Incubate at 37 °C for 48 h.

At one time, designated raw milks were expected to have plate counts less than 30 000 and pasteurized milks less than 100 000/ml.

For counts on pasteurized milk, incubate plate counts also at 25–28 °C. Some thermoduric organisms do not grow at temperatures above 30 °C.

Counts of thermophiles, incubated at 55 °C, are useful and also counts of psychrophiles incubated at 7 °C for seven days (but the time and temperature, 5–10 °C and five to ten days, is a matter of local choice and local standards).

Microbial content

Bacteria enter milk during milking and handling. Even with the most hygienic production, some bacteria gain access. The cooling that is normal practice after milking retards the multiplication of these bacteria. Pasteurization, intended to kill pathogenic bacteria, does not necessarily reduce the count of other organisms. It may increase the numbers of thermophiles.

The 'normal microflora' depends on temperature: at 15–30 °C *Streptococcus lactis* predominates and many streptococci and coryneform bacteria are present, but at 30–40 °C they are replaced with lactobacilli and coliform bacilli. All of these organisms ferment lactose and increase the lactic acid content, which causes souring. The increased acid content prevents the multiplication of putrefactive organisms. Spoilage during cold storage is due to psychrophilic pseudomonads and *Alcaligenes*, psychrotrophic coliforms, e.g. *Klebsiella aerogenes*, *Enterobacter liquefaciens* which are anaerogenic at 37 °C and in pasteurized milk, thermoduric coryneform organisms (*Microbacterium lactis*) may be significant. These coryneforms probably come from the animal skin or intestine and from utensils.

At temperatures above 45 °C, thermophilic lactobacilli (*Lactobacillus thermophilus*) rapidly increase in numbers.

Gram-negative bacilli are rarely found in quarter samples collected aseptically. These enter from the animal skin and dairy equipment during milking and handling. *Alcaligenes* species are very common in milk. *P. fluorescens* gels UHT milk.

Undesirable micro-organisms responsible for 'off flavours' and spoilage include psychrophilic *Pseudomonas*, *Achromobacter*, *Alcaligenes* and *Flavo-*

bacterium spp., which degrade fats and proteins and give peculiar flavours. Coliform bacilli produce gas from lactose and cause 'gassy milk'. *S. cremoris*, *Alcaligenes viscosus* and certain *Aerobacter* species, all capsulated, cause 'ropy milk'. *Oospora lactis* and yeasts are present in stale milk. *P. aeruginosa* is responsible for 'blue milk' and *Serratia marcescens* for 'red milk' (differentiate from bloody milk). *B. cereus* can cause rapid decoloration of methylene blue.

Pathogenic organisms which may be present include *S. aureus*, *S. pyogenes* and other streptococci from infected udders in mastitis. Tubercle bacilli may be found by guinea pig inoculation of centrifuged milk deposits and gravity cream. *B. abortus* is excreted in milk and can be isolated by animal inoculation, or antibodies can be demonstrated by the ring test or whey agglutination test. *Rickettsia burnettii*, the agent of Q fever, may be found by animal inoculation of milk from suspected animals.

Between 1970 and 1979 there were 29 outbreaks of Salmonella food poisoning affecting 2428 persons who had consumed raw milk[2].

Mastitis

Looking for evidence of mastitis in bulk milk is obviously unrewarding in any but in farm or other small samples. Centrifuge 50 ml of milk and make films of the deposit. Dry in air, treat with xylene to remove fat, dry, fix and stain with methylene blue. Examine for pus cells. For bacteriological examination and identification of causative organisms, *see* p. 332 and 336.

Examination of dairy equipment

See Chapter 22 (Sanitation control) for methods.

Microbial content of dairy equipment

The types of organisms found will be similar to those in milk. Soil organisms, e.g. *Bacillus* spp., may be present. *B. cereus* reduces methylene blue very rapidly and may also cause food poisoning. Organisms originating from faeces, e.g. various coliforms and *Pseudomonas* spp., are found. Coliform bacilli also rapidly reduce methylene blue. Pseudomonads may produce oily droplets in the milk.

Detection of antibiotics in milk and food

Milk from cows treated with antibiotics may contain these agents. Antibiotics are often added to feeding stuffs and may appear in food intended for human consumption.

If such antibacterial substances are present in milk used for cheese-making, the 'starters' may be inhibited and other, resistant organisms may cause spoilage or off-flavours.

To test use the AIM Test (Oxoid) or the Disc Assay Method (Difco).

For further reading on the Microbiology of milk *see* Robinson[3].

Natural cream

Viable counts

Do plate counts or Miles and Misra counts on serial dilutions ranging from 10^{-2} to 10^{-6} in a 0.1% peptone water solution, using yeast extract milk agar. For coliform counts, do Miles and Misra counts with these dilutions on MacConkey agar or use 10 ml of a 1:10 dilution in a pour plate as for milk examination (US method). Subculture colonies to identify as *E. coli* if required.

If the cream is solid or difficult to pipette, make the initial 1:10 dilution by weight as described on p. 131.

Methylene blue reduction test

This has been used for many years in the UK and gives a good correlation with the bacterial counts. This appears to be a useful guide for the hygienic production of cream.

Later workers stress the importance of psychotrophs and claim that these can be enumerated using a spiral plating technique and incubating for 25 h at 21 °C[4].

Culture

Plate the cream and serial dilutions on blood agar and incubate at 22 °C for 48 h and at 37 °C for 24 h. Inoculate salt meat medium with 0.1-ml amounts of cream. Incubate at 37 °C for 24 h, subculture on PPPA medium and examine for *S. aureus*. Add 25 ml or 25 g of cream to selenite medium and incubate at 35 °C for 24 h. Subculture on a suitable medium for salmonellas.

Microbial content

Fresh cream may contain large numbers of organisms, including coliforms, *E. coli*, *S. aureus*, other staphylococci, micrococci and streptococci. Most fresh cream has been pasteurized and therefore contamination by many of these organisms is due to faulty dairy hygiene.

There are no standards for cream, but it is not unreasonable to view with suspicion repeated samples that give counts of more than 100 000 ml and/or contain *E. coli*. *S. aureus* should not be present in 0.1 g. Cream should not be used in catering in circumstances where this organism can multiply rapidly[5].

Imitation cream

Examine for staphylococci only, as for natural cream.

Processed milks

Dried milk

This often has a high count and, if stored under damp conditions, is liable to mould spoilage. Spray-dried milk may contain *S. aureus*.

Examination

Reconstitute in pH 7.2 buffer in distilled water before testing. Examine direct films, do total counts as for fresh milk but prepare extra plates for incubation at 55 °C for 48 h for thermophiles. Do coliform counts to detect contamination after processing. Examine for yeasts and moulds as described for butter examination (p. 216).

The number of spore bearers is important in milk and milk powder intended for cheese making. Inoculate 10-ml volumes of bromcresol purple milk in triplicate with 10, 1.0 and 0.1-g amounts of milk. Heat at 80 °C for 10 min, overlay with 3 ml of 2% agar and incubate for seven days. Read by noting gas formation and use Jacobs and Gerstein's MPN tables to estimate the counts (p. 141).

Condensed milk

This contains about 40% of sugar and is sometimes attacked by yeasts and moulds that form 'buttons' in the product. Examine as for butter (p. 216).

Evaporated milk

This is liable to underprocessing and leaker spoilage. *Clostridium* spp. may cause hard swell and coagulation. Yeasts may cause swell. Open aseptically and examine as appropriate (*see* p. 235).

Fermented milk products

Yoghurt, leben, kefir, koumiss, etc., are made by fermentation (controlled in factory-made products) of milk with various lactic acid bacteria, and streptococci and/or yeasts. The pH is usually about 3.0–3.5 and only the intended bacteria are usually present, although other lactobacilli, moulds and yeasts may cause spoilage. The presence of large numbers of yeasts and moulds can be indicative of poor hygiene. Plate on potato glucose agar (pH adjusted to 3.5) and incubate at 23 ± 2 °C for five days.

Yoghurt is made from milk with *Lactobacillus bulgaricus* and *Streptococcus thermophilus*. In the finished product there should be more than 10^8 of each/g and they should be present in equal numbers[6]. Lower counts and unequal numbers predispose off-flavours and spoilage.

Make doubling dilutions of yoghurt in 0.1% peptone water and mix 5 ml of each with 5 ml of melted LS Differential Medium (*O*). Pour into plates, allow to set and incubate at 43 °C for 48 h. Both organisms produce red colonies, because the medium contains triphenyl tetrazolium chloride (TTC). Those of *L. bulgaricus* are irregular or rhizoid, surrounded by a white opaque zone; those of *S. thermophilus* small, round and surrounded by a clear zone.[7] Count the colonies and calculate the relative numbers of each organism (*see also* Tamine[8] and Robinson and Tamine[9]).

Buttermilk

Test for coliforms only. Use 10 ml of a 1:10 dilution in a pour-plate as for creams.

Human milk

Since the late 1940s interest in the storage and preservation of human milk has been growing. Milk banks have been set up to supply milk to babies of milk-deficient mothers or infants in hospital. Strict bacteriological control is necessary, particularly when raw milk is to be given.

Method

Use a micropipette. Drop 1 drop (0.1 ml) on each of two blood plates and one MacConkey plate. Make a 1:10 dilution of the milk (0.1 ml + 0.9 ml of 0.1% peptone water). Place one drop of this dilution on each of two blood plates. Spread, and incubate one of each pair of blood plates and the MacConkey plate aerobically and one of each pair of blood plates anaerobically all at 37 °C for 24 h. Re-incubate the anaerobic plate at this temperature for a further 24 h but keep the aerobic plate at 25–30 °C for a further 24 h. Make Gram films and do confirmatory tests as described in the appropriate chapters.

It is generally accepted that *S. aureus*, coliforms, β-haemolytic streptococci and enterococci are undesirable although milk containing these organisms and with a total count of lesss than 10 000 cfu/ml may be acceptable for pasteurization. Samples containing *P. aeruginosa* or spore-bearing aerobes or anaerobes are unacceptable.

Repeat tests after pasteurization should be sterile. For further information *see* Wright[10], Davidson *et al.*[11] and DHSS[12].

Butter

Soured cream is pasteurized and inoculated with the starter, usually *S. cremoris* or a *Leuconostoc* spp. Salted butter contains up to 2% of sodium chloride by weight.

Examination and culture

Emulsify 10 g of butter in 90 ml of warm (40–45 °C) diluent and do plate counts. Culture on a sugar-free nutrient medium and tributyrin agar.

Microbial content

Correct flavours are due to the starters, which produce volatile acids from the acids in the soured raw material.

Rancidity is due to *Pseudomonas*, *Achromobacter* and *Alcaligenes* spp., which degrade butyric acids. Anaerobic butyric organisms (*Clostridium*) cause gas pockets. Some lactic acid bacteria (*Leuconostoc*) cause slimes and others give cheese-like flavours. Coliforms, lipolytic psychrophiles and casein-digesting proteolytic organisms, which give a bitter flavour, may be found. To detect these, plate on Standard Methods agar containing 10% sterile milk and incubate at 23 ± 2 °C for 48 h. Flood the plates with 1% hydrochloric acid or 10% acetic acid. Clear zones appear round the proteolytic colonies.

Yeasts and moulds may be used as an index of cleanliness in butter. Adjust the pH of potato glucose agar to 3.5 with 10% tartaric acid and inoculate with the sample. Incubate at 23 ± 2 °C for five days. Report the yeast and mould count/ml or /g. (The International Dairy Federation uses glucose salt agar.) Various common moulds grow on the surface but often these grow only in water droplets or in pockets in the wrapper (*see* Murphy[13]).

Cheese

Pasteurized milk is inoculated with a culture of a starter, for example *S. lactis*, *S. cremoris*, *S. thermophilus* and rennet added at pH 6.2–6.4. To make hard cheeses, the curd is cut and squeezed free from whey, incubated for a short period, salted and pressed. The streptococci are replaced by lactobacilli (e.g. *L. casei*) naturally at this stage and ripening begins. Some cheeses are inoculated with *Penicillium* spp. (e.g. *P. roquefortii*), which give blue veining and characteristic flavours due to the formation under semi-anaerobic conditions of caproic and other alcohols.

Soft cheeses are not compressed, have a higher moisture content and are inoculated with fungi (usually *Penicillium* spp., which are proteolytic and flavour the cheese).

Types and flavours of cheeses are due to the use of different starters and ripeners and varying storage conditions.

Examination and culture

Use the sampling method described by Law *et al.*[14]. Take representative core samples aseptically, grate with a sterile Mouli grater, thoroughly mix and sub-sample. Fill the holes, left as a result of boring, with Hansen's paraffin cheese wax to prevent aeration, contamination and texture deterioration.

Make films, de-fat and stain. Bacteria may be present in colonies: examine thin slices under a lower-power microscope. Weigh 10 g and homogenize it in a blender with 90 ml of warm diluent. Do plate counts on yeast extract milk agar and coliform counts. Culture on whey and tributyrin agar and on media for staphylococci.

Microbial content

Apart from streptococci, lactobacilli and fungi that are deliberately inoculated or encouraged the following organisms may be found: contaminant moulds, *Penicillium*, *Scopulariopsis*, *Oospora*, *Mucor* and *Geotrichum* give colours and off-flavours. Putrefying anaerobes (*Clostridium* spp.) give undesirable flavours. *Rhodotorula* gives pink slime and *Torulopsis* yellow slime. Gassiness (unless deliberately encouraged by propionobacteria in Swiss cheeses) is usually due to *Enterobacter* spp. but these are not found if the milk is properly pasteurized.

Strains of *E. cloacae* and *E. aerogenes* have been shown to be inhibited by *S. lactis* and *S. cremoris*[15].

Gram-negative rods, some of which may hydrolyse tributyrin, may also be present.

Psychrophilic spoilage is common and is due to *Alcaligenes*, *Flavobacterium* and *Achromobacterium* spp. Counts may be very high.

Bacteriophages which attack the starters and ripeners can lead to spoilage.

Propionibacteria

Although deliberately introduced into some cheeses these can cause spoilage in others. Accurate counts are difficult to do because the organisms are microaerophilic.

Prepare doubling dilutions of cheese homogenate in 0.1% peptone water and add 1 ml of each to tubes of yeastrel agar containing 2% sodium lactate or acetate. Mix and seal the surface with 2% agar. Incubate anaerobically or under carbon dioxide at 30 °C for seven to ten days. Propionibacteria produce fissures and bubbles of gas in the medium. Choose a tube with a countable number of gas bubbles and calculate the numbers of presumptive propionibacteria (some other organisms may produce gas bubbles) (Harrigan and McCance[16] modified).

Staphylococcus aureus

Outbreaks of food poisoning caused by this product are occasionally reported. Cheddar cheeses may have a high staphylococcus count. In the USA cottage cheeses have been incriminated and in some states there is a statutory limit of 50 *S. aureus*/g. Enterotoxin may be produced during the long setting period, although the organisms themselves tend to die out on subsequent storage (*see also* Chapman and Sharp[17]).

Eggs

There are few bacteria in new-laid eggs and these are mostly in the yolk. Eggs are infected through the shell, which is normally impervious but may be damaged by rough handling. Removal of the 'bloom' by washing also permits the entry of micro-organisms.

Counts

Do counts on shell eggs separately or in batches of ten. Homogenize the egg and macerate the shell before making counts. Thaw frozen egg and reconstitute dried egg in warm peptone water diluent. Take frozen egg samples with a metal core-sampler.

Culture

Add 10–20 ml of egg to 100 ml of glucose tryptone broth. The egg must be well diluted because natural lysozyme prevents the growth of some bacteria which might be important when the eggs are used. Plate on glucose tryptone agar, blood agar, bile salt agar, inoculate cooked meat medium (anaerobic) and media for fungi if indicated.

Microbial content

Shell eggs

Externally, various salmonellas, coliforms, proteus, pseudomonas, *B. cereus*, yeasts, moulds and putrefactive anaerobes may be present. Internally, *S. gallinarum* and *S. pullorum*, coliforms, mycoplasmas, avian strains of mycobacteria and *Pasteurella anatipestifer* may be found. The degree of contamination is related to the rate of cooling and age of the egg, the porosity of the shell and humidity of the environment. Proteus causes coloured rots, e.g. 'black rot', in eggs.

Frozen egg

Total counts are high. Cultures may contain pseudomonas, aeromonas, achromobacter, alcaligenes, enterobacteria, serratia, micrococci and putrefactive anaerobes. Counts less than 50 000 cfu/g are rare; usually the count is several million and three million has been suggested as a reasonable maximum.

Frozen egg is invariably pasteurized nowadays, although this presents heat penetration problems and the α-amylase test is done on this by analytical chemists. Egg is incubated with a standard starch solution and the blue colour produced when iodine is added is measured. Adequate pasteurization, which reduces the bacterial content and destroys salmonellas, destroys most of the amylase. This test is less subject than bacteriological examination to sampling errors.

Dried egg

The total count is usually low (5000–10 000 cfu/g) in dried whole egg. Coliforms are rarely present in 1 g.

Dried albumen gives variable results. High-quality spray-dried material gives counts usually not exceeding 10 000 cfu/g but in some albumens counts may be as high as 10 million. Cross contamination may occur in tray drying of flake albumen.

Frozen and dried egg are recognized vehicles of salmonella. Culture replicate samples of up to 20 g in not less than 100 ml of medium (because of lysozyme). Use selenite broth (but good results can be obtained with ordinary broth). Incubate overnight at 35–37 °C and plate on DCA and/or bismuth sulphite medium (*see* Chapter 28).

Ice-cream and ice-lollies (water-ices)

The Food Standards (Ice-Cream) Regulations (1959) prescribe a standard of composition for ice-cream, dairy ice-cream (dairy cream ice or cream ice), 'parvex' (kosher) ice-cream and milk ice (including milk ice containing fruit, fruit pulp or fruit puree).

These regulations do not cover such articles of ice-lollies or fruit ices and, in deciding which tests are applicable to a product submitted for examination, it should be noted that if ice-cream is present as a core or covering, or if milk fat and milk solids other than fat are claimed to be present in the proportions of not less than 2.5 and 7%, respectively, the article qualifies as an ice-cream. A product containing no ice-cream or milk and consisting of fruit juices or pulp is treated as an ice-lolly.

Sampling

Send samples to the laboratory in a frozen condition, preferably packed in solid carbon dioxide. Remove the wrapper of small samples aseptically and transfer to a screw-capped jar. Sample loose ice-cream with the vendor's instruments (which can be examined separately if required, *see* Chapter 22). Keep larger samples in their original cartons or containers.

Hold ice-lollies by their sticks, remove the wrapper and place the lolly in a sterile screw-capped 500-ml jar. Break off the stick against the rim of the jar.

Reject samples received in a melting condition in their original wrappers.

Allow the samples to melt in the laboratory but do not allow their temperature to rise above 0–2 °C.

Methylene blue grading

This is used by the Public Health Laboratory Service.

Pipette 7 ml of quarter-strength Ringer solution into sterile 152 × 16 mm tubes calibrated at 10 ml (reductase tubes). Add 1 ml of the standard methylene blue solution used in milk testing. Pipette melted ice-cream into the tube until the level reaches the 10-ml mark. Stopper with a boiled rubber bung or Astell seal and mix by inversion. Check that the level is still at 10 ml and, if not, add more ice-cream. This product contains a variable amount of air which is trapped during manufacture.

Include boiled controls made in the same way and controls in which Ringer solution is substituted for methylene blue. These are necessary because of the variations in colour in ice-cream. It is difficult to apply this test to coloured and chocolate samples.

Mix by inversion. Place in a refrigerator at 0–4 °C until 5 p.m. and then transfer to a well insulated water-bath at 20 ± 0.5 °C with the level of the water above that of the ice-cream in the tubes (this temperature is near ambient; *see* Chapter 2).

At 10.0 a.m. the following day, remove the tubes from the bath and examine them. Report those in which the dye has been decolorized completely or to within half-an-inch of the top when compared with their controls as having decolorized methylene blue in 0 h.

Mix the contents of the tubes by inversion and place in a water-bath at

37 °C. Remove, examine, invert and replace at half-hourly intervals for 4 h and note the time taken to decolorize the methylene blue:

Methylene blue	*Grade*
Decolorized in 0 h	4
Decolorized in ½–2 h	3
Decolorized in 2½–4 h	2
Not decolorized in 4 h	1

Bacterial counts

It may be necessary to do counts as an alternative to the methylene blue test if the colour (usually chocolate) is such that blues are unreadable.

The difficulty with bacterial counts is the measurement of volumes of a product that contains a variable amount of air. For the best results, weigh about 10 ml of ice-cream into a screw-capped jar, multiply the weight to the nearest 0.1 g by nine and add this volume of 0.1% peptone water. Shake well. From this initial 1:10 dilution make two further tenfold serial dilutions and do plate counts on 1-ml samples with yeastrel agar. Incubate at 37 °C for 48 h.

Alternatively, and for routine work, make the initial 1:10 dilutions by adding 10 ml of melted ice-cream to 90 ml of Ringer solution. Smaller amounts are less reliable.

Miles and Misra counts (p. 133) are very convenient. Use the mixture prepared for the methylene blue test and drop three 0.02-ml amounts with a 50-dropper on blood agar plates. Allow the drops to dry, incubate at 37 °C overnight and count the colonies with a hand lens.

Multiply the average number of colonies per drop by 250 (i.e. dilution factor of 5 and volume factor of 50) and report as count/ml. If there are no colonies on any drops, the count can be reported as less than 100/ml. It is difficult to count more than 40 colonies per drop; therefore, report such counts as greater than 10 000 cfu/ml.

In counts, note the presence of *B. cereus*. This spore-bearer may be present in the raw materials. Its spores resist pasteurization and the organism is important in the finished product because it actively decolorizes methylene blue.

Counts at 0–5 °C for psychrophiles may be useful in soft ice-cream.

Coliform test (US)

To test for coliforms, dilute the sample 1:10 and use the pour-plate technique described for milks (p. 210).

Ice-lollies

Measure the pH of the melted product. If it is 4.5 or less, no bacteriological examination is necessary. A few yeasts may be present. When the pH is higher, do plate or Miles and Misra counts and coliform counts by the MPN or drop count methods.

Aseptically separate the components of lollies that have ice-cream cores while still frozen and treat separately as above.

Ingredients and other products

Do plate counts at 5, 20 and 37 °C. Although ice-cream is pasteurized and then stored at a low temperature, only high-quality ingredients with low counts will give a satisfactory product.

In cases of poor grading or high counts on the finished ice-cream, test the ingredients and also the mix at various stages of production. Swab-rinses of the plant are also useful (*see* Chapter 22 and Rothwell[18]).

References

1. *Standard Methods for the Examination of Dairy Products* (1967) 12th edn. New York: American Public Health Association
2. SHARPE, J.M.C., PATERSON, G.M. and FORBES, G.I. (1980) *Journal of Infectious Diseases*, **2,** 333
3. ROBINSON, R.K. (Editor) (1981) *Dairy Microbiology, Volume I The Microbiology of Milk.* London: Applied Science Publishers
4. GRIFFITHS, M.W., PHILLIPS, J.D. and MUIR, D.D. (1980) *Journal of the Society of Dairy Technology*, **33,** 8
5. DAVIS, J.G. (1981) Microbiology of cream and dairy desserts. In *Dairy Microbiology, Vol. 2.* Edited by R. K. Robinson, pp. 31–89. London: Applied Science Publishers
6. DAVIS, J.G., ASHTON, T.F. and McCASKILL, M. (1971) *Dairy Industry*, **36,** 569
7. *Oxoid Manual* (1982) 5th edn. Basingstoke: Oxoid Ltd
8. TAMIME, A.Y. (1981) Microbiology of starter cultures. In *Dairy Microbiology, Vol. 2.* Edited by R.K. Robinson, pp. 113–156. London: Applied Science Publishers
9. ROBINSON, R.K. and TAMIME, A.Y. (1981) Microbiology of fermented milks. In *Dairy Microbiology, Vol. 2.* Edited by R.K. Robinson, pp. 245–278. London: Applied Science Publishers
10. WRIGHT, J. (1947) *Lancet*, **ii,** 121
11. DAVIDSON, D.C., POLL, R.A. and ROBERTS, C. (1979) *Archives of the Diseases of Children*, **54,** 760
12. DHSS (1982) *Hazard Notice* HN 82/11, 20th July, London
13. MURPHY, M.F. (1981) Microbiology of butter. In *Dairy Microbiology, Vol. 2.* Edited by R.K. Robinson, pp. 91–111. London: Applied Science Publishers
14. LAW, B.A., SHARPE, M.E., MABBETT, L.A. and COLE, C.B. (1973) Microflora of cheddar cheese. In *Sampling—Microbiological Monitoring of Environments.* Edited by R. G. Board and D.W. Lovelock, pp. 1–7. London: Academic Press
15. RUTZINSKI, J.L. and MARTH, E.H. (1980) *Journal of Food Protection*, **43,** 720
16. HARRINGTON, W.F. and McCANCE, M.E. (1966) *Laboratory Methods in Microbiology*, p. 176. London: Academic Press
17. CHAPMAN, H.R. and SHARPE, M.E. (1981) Microbiology of cheese. In *Dairy Microbiology, Vol. 2.* Edited by R.K. Robinson, pp. 157–243. London: Applied Science Publishers
18. ROTHWELL, J (1981) Microbiology of ice cream and related products. In *Dairy Microbiology, Vol. 2.* Edited by R.K. Robinson, pp. 1–30. London: Applied Science Publishers

Chapter 19

Miscellaneous foods

Frozen foods

Some frozen convenience foods are mentioned here, others—frozen meat, fish and ice-cream—are included under their appropriate headings in Chapters 16, 17 and 18. For frozen vegetables *see* (p. 225).

Storage

Unless the cabinets are maintained at −18 to −20 °C there will be difficulties due to the different melting points of the stored products, the presence of water films and the variety of microclimates. Two kinds of spoilage occur:

(1) low-temperature spoilage due to enzymes and, less often, to psychrophiles if the temperature is at or about the freezing point of water;
(2) unfreeze spoilage, when bacteria can grow because of gross temperature fluctuations. Off-odours and off-flavours and spoilage losses are likely to be doubled for each 2–3 °C rise in temperature.

Counts

Count total bacteria growing on at 5–7 °C in 5–7 days and 20–30 °C in 2–3 days. Use enriched media because organisms will be cold shocked. Count lactic acid bacteria (except in fish), moulds and yeasts. Estimate coliforms by the method described on p. 193.

Culture

For total counts, inoculate one of the media for fastidious organisms mentioned on p. 57 and on media for lactobacilli and fungi, for staphylococci, salmonellas and enterococci. There is evidence that enterococci survive longer than coliforms in frozen foods.

Frozen pies and complete meals

Frozen pies often have low counts (5000–30 000 cfu/g) but may contain enterobacteria and staphylococci. 'Complete meals' vary enormously in their counts and flora, usually reflecting factory conditions.

The practice of keeping complete meals in cold storage for several months appears to reduce the counts but this may be a false effect and reflect the failure to resuscitate cold-shocked organisms.

Storage life

This is usually controlled by the better manufacturers, who date-stamp their products. Chicken, pies and complete meals keep for two to six months. Storage life depends on the maintenance of a constant low temperature.

Arbitrary standards

Standards that have no legal status are used by the trade and public health authorities as a guide in the frozen food industry:

Plate count at 35 °C in 48 h—not more than 100 000 cfu/g

Coliform bacilli NOT present in 0.1 g.

S. aureus NOT present in 0.01 g.

For further information on the microbiology of frozen foods, *see* Roberts *et al.*[1].

Fruit and vegetables

Fresh fruit and vegetables

Spoilage of fresh fruit is mostly caused by moulds. Mould and yeast counts may be indicated in fruit intended for jam making or preserving. Orange serum agar is useful for culturing citrus fruit.

Washed, peeled and chopped vegetables are common in supermarkets in the rest of Europe and are gaining acceptance in the UK. Spoilage is caused by pseudomonads, lactobacilli, Lancefield Group D streptococci and leuconostoc. Salad crops, e.g. watercress, which may be grown in polluted streams, should be tested for *E. coli*.

Dehydrated fruit and vegetables

Counts

Do total, mould and yeast counts. Counts of lactic acid bacteria may be useful: use Rogosa agar medium and incubate at 30–32 °C for three days.

Culture

Inoculate glucose tryptone agar, tomato juice, or Rogosa medium and incubate at 30–32 °C. Incubate glucose tryptone cultures also at 55–60 °C aerobically and anaerobically.

Microbial content

Counts should be low. Lactic acid bacteria, flat sour thermophiles (*B. stearothermophilus*) and hydrogen sulphide-producing anaerobes may be present.

Blakey and Priest[2] examined red and brown lentils, yellow and green peas, black-eyed, kidney, mung and soya beans, scotch broth mix, rice, pearled barley and chapatti flour. They found *B. cereus* at levels ranging from 1×10^2 to 6×10^4/g.

Frozen vegetables and fruit

Frozen peas offer the greatest problem as they deteriorate rapidly on thawing. Slime usually contains large numbers of leuconostocs, which imparts a yellow appearance (acid pH) but sometimes a heavy growth of coryneforms produces ammonia, which neutralizes the acid formed by the leuconostocs. If sucrose is used in the medium instead of glucose, a levan is produced by leuconostocs.

Counts on other frozen vegetables are usually low (*ca.* 100 000 cfu/g). Coliforms and enterococci are commonly found on vegetables, *E. coli* type 1 may have public health significance.

Fruits may have low yeast and lactobacilli counts. In general, they keep well at 0 °F (−15 °C) for two to three years. Vegetables remain in good condition at this temperature for 6–12 months.

Pickles and sauces

Pickles

Vegetables are first pickled in brine. The salt is then leached out with water and they are immersed in vinegar (for sour pickles) or vinegar and sugar (for sweet pickles). Some products are pasteurized. Spoilage is due to low salt content of brine, poor quality vinegar, underprocessing and poor closures.

Counts

Use glucose tryptone agar at pH 6.8 and 4.5 and do total counts and also counts of acid-producing colonies (these have a yellow halo due to a colour change of indicator). Count lactic acid bacteria on Rogosa or other suitable medium. Estimate yeasts either by the counting chamber method (stain with 1:5000 erythrosin) or on malt agar or Mycophil agar.

Culture

Use glucose tryptone and Rogosa or MRS. Grow suspected film-yeasts in a liquid mycological medium, containing 5 and 10% of sodium chloride for three days at 30 °C. For obligate halophiles, use a broth medium containing 15% of sodium chloride.

Microbial content

Pickles are high-acid foods. Counts are usually low, for example 1000 cfu/g. Yeasts are a frequent source of spoilage and may be either gas-producing or film-producing. In the former case, enough gas may be generated to burst the container. Bacterial spoilage may be due to acid-producing or acid-tolerant bacteria such as acetic acid bacteria, lactic acid bacteria and aerobic spore-bearers. Infected pickles are often soft and slimy.

Fermented pickles of the kraut type contain large numbers of lactic acid bacteria (*Lactobacillus* and *Leuconostoc*) which are responsible for their texture and flavour.

Acid-forming bacteria are active at salt concentrations below 15%. Above this concentration, obligate halophiles are found.

Ketchups and sauces

The most common cause of spoilage is *Zygosaccharomyces* (*Saccharomyces*) *bailii*. This organism grows at pH 2, at $<5\,^{\circ}C$ up to 37 °C, in 50–60% glucose and is heat resistant to 65 °C.

Salad creams

These contain olive oil and spoilage may be due to lipolytic bacteria. Test for these with tributyrin agar. Examine for coliform bacilli and thermophiles. All these should be absent. There should not be more than five yeasts or moulds/g.

Sugar and confectionery

Sugars, molasses and syrups

The importance of micro-organisms is related to the use of the products, for example flat-sour bacteria are less important in bakery than in canning. Do total counts, examine for thermophiles (*B. stearothermophilus*, *C. nigrificans*, *C. thermosaccharolyticum*) and for yeasts and moulds. Osmophilic yeasts, aspergillus, penicillium, etc., may cause inversion.

Chocolate

This has been shown to be the vehicle of salmonella infection in a number of cases. A review of this subject was published in 1977.[3]

Shave the chocolate into nutrient broth and selenite, incubate overnight and plate out on DCA medium. Do not place large lumps of chocolate into liquid medium.

Cereals and protein additives

It is important that these materials should not contribute unduly to the bacterial load of the product. Do total counts, and, if the material is to be used in canned foods, do spore counts. The total count at 30 °C should not exceed 20 000 cfu/g. The spore counts at 30 and 55 °C should not be greater than 100/g.

Flour

Contaminated flour may be responsible for the spread of infection by spoilage organisms in kitchens as well as spoilage of the bread, pastry, etc., for which it is used. Grain is naturally infected by soil, dust and rodent and bird faeces during ripening, havesting and storing. During transport and handling, this contamination is distributed throughout the bulk. Before milling, the grain is washed, sometimes with polluted water.

Counts

Weigh 10 g of flour into a sterile jar containing coarse sand or small glass beads. Add 100 ml of 0.1 % peptone water diluent and shake mechanically for 20 min. If moisture absorption is great this dilution may need to be adjusted.

Do total bacterial counts on serial dilutions using glucose tryptone agar. Because of turbidity, it may be necessary to use the MPN method, in which case test 5 × 10, 5 × 1 and 5 × 0.1 ml samples of tenfold dilutions of the above homogenate in glucose tryptone broth plus bromocresol purple and record as positive any tube that shows acid production. Incubate at 32 °C for three days.

Count anaerobes in the same way with thioglycollate broth, and coliform bacilli by the MPN method.

To count 'rope spores', heat the homogenate for 20 min at 90 °C to kill vegetative bacteria, make dilutions and do MPN counts in glucose tryptone broth. Alternatively, inoculate two tubes of glucose tryptone broth each with 1 ml of the serial dilutions of heated homogenate. Incubate at 30 °C for three days and record as positive tubes that show a pellicle. Examine the pellicle to identify *B. mesentericus* or *B. subtilis*. Record as rope spores/g or as present in so much of 1 g.

Culture

Use glucose tryptone agar for aerobic and glucose tryptone agar and iron sulphite medium for anaerobic culture. Inoculate malt extract or a similar medium.

Microbial content

Counts of 5000–500 000 are usual and *E. coli* is often present in 1 g or less depending on processing. Flat-sour bacteria (*B. stearothermophilus*) and hydrogen sulphide-producing clostridia may be found in varying numbers. 'Rope organisms' ('*B. mesentericus*', but *see* p. 361) are important and cause 'ropy bread'. Mould counts may be 2000 or more per g.

Pastry

Uncooked pastry, prepared for factory use or for sale to the public, may suffer spoilage due to lactobacilli. Do counts on Rogosa agar and incubate at 30 °C for four days. Counts of up to 10 000/g are not unreasonable.

Baby food (infant formulae)

The importance of testing this type of product is emphasized by Collins-Thompson *et al.*[4]. They review the existing literature and suggest appropriate standards.

Gelatin

Gelatin is often used to top-up pastry cases of cold-eating pies, in canned ham production and in ice-cream manufacture. It should be free from spores and coliforms.

Laboratory examination

Prepare an initial 20% solution. Weigh 5 g of gelatin into a bottle containing 100 ml of sterile water and allow to stand at 0–4 °C for 2 h. Place the bottle in a water-bath at 50 °C for 15 min and then shake well. Mix 20 ml of this 20% gelatin solution with 80 ml of sterile water. This gives a 1:100 dilution. Use 1.0 and 0.1 for total counts by the pour-plate method. Incubate at 35 °C for 48 h.

Gelatin for ice-cream manufacture

Do a semi-quantitative coliform estimation using MacConkey or similar broth. Add 10 ml of 1:100 gelatin to 10 ml of double-strength broth; add 1 ml of 1:100 gelatin to 5 ml of single-strength broth and 0.1 ml of 1:100 gelatin to 5 ml of single-strength broth. Incubate at 35 °C for up to 48 h and do confirmatory tests where indicated (*see* p. 285). Thus the presence or absence of coliforms and *E. coli* in 0.1, 0.01 or 0.001 g of the original material can be determined (*see* p. 193). It is desirable that coliforms should be absent from 0.01 g and *E. coli* absent from 0.1 g. The total count in gelatin to be used for ice-cream manufacture should not exceed 10 000 cfu/g.

Gelatin for canned ham production

This should have a low spore count. After doing the total counts, heat the remaining 1:100 solution of gelatin at 80 °C for 10 min. Plate 4 × 1 ml of this solution on standard plate count medium. Incubate two plates at 35 °C and two plates at 55 °C for 48 h. There should be not more than one colony per plate, i.e. 100 g of the original gelatin. The total count should not exceed 10 000 cfu/g.

Spices and onion powder

A plastic bag inverted over the hand is a satisfactory way of sampling spices. Some are toxic to bacteria and the initial dilution should be 1/100; for cloves use 1/1000, in broth.

The total counts on these materials vary widely; white peppers usually have low counts and black peppers often have very high counts. Twenty-five samples each of unprocessed black pepper, red pepper, ginger and turmeric were examined by Seenappa and Kempton[5]. Eleven per cent of samples had a total count of less than 10^4 cfu/g. *B. cereus* and *B. coagulans* were the most commonly isolated members of the genus *Bacillus*.

Low spore counts are important if these materials are to be used for canned foods: an acceptable level at 30 and 55 °C is less than 100/g.

Oily material

Mix 1.5 g of tragacanth with 3 ml of ethanol and add 10 g of glucose, 1 ml of 10% sodium tauroglycocholate and 96 ml of distilled water. Autoclave at 115 °C for 10 min. Add 25 g of the oily material under test and shake well. Make dilutions for examination in warm diluent.

References

1. ROBERTS, T.A., HOBBS, G., CHRISTIAN, J.H.B. and SKOVGAARD, N. (Editors) (1981) *Psychrotrophic Micro-organisms in Spoilage and Pathogenicity*. London: Academic Press
2. BLAKEY, L.J. and PRIEST, F.G. (1980) *Journal of Applied Bacteriology*, **48,** 297
3. D'AOUST, J.Y. (1977) *Journal of Food Protection*, **40,** 718
4. COLLINS-THOMPSON, D.L., WEISS, K.F., RIEDEL, G.W. and CHARBONNEAU, S. (1980) *Journal of Food Protection*, **43,** 613.
5. SEENAPPA, M. and KEMPTON, A.G. (1981) *Journal of Applied Bacteriology*, **50,** 225

Chapter 20

Thermally processed foods

The practice of prolonging the shelf-life of foods by sealing them in metal or glass containers and heat processing them has long been established. In recent years, it has been possible to extend the range of packaging to include semirigid aluminium trays and foil containers laminated with polypropylene. The basic principles are the same. In simple terms, successful processing depends upon clean food being packed into sound containers which are hermetically sealed. Sufficient heating is then given to eliminate *Clostridium botulinum* and spoilage organisms. After heating, the containers are rapidly and hygienically cooled and dried. The technical expertise required to achieve this is beyond the scope of this book. It is described in detail by Hersom and Hulland.[1]

Routine control of the product in the factory is the responsibility of the quality assurance and laboratory departments of the manufacturer but general laboratories may be asked to help when defects or spoilage have developed after the products have left the factory or where the product is suspected of having caused food poisoning or enteric fever. Occasionally, arbitration is needed between suppliers of ingredients and manufacturers of finished goods.

Types of containers

Cans

Tin-coated steel, sometimes lacquered, aluminium easy-open ends (EOE) used for beverages.

Jars and bottles

Glass closed with metal cap and resilient seal, sometimes under vacuum.

Trays

Aluminium or aluminium/polypropylene laminated, with lids.

Flexible pouches

Aluminium/polypropylene or ethylene, are slowly gaining acceptance.

Types of cans

There are four main types of cans: three-piece (sanitary), two-piece, non-soldered and easy-open ends (EOE).

Three-piece cans

The side or body seam is made by interlocking and an external soldering process. This seam is rather thick and to make the covers sit well upon the body, the last $\frac{1}{4}$ in at each end is lapped, not interlocked. The ends of the body are opened slightly to form a flange. The rims of the covers are recessed

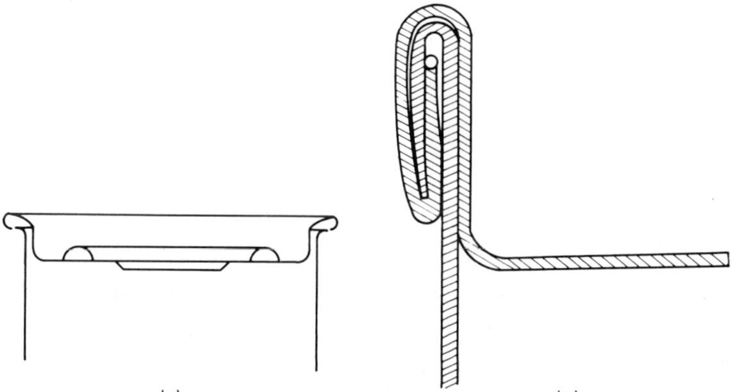

(*a*) (*b*)

Figure 20.*1* Assembly of compound lined end-to-can body, with section, through finished double seam. (From T.E. Bashford and D.A. Herbert. Reproduced by permission of the Royal Society of Health)

(*Figure 20.1a*) and contain a rubber compound. The covers are machined on to the body ends to form the double seal (*Figure 20.1b*). The rubber compound fills and seals the seam. These cans are made by specialist manufacturers to a high degree of mechanical precision and are supplied to the canners with one cover only. The other cover is put on by the canner after filling, usually with machinery supplied by the can manufacturer.

Two-piece cans

A single metal disc forms the body of the can. This has several advantages: less metal is required because there are fewer overlaps; fewer overlaps reduce the chances of leaks; absence of solder removes the risk of lead contamination. This type of can has been in use for fish products for many years. Recent work has made possible the production of deeper cans of this design.

Non-soldered cans

Beverage manufacturers favour cemented seams. These cans are made of tin-free steel. Nylon cement is used both as a side seam seal and to avoid iron pick-up. A lacquer coating is applied to the internal surface as with conventional cans.

Welded side seams are now technically possible. These are used in the UK for low carbonation (draught) beer in 5-pint containers.

Easy-open ends

These cans are made of aluminium and used for beverages. They invariably have ring pull devices for ease of opening. As yet, they are of limited use for foods.

Jars and bottles

Closures are made of metal and coated with a suitable lacquer to prevent corrosion. A resilient compound makes an airtight seal between the glass and cap. A vacuum is often present. The type of closure used is determined by the product.

Flexible pouches

Although it is technically possible to produce flexible pouches using combinations of aluminium and thermoplastic materials, these have not gained worldwide acceptance. A common problem is seam leakage occasioned by the pumping action of the contents when handling or when pressure changes occur.

Trays

Aluminium/polypropylene laminates are commonly used. They may be formed on the food manufacturer's premises or supplied to him complete with lids of the same material. Filled and heat-sealed trays may be processed at up to 121 °C. These packs are gaining in popularity because, although material costs are higher and line speeds slower than for conventional cans, heat penetration is faster and consumer appeal is considerable.

Physical defects of cans

These defects may be due to improper processing; during exhausting and autoclaving incorrect stresses may be imposed and cause 'peaking' or 'panelling', which are distortions of the ends (distinct from 'swelling') and of the body, respectively. The ends of the cans have concentric rings impressed in them to absorb the normal strain and to permit swelling in normal stresses.

Rusting of cans, causing pinholes and consequent spoilage, is revealed by inspection. Hydrogen swell is caused by hydrogen formed when acidic foods attack the metal in places where the lacquer is defective.

Faulty can manufacture and improper closure of seams, either side or lid, can lead to spoilage.

Inadequate drying may result in contamination of the can contents by bacteria which enter in water droplets through minute pinholes in the seams. These holes are usually self-sealing when the can is dry. Wet cans also tend to rust.

Spoilage due to micro-organisms

Leaker spoilage

When this occurs, usually only a small number of cans in each batch is affected. Minute faults may be present in some cans (*see above*), particularly where the end seams cross the body seams. After autoclaving and during cooling there is a negative pressure in the can and cooling water containing bacteria may be drawn in, and canned foods may thus be infected with pathogens. Only very few organisms, pathogens or spoilers need be drawn into a can, as they will multiply rapidly. Gas producers will manifest themselves by 'blowing' the can but organisms that produce gas in normal culture may not do so in cans. Other spoilage will be obvious when the can is opened but food infected with pathogens, e.g. typhoid bacilli, may appear sound and wholesome. Care must also be taken that there is not a build-up of organic matter in the line which can overcome the disinfectant and lead to bacterial multiplication in the water remaining on the cans. Personnel with septic conditions must be excluded from handling cans. Dirty or infected hands can infect the surface water on the cans before it is drawn into the cans through these defective seams or pinholes. Water used for cooling is usually chlorinated but some organisms are relatively resistant to the process. When the seams are dry, the chances of contamination are slight.

Gross underprocessing will usually have been found by the manufacturer's tests. If this is not the case, it is common to find only one type of organism in this situation.

Some cured meats are deliberately underprocessed because they are rendered less palatable by autoclaving. Ham and mixtures of ham and other meat are usually salted and spiced and given the minimum of heat treatment. The manufacturers do not claim sterility and the label on the can invariably recommends cold or cool storage. The conditions of pH, salt content and storage temperature should be such that the bacteria in the can are prevented from multiplying. There is evidence that some species of enterococci produce antibacterial substances in canned hams, which act antagonistically on some species of clostridia, lactobacilli and members of the genus *Bacillus*. It is suggested that this is a factor in the successful preservation of commercial products of this nature[3].

Poor plant hygiene, faulty design or careless operation of equipment, e.g. bad stacking of retorts, may be a contributory cause of underprocessing. Vegetative organisms are killed, except in very rare cases of gross underprocessing, but spores are unaffected; they subsequently germinate and cause spoilage. The spores of *C. botulinum* are very heat resistant and they may germinate and produce toxin. Spores of *Bacillus stearothermophilus* are even more resistant than those of *C. botulinum*, and fortunately, *C. botulinum* is a poor competitor.

Gas-producing organisms cause the can to swell. The first stage is the 'flipper' when the end of the can flips outward if the can is struck sharply. A 'springer' is caused by more gas formation. Pressing the end of the can causes the other end to spring out in a bulge. The next stage is the 'swell' or 'blower' when both ends bulge. A 'soft swell' can be pressed back but bulges again when the pressure is released. 'Hard swells' cannot be compressed.

Spoilage may not result in gas formation and may be apparent only when the can is opened. This kind of spoilage includes the 'flat-sour' defect.

Inadequate cooling

If cooling is too slow or inadequate there may be sufficient time for the highly resistant spores of *B. stearothermophilus* to outgrow and multiply. The optimum growth temperature for this organism is between 59 and 65 °C. It will not grow at 28 °C or at a pH of less than 5[4]. Together with *B. coagulans* it is the chief cause of 'flat-sour' spoilage in canned foods. Most strains of *B. coagulans* will grow at 50–55 °C. They will also grow in more acid conditions. Pederson and Beaker[5] showed that growth at pH 4.02 was possible.

Preprocessing spoilage

This can occur when the material to be canned is mishandled before processing, e.g. if precooked meats with a large number of surviving spores are kept for too long at a high ambient temperature this may result in the outgrowth of spores and the production of gas. When canned the organisms will be killed leaving the gas to give the appearance of a blown can.

Type of food, and organisms causing spoilage

High-acid foods

Spoilage is rare in processed foods with a pH of 3.7 or less, for example pickles and citrus fruits. Yeasts may occur in serious underprocessing. These foods are usually not pressurized during heating.

Acid foods

When the pH is 3.7–4.5, as in most canned fruits, aerobic and anaerobic spore-bearers may cause spoilage but this is not common. Lactobacilli and *Leuconostoc* have been reported. Osmophilic yeasts and the mould *Byssochlamys* are sometimes found. These foods are usually not pressurized. They are too acidic for the growth of most bacteria and leaker spoilage is uncommon.

Low-acid foods

If the pH is 4.5 or above, as in canned soups, meat, vegetables and fish (usually about pH 5.0), one of the following thermophiles is usually found in spoilage due to underprocessing:

(1) *B. stearothermophilus*, causing 'flat-sour' spoilage.
(2) *C. thermosaccharolyticum*, causing 'hard swell'.
(3) *C. nigrificans*, causing 'sulphur stinkers'.
(4) Mesophilic spore-bearers, obligate or facultative anaerobes, causing putrefaction.

Leaker spoilage may be due to a variety of organisms; aerobic and anaerobic spore-bearers, Gram-negative non-sporing rods and various cocci, including *Leuconostoc* and *Micrococcus* may be found. *S. aureus* (food poisoning type) has been isolated.

Methods of examination

Sampling

If packs are blown or swollen, examine six and take six normal ones from another batch as controls. In suspected underprocessing, examine 6–12 packs from each batch. Leaker-spoilage is likely to occur in only a very small number of packs in a batch; therefore, examine as many as possible.

Physical examination

Inspect the seams and can surfaces. A jeweller's saw is useful for cutting across seams. Note the batch or code numbers printed on the labels or stamped on the lid.

Jars

Examine cap for perforations and note code.

Trays and pouches

Inspect the seals. These are formed from a continuous weld and less likely to leak than the double seams of cans. Furthermore there is no headspace or vacuum which could cause organisms to be sucked in if there was a small hole in the seal. Note code.

Pre-incubation

Incubate apparently sound packs at 35–37°C for seven days. This encourages the multiplication of small numbers of organisms which might otherwise be missed in sampling the contents.

Sampling contents: cans normal in appearance

Swab the top with cotton wool and methylated spirit. Pour 1 ml of spirit on the swabbed area and flame it. Allow the spirit to burn out.

If the contents of the can are liquid, puncture the flamed surface with a 100-mm wire nail (sterilized in tins containing 10–12 nails in a hot air oven) by a sharp blow with a hammer. Remove a sample of the contents with a pasteur pipette into culture media and into a screw-capped bottle for viable counts if required.

If the contents are solid, use a punch made of brass rod 9–10 mm in diameter with one end drawn to a point. Sterilize these individually. Drive the punch well in to make a large hole. Remove a core sample with a length of glass tubing of 7–8 mm outside diameter by pushing it tight to the bottom of the can. Push the core sample from the glass tube into a screw-capped bottle with a piece of glass rod of suitable thickness. These core and rod samplers can be sterilized together in copper pipette drums. In addition to taking core samples, it is desirable to sample jelly adjacent to the seam of the can. To do this, remove the end of the can, previously punctured, with a sterile domestic can opener and tip the contents on to a sterile tray. Take care not to disturb the material under the seam. Note the appearance of this material and sample with a cotton wool swab.

Jars normal in appearance

Sterilize as for cans and pierce with a sterile nail. It is very important to release any vacuum. Carefully remove the cap so that the sealing gasket is undisturbed. This should be examined thoroughly for evidence of improper seating or twisting. Liners are sometimes misplaced causing inadequate closure of the jar. Look for damage to the rim of the jar. Examine contents as for cans.

Flexible pouches and trays normal in appearance

Support in suitable racks. Sterilize with 50:50 alcohol/ether and allow to dry. Open with sterile scissors and sample with a sterile spoon. Examine contents as for cans.

Sampling blown, swollen or leaking packs

These contain gas under pressure and the contents may be offensive. Chill before opening. Place the container on a metal tray with the seam facing away from the operator. Swab with 4% iodine in 70% alcohol, allow to stand for a few minutes then dry with a sterile towel. Do not flame.

Invert a previously sterilized metal funnel or new plastic bag over pack. The diameter of the funnel should be slightly larger than that of the can. Pass a sterile brass rod with a point at one end down the funnel spout until it rests on the can; hold both firmly and puncture the can by tapping the rod with a hammer, then withdraw the rod slightly. The contents of the can may be ejected with some force but the funnel and tray will prevent broadcast. Before removing the funnel and brass rod, push the latter in and out of the hole in the can several times. Sometimes a piece of food is forced against the hole by internal gas pressure and when a sampler is inserted more gas and food are ejected.

Take samples with a pasteur pipette or core sampler as described above. After sampling, open the can with a domestic can opener and inspect the contents.

Direct film examination

Make Gram films of sample. The presence of Gram-positive rods may suggest underprocessing while the presence of cocci, yeasts, etc., indicates leaker spoilage.

Note: The organisms seen may be dead (killed during processing), so too much reliance must not be placed on this examination.

Autosterilization may also account for this phenomenon. In this instance the organisms die out during storage. When this has occurred the organisms appear degenerate and poorly stained (*see* Preprocessing spoilage p. 234).

Culture

For general examination inoculate glucose tryptone agar (with bromocresol purple indicator) and incubate aerobically and anaerobically at 22–25, 35–37 and 55–60 °C for 24–36 h.

For high-acid foods, i.e. pH 4.6 or less, inoculate four tubes of acid broth. Incubate two tubes at 55 °C for 48 h and two tubes at 30 °C for 96 h. Inoculate two tubes of malt extract broth and incubate these at 30 °C for 96 h. Subculture and make Gram-stained smears as necessary.

Also inoculate the following media if indicated: iron sulphite medium (sulphur stinkers), blood agar and MacConkey (for putrefactive organisms, micrococci, leuconostoc, etc.), Crossley Milk Medium (putrefactive aerobic or anaerobic spore-bearers). Clostrisel agar for clostridia, malt extract or other mycological medium (for yeasts and fungi).

Make Gram films of colonies.

Microbial content

Identify as follows:

(1) Gram-positive rods
- (a) Thermophiles:
  - (i) Aerobic: *B. stearothermophilus* (flat-sour). *See* p. 361.
  - (ii) Anaerobic: *C. thermosaccharolyticum* (hard swell). *See* p. 370.
  - (iii) Anaerobic: black colonies in iron sulphite medium: *C. nigrificans* (sulphur stinkers). *See* p. 370.
- (b) Mesophiles:
  - (i) Aerobic: *Bacillus*. *See* Chapter 34.
  - (ii) Anaerobic: *Clostridium*. *See* Chapter 35.

(2) Gram-negative rods

Pseudomonas–Achromobacter–Alcaligenes Groups or Enterobacteria. *See* Chapters 26 and 28.

(3) Gram-positive cocci

Micrococci, Leuconostoc. *See* Chapters 31 and 32.

(4) Yeasts and Moulds

See Chapters 39 and 40.

Pathogens in canned food

Outbreaks of enteric (typhoid) fever and staphylococcal disease have caused food bacteriologists, public health authorities and canners to revise their opinions on the safety of canned foods, although these outbreaks are very few in proportion to the enormous amount of canned food consumed. Random or 'routine' sampling of canned foods for pathogens is an unrewarding procedure. Only low-acid foods, meat and dairy products and certain canned vegetables can support the growth of enteric organisms, staphylococci and botulinus.

It has been shown that certain moulds and other organisms can raise the pH of some acid fruits, e.g. tomato juice and pears to a level at which *C. botulinus* will grow.[6]

Examination for pathogens

When this is indicated, open the cans with sterile can openers and if the food is solid take samples from opposite the seams, particularly where the end seams cross the side seam. Culture in selenite medium for salmonellas (continue as on p. 291), in salt meat for staphylococci (continue as on p. 325). For botulism, *see* p. 362.

References

1. HERSOM, A.C. and HULLAND, E.D. (1980) *Canned Foods: Thermal Processing and Microbiology*. 7th edn. Edinburgh: Churchill-Livingstone
2. *Report of the Conference on the Safety of Canned Foods* (1965) London: Royal Society of Health
3. KAFEL, S. and AYRES, J.C. (1969) *Journal of Applied Bacteriology*, **32,** 217
4. ALLEN, M.B. (1953) *Bacteriological Review*, **17,** 125
5. PEDERSON, C.S. and BECKER, M.E. (1949) *New York State Agricultural Experimental Station Technical Bulletin*, No. 150
6. MUNDT, J.O. (1978) *Journal of Food Protection*, **41,** 267

Chapter 21

Soft drinks, fruit juices and alcoholic beverages

Soft drinks

Total counts should be low and coliform bacilli absent, as in drinking water. Membrane filters can be used to test water intended for soft drink production. Millipore[1] publish a very useful booklet on the microbiological examination of soft drinks. Yeast and mould spoilages are not uncommon and they may raise the pH and allow other less acid tolerant organisms to grow.

In non-carbonated fruit drinks, yeasts are not inhibited by the amounts of preservatives that are permitted by law. Spoilage is generally controlled by acidity (except *Z. baillii*) (*see* Ketchups, (p. 226). In both these and carbonated drinks the microbial count diminishes with time.

Lactic acid and acetic acid bacteria may grow at pH 4.0 or less in some fruit juices.

Fruit juices

Heat-resistant fungi can cause problems in concentrated juices. Screen by heating at 77 °C for 30 min. Cool and pour plates with 2% agar. Incubate for up to 30 days.[2]

Alcoholic beverages

In the cider and perry industry problems can be caused by modern methods of husbandry and harvesting. Soil and rotting fruit are not infrequently gathered and this introduces a heavy load of spoilage organisms. Cider, beer and wine may all be spoiled by acetic acid bacteria which convert ethanol into acetic acid[3] (p. 269). Beer, especially if unhopped, may be spoiled by *Pediococcus* (sarcina-sickness, *see* p. 342).

References

1. MILLIPORE CORPORATION (1973) *Microbiological Analysis of Soft Drinks*. Application Manual, AM 601. Bedford, Mass
2. MURDOCK, D.I. and HATCHER, W.S. (1978) *Journal of Food Protection*, **41**, 254
3. CARR, J.G. and PASSMORE, S.M. (1979) Methods for identifying acetic acid bacteria. In *Identification Methods for Microbiologists*, 2nd edn. Edited by F.A. Skinner and D.W. Lovelock. pp. 33–45. London: Academic Press

Chapter 22

Sanitation control

An estimate of the numbers of bacteria in the environment may be necessary for a variety of reasons:

(1) to test the standard of hygiene and efficiency of cleaning procedures in hospital wards, kitchens, canteens, operating theatres, food factories, dairies, shops, restaurants, offices or schools;
(2) to assess the level of bacteria in 'sterile' environments, e.g. in the pharmaceutical industry;
(3) to trace the route of contamination from dirty to clean situations;
(4) to educate staff.

There are four main methods of sampling:

(1) contact plates or slides using appropriate media,
(2) swabbing,
(3) rinsing,
(4) air sampling.

The choice of method will depend upon the situation. It is fundamental to the success of any investigation that the microbiologist is fully conversant with the problem and is prepared to go and look (and smell) at the site and talk to the personnel concerned. For background reading *see* Howie[1].

Control procedures

Agar contact for flat or nearly flat surfaces

Agar-filled contact plates are made by Medical Wire, Partome, Germ Control and BBL. Agar-covered slides are marketed by BBL, Orion Diagnostica, Biotest, Oxoid and Tillomed. The manufacturer's instructions should be followed. It is possible to estimate total numbers, enterobacteria, staphylococci, yeasts and fungi using an agar contact method. Apart from being very easy to use they have the additional merit of encouraging cleaning staff. Before showing agar 'contacts' to staff the cultures should be sprayed with domestic hair spray (vinyl acetate in methylated spirit)[2] to minimize the risk of infection. Cultures should always be returned to the laboratory for autoclaving or incineration.

Swab counts

It is generally accepted that the swabbing technique gives a count approximately ten times higher than that obtained by agar surface contact when sampling smooth surfaces.

Cut card or cellophane squares with 10-cm sides and cut squares in them with 5-cm sides. Sterilize these templates in envelopes. Place a template on the surface to be examined and swab the area within the 5 × 5 cm square with a cotton wool or alginate swab. Treat these swabs as described below. The count/25 cm^2 is given by the number of colonies/ml of rinse or solvent multiplied by 10. With Miles and Misra counts, it is given by the number of colonies/five drops multiplied by 100.

The swab rinse method for crockery cartons and containers

Dip a swab in sterile 0.1% peptone water and rub over the surface to be tested, for example the whole of the inside of the cup or glass or the whole surface of a plate. Use one swab for five such articles. Use one swab for a predetermined area of a cutting table, chopping board, etc. Swab both sides of knives, ladles, etc. Return this swab to the tube and swab the same surfaces again with another, dry swab.

To the tube containing both swabs add 10 ml of 0.1% peptone water. Shake and stand for 20–30 min. Do plate counts with 1.0 and 0.1 ml amounts using yeast extract agar. Divide the count/ml by 5 so as to obtain the count per article. Inoculate three tubes of single-strength MacConkey broth with 0.1-ml amounts.

There are no standards for crockery and utensils in the UK but in the USA the Public Health Service[3, 4] requires counts on crockery and utensils used by the public in restaurants, bars, etc., to be less than 100 per article. Coliform organisms should be absent.

Alginate swabs and drop counts

This technique was originally developed by Higgins[5].

Buy or make the swabs of alginate wool and sterilize by autoclaving. Proceed as above but add 9 ml of Calgon Ringer's solution (*O*). Shake gently; the swabs will dissolve in 1 or 2 min. Do Miles and Misra counts (p. 133) with six drops from each pair of swabs on one blood agar and one MacConkey plate. Two colonies or less per six drops on the blood agar plate is within the US Public Health Service limits. This method also allows the organisms to be identified. Apart from coliform bacilli, the presence of respiratory organisms such as viridans and salivary streptococci, staphylococci and neisseria is evidence of inefficient sanitation.

Alginate swabs are considered by some workers to give less efficient recovery of sub-lethally damaged organisms.

Rinse method

For churns, bins and large utensils

Add 500 ml of 0.1% peptone water to the vessel. Rotate the vessel to wash the whole of the inner surface and then tip the rinse into a screw-capped jar.

Do counts with 1.0- and 0.1-ml amounts of the rinse in duplicate. Incubate one pair at 37 °C and the other at 22 °C for 48 h.

Take the mean of the 37 °C and 22 °C counts/ml and multiply by 500 to give the count per container.

For milk churns, counts of not more than 50 000 cfu per container are regarded as satisfactory, between 50 000 and 250 000 as fairly satisfactory and over 250 000 as unsatisfactory.

For milk, soft drink bottles and jars

If bottles are sampled after they have been through a washing plant where a row of bottles travels abreast through the machine, test all those in one such row. This shows if one set of jets or carriers is out of alignment. In any event, examine not less than six bottles. Cap or stopper them immediately. To each bottle add 20 ml of 0.1% peptone water. Close with sterile rubber bungs and roll the bottles on their sides so that all of the internal surface is rinsed. Leave them on their sides and roll at intervals for half an hour.

Pipette 5 ml from each bottle into each of two petri dishes. Add 15–20 ml of yeast extract agar, mix and incubate one plate from each bottle at 37 °C and one at 22 °C for 48 h. Pipette 5 ml from each bottle into 10 ml of double-strength MacConkey broth and incubate at 37 °C for 48 h.

Take the mean of the 37 °C and 22 °C plate count and multiply by 4 to give the count per bottle. Find the average of the counts, omitting any figure that is 25 times greater than the others (indicating a possible fault in that particular line).

The figure obtained is the Average Colony Count per container. In the Ministry of Agriculture and Fisheries (1947) classification, milk bottles giving average colony counts of 200 cfu or less are satisfactory; counts from 200 to 600 cfu are regarded as fairly satisfactory, but over 600 as unsatisfactory.

Coliform bacilli should not be present in 5 ml of the rinse.

Membrane filter method

Examine clean bottles and jars by passing the rinse through a membrane filter. Place the membrane on a pad saturated with an appropriate medium, e.g. double-strength tryptone soya broth for total count. Incubate at 35 °C for 18–20 h. Stain the membrane with methylene blue so as to assist with counting the colonies under a low-power lens. Examine another membrane for coliforms using MacConkey membrane broth or membrane enriched lauryl sulphate broth. Incubate at 35 °C for 18–24 h. Staining to reveal colonies is unnecessary. Subculture suspect colonies into lactose peptone water and incubate at 37 °C for 48 h to confirm gas production.

Roll-tube method for bottles

Use nutrient or similar agar for most bacteria, MRS medium for lactobacilli, RCM agar for clostridia and buffered yeast agar for yeasts. Increase the agar concentration by 0.5%. Use the roll-tube MacConkey agar for coliforms.

Melt the medium and cool it to 55 °C. To 1-quart or 1-litre bottles add 100 ml of medium, and to smaller bottles proportionally less. Stopper the bottles with sterile rubber bungs and roll them under a cold water tap to form a film of agar over all the inner surface (*see* Roll-tube counts, p. 132). Incubate vertically and count the colonies. For lactobacilli and clostridia, replace the bung with a cotton wool plug and incubate anaerobically.

Vats, hoppers and pipework

Large pieces of equipment are usually cleaned in place (CIP). Test flat areas by agar contact or swabbing. Take swab samples of dead ends of pipework and crevices. The dairy industry has devised a simple and effective test for testing filling equipment. Sample the first, 100th and 200th container. The first will contain any residual bacteria not killed by CIP treatment. If this has not been effective, sample 1 will give a higher count than samples 100 and 200. If all three samples are satisfactory then the cleaning was efficient[6].

Examination of sink waters, cloths, towels, etc.

To demonstrate unhygienic conditions and the necessity for frequent changes of washing water and cloths, examine by agar contact or sample and examine as follows.

Take 100 ml of washing or rinse water by immersing a water sample bottle (containing sodium thiosulphate in case hypochlorites are used in washing-up) into the sink and allow it to fill. Stopper and cool under a tap. Do plate and coliform counts. Test at the beginning and at intervals during the washing-up.

Spread a wash-cloth or drying cloth over the top of a screw-capped jar of known diameter. Pipette 10 ml of 0.1% peptone water on the cloth so that the area over the jar is rinsed into it. Do plate counts and compare with freshly laundered cloths. If a destructive technique is possible, cut portions of cloths, sponges or brushes with sterile scissors and add to diluent.

Mincers, grinders, etc.

After cleaning, rinse with 500 ml of 0.1% peptone water. Treat removable parts separately by rinsing them in a plastic bag in diluent. Do colony count on rinsings as for milk bottles.

Counts on working and food surfaces, floors, walls, etc.

These are used to assess hygienic conditions.

Chopping blocks

Organisms are sometimes deeply embedded in these blocks. Sample by taking scrapings from representative areas, e.g. 100 cm^2, with a sterile scalpel. Disperse the scrapings in warm diluent and do counts.

Air sampling

Settle plates

Several plates containing appropriate media are exposed for a given time and incubated. The colonies are counted. Settle plates are favoured by those who need to monitor the air over long periods, e.g. in hospital cross-infection work. Expose blood agar to test for the presence of 'presumptive' *S. aureus*. This method is less satisfactory than using slit samplers for testing for very small suspended particles.

Equipment for air sampling has been described by various workers. Bourdillon[7] used a slit sampler beneath which an agar plate was exposed. Andersen[8] and Andersen and Andersen[9] suggested an agar drum sampler that monitored continuously the number of viable bacteria per unit volume of air at the time and point of sampling. This device has an impaction line of 484 in. Tryptone agar is used to coat a revolving drum that is capable of sampling 3 litres of air/min for a period of 24 h. This equipment is useful in monitoring airborne bacteria in hospitals and public places and in food, pharmaceutical and industrial plants. It is favoured for its reliability and is the equipment of choice for many tasks which do not require extreme mobility.

The Casella (UK) and Reynier (USA) models are both slit samplers. The Fisons model (UK) draws air through a membrane filter. Kitchell, Ingram and Hudson[10] tested all three and found the Casella to be the most flexible and efficient, the Reynier smaller and more readily transported than the Casella, while the Fisons model was the easiest to use in field studies.

Recently, hand-held air samplers have appeared on the market. A surface air system (SAS)[11] sold by Cherwell. This model uses contact plates and is portable. Folex-Biotest also market a hand-held model weighing 2.5 lb which looks like an electric torch. This is called a Reuter Centrifugal Sampler (RCS). Air is subjected to centrifugal acceleration and particles are impacted on to an agar coated strip which fits round a drum. Nakla and Cummings[12] have compared the performance of the RCS with the more conventional slit air sampler in hospital use. Clark *et al.*[11] have also studied the RCS system. Millipore[13, 14] publish booklets describing their Sterifil system. Air is streamed into a special broth which is passed through a membrane filter. The filter is then transferred to a pad impregnated with suitable medium and incubated appropriately.

References

1. HOWIE, J.W. (1979) *Journal of Applied Bacteriology*, **47**, 233
2. LINE, S.J. (1980) *The Medical Technologist*, October, p. 5
3. Report on Ordinance and Code Regulating Eating and Drinking Establishments (1943) *Public Health Bulletin No. 280*
4. Report on Progress Report on Food Utensil Sanitation (1944) *American Journal of Public Health*, **34**, 255
5. HIGGINS, M. (1950) *Monthly Bulletin Ministry of Health and Public Health Laboratory Service*, **9**, 50
6. DAVIS, J.G. (1979) *Culture*, September, p. 1. Basingstoke: Oxoid Ltd
7. BOURDILLON, R. B., LIDWELL, O.M. and THOMAS, J.C. (1941) *Journal of Hygiene, Cambridge*, **41**, 197
8. ANDERSEN, A.A. (1958) *Journal of Bacteriology*, **76**, 471

9. ANDERSEN, A.A. and ANDERSEN, M.R. (1962) *Applied Microbiology*, **10,** 181
10. KITCHELL, A.G., INGRAM, C.G. and HUDSON, W.R. (1973) Microbiological Sampling in Abattoirs. In *Sampling—Microbiological Monitoring of Environments*. Edited by R.G. Board and D.W. Lovelock, pp. 43–59. London: Academic Press
11. CLARK, S., LACH, V. and LIDWELL, O.M. (1981) *Journal of Hospital Infection*, **2,** 181
12. NIKHLA, L.S. and CUMMINGS, R.F. (1981) *Journal of Hospital Infection*, **2,** 261
13. MILLIPORE CORPORATION (1972) *Detecting Micro-organisms in Air*, Application Procedure, AP 309. Bedford, Mass
14. MILLIPORE CORPORATION (1973) *Sterility Testing with the Membrane Filter*, Application Manual, AM 201. Bedford, Mass

Chapter 23

Pharmaceuticals and cosmetics

There are no universally approved standards for pharmaceutical products. Acceptable levels of contamination vary between products to be swallowed, used on broken skin, eyes or on neonates and those to be used as lotions or ointments on intact skin surfaces. In guidelines published by the Council of the Society of Cosmetic Scientists of Great Britain in 1970[1] it was recommended that relevant known pathogenic micro-organisms be absent from preparations intended for the former categories. The genera mentioned were *Clostridia*, *Salmonella*, *Shigella*, *Pseudomonas*, *Escherichia*, *Klebsiella*, *Proteus*, *Staphylococcus* (*S. aureus*) and *Streptococcus*. In 1973 the Cosmetic Toiletry and Fragrance Association of America defined a maximum number of organisms permitted as 1000 cfu/g for cosmetics and 500 cfu/g for products for ophthalmic or infant use[2].

The Food and Drugs Administration (FDA) Manual of the USA states[3] that 'injury to the user is the basis of the cosmetic provisions of the Food Drug and Cosmetic Act as amended in 1976'.

In 1977 Hugo and Russell[4] pointed out that the complexity of many products intended for pharmaceutical or cosmetic use invited contamination by and multiplication of a variety of organisms. They emphasized that control is largely a matter of identifying the hazards and employing a good manufacturing process (GMP). Their book, *Pharmaceutical Microbiology*[4], is a useful reference work on the subject.

Methods of examination

Visual

Examine the pack and note any defects.

Opening

Disinfect the outside with 1% HCl in 80% ethanol and wipe dry with a sterile swab.

Neutralization and culture of antibacterial agents

Add 1 g or 1 ml to 20 ml of nutrient broth containing the following:

Halogens	1% Sodium thiosulphate
Aldehydes	2% Sodium sulphite
Hexachlorophenes and QACs	3% Tween 80

For phenols and alcohols add 1 g or 1 ml to 100 ml of nutrient broth. Homogenize in a blender or Stomacher and incubate aerobically for 24 h at 37 °C. Subculture on blood agar, mannitol salt agar and 0.03% cetrimide agar. Incubate for 24–48 h at 37 °C[5]. The following methods have been adapted from the FDA Bacteriological Analytical Manual[3].

Pretreatment of oily substances and powders

Add 1 g or 1 ml to 1 ml of sterile Tween 80, disperse and make up the volume to 10 ml with modified Letheen broth, *see* Chapter 4.

Screening test for aerobes

(1) Add 1 ml or 1 g to 90 ml Letheen broth.
(2) Add 1 ml or 1 g to 90 ml Sabouraud dextrose broth.

Treat oils and powders as above before adding to broth. Incubate at 30 °C for seven days. Subculture 0.5 ml of solution (1) to modified Letheen agar (*see* Chapter 4) and 0.5 ml of solution (2) to malt extract agar containing 40 ppm chlortetracycline after two and seven days or when there is growth. Spread the inoculum with a glass spreader and allow the fluid to soak into the agar before incubating the plates at 30 °C for four days.

Screening test for anaerobes

All powders should be tested. Prepare the sample as above and inoculate 90 ml thioglycollate broth 135c (*see* Chapter 4). Do not shake the bottle. Incubate at 35 °C for up to seven days. Plate 0.1 ml of broth on pre-reduced anaerobic agar (*see* Chapter 5) and incubate immediately under strict anaerobic conditions at 35 °C for four days. Inoculate 0.1 ml of broth on Letheen agar and incubate aerobically at 35 °C for four days.

Counts

Aerobic

Make dilutions in Letheen broth. Use spread plate method on Letheen agar in duplicate and incubate at 30 °C for 48 h.

Anaerobic

Do as for aerobic count but plate 0.1 ml on anaerobic agar and 0.1 ml on 5% sheep blood agar. Incubate the anaerobic plate in a Gas Pak or similar jar and the blood agar in 5–10% CO_2 both at 35 °C for 48 h.

S. aureus

Spread 0.5 ml of dilution broths on Baird-Parker agar. Proceed as for total counts but incubate at 35 °C.

Yeasts and moulds

Count as for *S. aureus* (above) but use Sabouraud's medium. Incubate at 30 °C for seven days.

Enrichment cultures

Incubate all dilution bottles at 30 °C for seven days then subculture on Letheen and MacConkey agar.

Identification

Prepare Gram-stained films of any significant growth and proceed as described in the appropriate chapter. If Gram-positive rods are seen, test for spores using Ellner's medium and a spore staining technique.

Intravenous fluids

Intravenous infusion fluids are occasionally contaminated[6]. Pass the fluid remaining in the container after the drip has been disconnected, or the contents of suspect or unused containers, through a membrane filter (p. 134). Cut the membrane in two; place one-half on a nutrient medium for bacteria and the other on a mycological medium. Incubate and examine the filters for colonies.

If there is very little infusion left in the container, add 100 ml of nutrient broth aseptically to it (use a laminar flow cabinet if possible). Incubate overnight. If the broth is turbid, subculture on blood agar. If not pass it through a membrane filter as above.

Incubate the membrane filter at room temperature and if there is no growth after 24 h prolong the incubation. Psychotrophs are not uncommon contaminants. For information on this problem *see* Phillips *et al.*[6] and Denyer[7]. For details about the efficiency of preservatives in pharmaceutical products see the British[8] and United States[9] Pharmacopoeias.

References

1. THE SOCIETY OF COSMETIC CHEMISTS (1970) *Journal of the Society of Cosmetic Chemists*, **21**, 719
2. PARKER, M.S. (1982) *Culture* (Oxoid) 3rd September, **3**, 3
3. *Bacteriological Analytical Manual* (1978) 5th edn. Washington DC: Association of Official Analytical Chemists
4. HUGO, W.B. and RUSSELL, A.D. (1977) *Pharmaceutical Microbiology*, Oxford: Blackwell Scientific Publications
5. BAIRD, R.M., BROWN, W.R.L. and SHOOTER, R.A. (1976) *British Medical Journal*, **1**, 511
6. PHILLIPS, I., MEERS, P.D. and D'ARCY, P.F. (1976) *Microbiological Hazards of Infusion Therapy*. Lancaster: MTP Press

7. DENYER, S.P. (1982) In-use contamination in intravenous therapy—the scale of the problem. In *Infusions and Infections? The Hazards of In-Use Contamination in Intravenous Therapy*. Edited by P.F D'Arcy, pp. 1–16. Oxford, UK: The Medicine Publishing Foundation
8. *British Pharmacopoeia* (1980) London: HMSO
9. *United States Pharmacopoeia* (1980) 20th edn., Bethesda MD: Government Printing Office

Chapter 24

Animal by-products, grass and silage

Animal by-products

It is convenient to consider animal feeds and fertilizers together, as the bacteriological problems are similar. The microbiological quality of these products is very important because it is directly or indirectly related to the level of contamination in human and pet-foods (*Figure 24.1*). In recent years, efforts have been made to improve methods of producing salmonella-free materials but even if sterilization of the end-product is thorough and efficient there is always the possibility that re-contamination will occur. Some animals suffer only a mild salmonella infection and may pass unnoticed through the slaughterhouse to the food factory.

The product may be relatively sterile (only 0.3% of re-sterilized imported meat and bone meal from Denmark contained salmonellas[1]) or may be highly contaminated with a variety of organisms, some of which may be salmonellas or *B. anthracis* (*see* p. 356). If salmonellas are present, they may be there only in small numbers and may be heat shocked. Williams-Smith *et al.*[2] reported that more than 50% of untreated and 25% of treated garden fertilizer

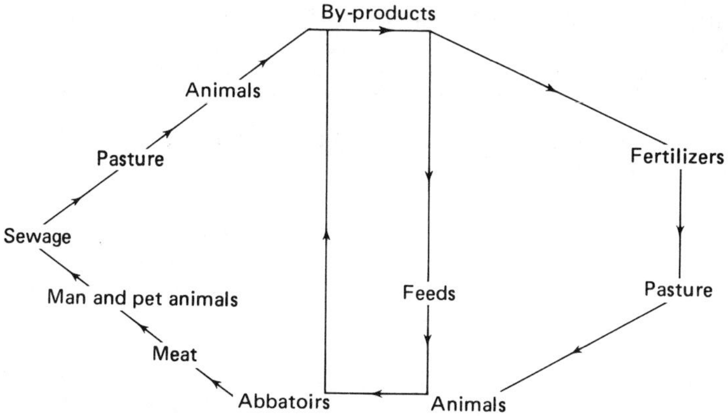

Figure 24.1 Animal by-products. Recirculation of an infection man and animals

contained salmonellas[2]. There is evidence that fish meal also contributes to this problem[3].

Tompkin and Kueper[4] suggest that a linear relationship exists between the detection of salmonellas and the total plate count. They say that salmonellas are most likely to be found in samples that have a plate count between 10^3 and 10^7.

Counts

When the microbiological quality of the sample is unknown, do a preliminary plate count using dilutions of 10^2, 10^4 and 10^6 and incubate at 37 °C for 48 h. Then use narrower limits to find a more exact range.

Examination for salmonellas

If facilities permit, pre-enrich four 25-g aliquots in 75-ml volumes of nutrient broth and incubate at 37 °C for 4 h. Add to each sample 75 ml double-strength selenite F broth and incubate at 43 °C for 24 h. Subculture to deoxycholate citrate, Wilson and Blair and brilliant green MacConkey medium, because lactose-fermenting salmonellas belonging to subgenus III may be present[5].

Anthrax bacilli may also be present (*see* Chapter 34).

Grass and silage

Botulism in cows has resulted in contaminated grass. It has been shown that silage made from this material has supported the growth of *C. botulinus* and contained toxin[6].

References

1. PUBLIC HEALTH LABORATORY SERVICE WORKING GROUP and SKOVGAARD, N. and NEILSEN, B.B. (1972) *Journal of Hygiene, Cambridge*, **70,** 127
2. WILLIAMS-SMITH, H., TUCKER, J.F., HALL, M.L.M. and Rowe, B. (1982) *Journal of Hygiene, Cambridge*, **89,** 25
3. LEE, J.A., GHOSH, A.C., MANN, P.G. and TEE, G.H. (1972) *Journal of Hygiene, Cambridge*, **70,** 141
4. TOMPKIN, R.B. and KUEPER, T.V. (1973) *Applied Microbiology*, **25,** 485
5. HARVEY, R.W.S., PRICE, T.H. and HALL, L.M. (1973) *Journal of Hygiene, Cambridge*, **71,** 481
6. NOTEMANS, S., KOZAKI, S. and VAN SCHOTHORST, M. (1979) *Applied and Experimental Microbiology*, **38,** 767

Chapter 25

Water

Stored and river water may contain a wide variety of organisms, including pseudomonads, achromobacteria, flavobacteria, micrococci, aerobic spore-bearers, enterobacteria and streptomycetes. Piped water may, in addition, contain iron bacteria and river waters both iron bacteria and sulphate reducers.

From the public health point of view, the coliform test is the most important as the presence of these organisms, particularly of *E. coli*, indicates if not actual pollution then a less than satisfactory supply. Tests for enterococci and *C. perfringens* are also useful.

As water is not the natural habitat of the enterobacteria and these organisms do not multiply in reasonably clean water, care must be taken in selecting media for their isolation. The organisms may be still viable but damaged and therefore media should not contain inhibitory substances.

In the UK, water is examined for coliforms by the multiple tube (MPN) method or the membrane filter technique. Minerals Modified Glutamate is the medium of choice for the MPN method and lauryl sulphate has now replaced Teepol in membrane culture because the latter is difficult to obtain.

Officially approved methods for examining water and for interpreting the results are published in the UK by DHSS[1] and in the US American Public Health Association[2].

Sampling

Samples are collected by health inspectors and water engineers. The laboratory should supply 120-ml glass bottles with glass, dust-proof stoppers, covered with kraft paper tied at the neck and sterilized in a hot air oven or autoclave. For samples of chlorinated waters, the bottles must contain sodium thiosulphate (0.1 ml of a 3% solution) to neutralize residual chlorine. These should be freshly prepared every three days.

Samples must be delivered to the laboratory within 6 h of sampling to ensure a meaningful report.

Plate counts

These are not usually done as a routine in the UK, except on new supplies of raw water. In the USA, plate counts are standard procedure[2].

Appropriate dilutions are prepared in phosphate buffer, mixed with Standard Methods agar and incubated at either 20 °C for 48 h or 35 °C for 24 h. Plates showing between 30 and 300 colonies are counted (*see* Chapter 9).

Coliform test: MPN method with Minerals Modified Glutamate broth

Select the range according to the expected purity of the water:

Mains chlorinated water	*A* and *B*
Piped water, not chlorinated	*A*, *B* and *C*
Deep well or borehole	*A*, *B* and *C*
Shallow well	*B*, *C* and *D*
No information	*A*, *B*, *C* and *D*

A: 50 ml of water to 50 ml of double-strength broth.
B: 10 ml of water to each of five tubes of 10 ml of double-strength broth.
C:1 ml of water to each of five tubes of 5 ml of single-strength broth.
D: 0.1 ml of water to each of five tubes of 5 ml of single-strength broth.

Incubate at 35–37 °C and note the numbers of tubes showing acid and gas at 48 h. Tap tubes showing no gas. A bubble may then form in the Durham's tube. Consult the MPN tables (*Tables 9.1–9.3*) and read the most probable number of *presumptive* coliform bacilli/100 ml of water. Small amounts of gas occurring after 48 h in *presumptive* tubes are disregarded unless the presence of coliform bacilli is confirmed by plating.

From each tube showing acid and gas, inoculate a tube of 1% ricinoleate broth or a tube of brilliant green bile salt broth and a tube of peptone water. Incubate these at 44 °C for 24 h in a reliable water-bath (Eijkman test) along with control of known strains of *E. coli* (which grows at 44 °C) and *K. aerogenes* (which does not). Plate also from positive tubes on MacConkey agar.

Observe gas formation at 44 °C and test the peptone-water culture for indole. Only *E. coli* produces gas *and* indole at 44 °C.

Read the most probable numbers of *E. coli* ('faecal coli') from *Tables 9.1–9.3*.

Translation from the MPN tables of the 44 °C positive tubes sometimes causes difficulty. Remember that the organisms cultured from any positive 37 °C tube and grown at 44 °C represent coliforms cultured from the volume of water placed in the 37 °C tube. For example:

	50 ml	*10 ml*	*1 ml*	*MPN/100 ml*
Tubes positive at 37 °C	1	2	2	10 'presumptive coli'
Tubes positive at 44 °C	1	1	0	3 *E. coli*

For further investigation, pick colonies from the MacConkey plate into peptone water, glucose phosphate medium and citrate medium for indole, MR, VP and citrate utilization tests (Chapter 7).

Acid and gas in formate lactose glutamate medium, as in MacConkey broth, may occasionally be due to spore bearers, e.g. *C. perfringens* at both 37 and 44 °C. These organisms do not grow in brilliant green or ricinoleate broths or on the MacConkey plate.

Most raw waters in the UK showing acid and gas do in fact contain coliform bacilli but in about 5% of chlorinated waters acid and gas is due to *C. perfringens*.

Coliform test: US method[2]

In the US Standard Method, 15 fermentation tubes, each containing 20 ml of 0.5% lactose broth are used. They are inoculated with 5 × 10 ml, 5 × 1 ml and 5 × 0.1 ml of water sample. These are incubated at 35 °C and examined for gas production after 24 and 48 h. Gas within 48 h is presumptive evidence of coliform bacilli.

Confirmatory test

Tubes showing gas are subcultured on eosin methylene blue (EMB) agar, incubated at 35 °C for 24 h and examined for typical colonies of *E. coli*. If atypical colonies are seen, the 'completed test' is carried out.

Completed test

Several colonies from the EMB plate are subcultured into lactose broth fermentation tubes and on a nutrient agar slope. Both are incubated at 35 °C for 24 h. Gas in the broth and a Gram-negative non-sporing rod on the slope is evidence of coliform bacilli.

Faecal coli test

Tubes showing gas are subcultured into EC or similar medium and incubated at 44.5 ± 0.2 °C for 48 h to test for gas production, which indicates faecal coli. Colonies from solid medium are subcultured to lactose broth and if gas is produced in these they are further subcultured into EC broth as above. For full technical methods and directions, *see* Standard Methods[2].

Coliform test: membrane filter method

Advantages of using membrane filter techniques for waters

(1) Speed of obtaining results.
(2) Saving of labour, media, glass and cost of materials if the filter is washed and re-used.
(3) Sample can be filtered on site, if the filter is placed on transport medium and posted to the laboratory, thus avoiding delay in transporting the sample.
(4) Organisms can very easily be exposed to pre-enrichment media for a short time at an advantageous temperature.

Disadvantages of using membrane filter techniques for waters

(1) There is no indication of gas production (some waters contain large numbers of non-gas-producing lactose fermenters capable of growth in the medium).
(2) Membrane filtration is unsuitable for waters with high turbidity and low count because the filter will become blocked before it can pass sufficient water.
(3) Large numbers of non-coliform organisms capable of growing on the medium may interfere with coliform growth.

If large numbers of water samples are to be examined and much field work is involved the membrane method is undoubtedly the most convenient. The booklets published by Millipore[3] and Gelman[4] give valuable advice and information about all aspects of the application of this technique.

Pass two separate 100-ml volumes of the water through 47-mm membrane filters. If the supply is known or is expected to contain more than 100 coliform bacilli/100 ml, use 10 ml of water diluted with 90 ml of quarter-strength Ringer's solution.

Place sterile Whatman No. 17 absorbent pads in sterile petri dishes and pipette 2.5–3 ml of enriched lauryl sulphate broth over the surface. Place a membrane face up on each pad and incubate as follows.

Chlorinated samples

One membrane at 25 °C for 6 h followed by 35 °C for 18 h for the *presumptive* count and one membrane at 25 °C for 6 h followed by 44 ± 0.2 °C for 18 h for the *E. coli* count.

Unchlorinated samples

One membrane at 30 °C for 4 h followed by 35 °C for 14 h for the *presumptive* count and one membrane at 30 °C for 4 h followed by 44 ± 0.2 °C for 14 h for the *E. coli* count.

These times allow for 'resuscitation' of coliforms.

For incubation in water-baths at 44 °C, waterproof submersible boxes are required. Several laboratory suppliers sell these. Methods of automatic changes in temperature in incubators and water-baths are mentioned on p. 12.

Counting

Count the yellow colonies only and report as *presumptive* coliform and *E. coli* count/100 ml of water. Membrane counts may be higher than MPN counts because they include all organisms producing acid, not only those producing acid and gas. *C. perfringens* does not grow.

Confirmatory test

Subculture colonies to lauryl sulphate tryptose broth for gas production and peptone water for indole test. Incubate both at 44 ± 0.2 °C overnight.

Smith and Rockcliff[5] describe a single tube confirmatory test which can be read in 4 h. They claim that it gives 99% agreement with traditional methods.

Membrane filter: US method

The methods are very similar to those above, but the filter is rinsed with three volumes of 20–30 ml of sterile buffered water before the sample is passed. Membrane agar medium (EMB) or membrane fluid (EMB) media are used. Pre-enrichment, where necessary, is effected on pads saturated with lactose broth at 35 °C for 2 h.

Full details are given in the Standard Methods[2].

Faecal streptococci in water

These organisms are useful indicators when doubtful results are obtained in the coliform test. They are more resistant than *E. coli* to chlorine and are therefore useful when testing repaired mains. Group D organisms only are significant.

MPN method

Use one of the azide broths, e.g. azide glucose broth, Enterococcus Presumptive Broth or Slanetz and Bartley Broth.

Add 50 ml of water to 50 ml of double-strength medium.

Add 10 ml of water to each of five tubes of 10 ml of double-strength broth.

Add 1 ml of water to each of five tubes of 5 ml of single-strength broth.

Incubate at 37 °C for 72 h. Subculture any tubes showing acid production to tubes of single-strength medium and incubate at 44–45 °C for 18 h. Record tubes showing acid and consult the MPN tables (*Tables 9.1–9.3*). Confirm by microscopic examination for short-chain streptococci. Subculture each presumptive positive tube to ethyl violet azide broth and incubate at 37 °C for 24–48 h. Turbidity and a purple-stained button of growth at the bottom of the tube indicate enterococci.

Membrane method

Always use a new membrane when testing for faecal streptococci because these organisms are not always removed when membranes are cleaned. Pass 100 ml of water through a membrane filter and place the filter on a pad of one of the enterococcus membrane broths or a plate of membrane enterococcus agar. Incubate at 37 °C for 4 h and then at 44–45 °C for 44 h. All red or maroon colonies are presumptive positives. Carefully remove the filter and place face downwards on a plate of Mead's medium to imprint the colonies.

Remove the membrane and incubate the plate at 37 °C for 18 h. *S. faecalis sensu strictu* gives maroon colonies surrounded by a clear zone.

US methods for faecal streptococci

These methods[2] are very similar to the above. In the membrane method, suspected colonies are subcultured for catalase tests and into broth incubated at 45 °C for confirmation.

C. perfringens in water

MPN method

The litmus milk method is not now regarded as satisfactory for isolating this organism. The 'black tube' method is better. This is done in reinforced clostridial medium (RCM) containing 70 μg/ml of polymyxin to inhibit facultative anaerobes.

Add 50 ml of water to 50 ml of double-strength medium.

Add 10 ml of water to each of five 10-ml amounts of double-strength medium.

Add 1 ml of water to each of five 5-ml amounts of single-strength medium.

Fill the bottles almost to the neck with single-strength medium to exclude most of the air. Replace the caps and incubate at 37 °C for 48 h. Tubes showing blackening are presumptive positives, but other clostridia also give this reaction. Confirm *C. perfringens* by subculturing into purple or litmus milk. Incubate at 37 °C for 24 h and record as positive tubes that show stormy fermentation. Consult the MPN tables (*Tables 9.1–9.3*).

Alternatively, plate out and identify as on p. 106.

Membrane method

Pass 100 ml of water, or an amount suggested by experience, through a membrane filter and place the membrane face down on a plate of iron sulphite agar or Wilson and Blair medium. Pour 20 ml of the same medium, cooled to 50 °C, on top. Incubate at 44 °C for 48 h and count the black colonies with haloes. These are probably *C. perfringens*. If too many clostridia are present, the whole medium will be blackened.

Other clostridia

MPN method

Heat the black tubes (above) at 75 °C for 20 min to kill vegetative organisms (including *C. perfringens*, which does not form spores in the medium) and subculture into other tubes of RCM medium. This will give the MPN count of other clostridia, which can then be identified if necesssary.

Microfungi and streptomycetes

These organisms cause odours and taints and often grow in scarcely used water pipes, particularly in warm situations, e.g. basements of large buildings where the drinking water pipes run near to the heating pipes.

Sample the water when it has stood in the pipes for several days. Regular running or occasional long running may clear the growth.

Centrifuge 50 ml of the sample and plate on rose bengal agar and malt agar medium. Add 100 μg/ml of kanamycin to the malt medium to suppress most eubacteria.

Alternatively, pass 100 ml or more of water through a membrane filter and apply this to a pad soaked in liquid media of similar composition.

Pathogens

Pass a large volume of the sample through membrane filters and add the filters to bottles of the appropriate enrichment media.

For salmonellas use selenite broth, incubate at 43 °C (except for *S. typhi*, when 37 °C may be better) and subculture every 12 h for four days on bismuth sulphite and DC agars.

For *V. cholerae* and other vibrios, use alkaline peptone water and subculture at 4, 8 and 12 h on TCBS medim.

For *Clostridium perfringens*, place the membrane in the bottom of a petri dish and pour over it reinforced clostridial medium, melted and cooled to 50 °C, to a depth of at least 5 mm.

Swimming pools

Test the chlorine content with the portable Lovibond device at the pool side. This is often more useful than bacteriological examination, but if this is required sample from below the surface at both ends, using thiosulphate to destroy chlorine (*see above*). Do plate counts on 1-ml amounts and presumptive coliform tests. The plate count should be less than ten in 80% of samples and coliforms should be absent.

Staphylococci are more resistant than coliforms to chlorination. They tend to accumulate on the surface of the water in the 'grease film'. To find staphylococci, take 'skin samples'. Open the sample bottle so that the surface water flows into it. Add 20-ml aliquots to each of five tubes containing 20 ml of Robertson's meat medium plus 10% of sodium chloride. Incubate overnight and plate on one of the staphylococcal media.

Interpretation

It is not possible to assess the potability of any water supply by a single examination.

Presumptive coliforms and *E. coli* should be absent from 100 ml. Enterococci and *C. perfringens*, in the absence of coliforms, suggest contamination at a remote time. These organisms persist longer than coliforms in water.

Interpretation requires consideration of geographical and engineering factors as well as a laboratory report. Guidance is given in the Report of the Department of Health and Social Security[1] and Standard Methods[2].

Bordner[6] has reviewed attempts to standardize microbiological methods for potable water and water used for various industrial processes.

Sea water

The sampling techniques and bacteriological examination of sea water is discussed by Collins *et al.*[7]. *See also* the booklets by Millipore[3] and Gelman[4].

References

1. DEPARTMENT OF HEALTH AND SOCIAL SECURITY (1983) *The Bacteriological Examination of Water Supplies*, Report No. 71. London: HMSO
2. *Standard Methods for the Examination of Water and Wastewater* (1975) 14th edn. Washington DC: American Public Health Association
3. MILLIPORE CORPORATION (1973) *Biological Analysis of Water and Wastewater*. Application Manual, AM 302, Total coliform analysis and Faecal coliform analysis, AB 311. Bedford, Mass
4. GELMAN INSTRUMENT COMPANY (1978) *Microbiological Analysis of Water*. Ann Arbor
5. SMITH, J.L. and ROCKCLIFF, S. (1982) *Journal of Hygiene, Cambridge*, **89,** 149
6. BORDNER, R.H. (1978) *Journal of Food Protection*, **41,** 314
7. COLLINS, V.G., JONES, J.G., HENDRIE, M.S., SHEWAN, J.M., WYNN-WILLIAMS, D.D. and RHODES, M.E. (1973) Sampling and estimation of bacterial populations in the aquatic environment. In *Sampling—Microbiological Monitoring of Environments*. Edited by R.G. Board and D.W. Lovelock, pp. 77–107. London: Academic Press

Chapter 26

Pseudomonas, acinetobacter, alkaligenes, achromobacter, flavobacterium, chromobacterium and acetobacter

Aerobic Gram-negative non-sporing bacilli that grow on nutrient and usually on MacConkey agars are frequently isolated from human and animal material, from food and from environmental samples. Some of these bacilli are confirmed pathogens; others, formerly regarded as non-pathogenic, are now known to be capable of causing human disease under certain circumstances, e.g. in 'hospital infections', after chemotherapy or treatment with immunosuppressive drugs; others are commensals; many are of economic importance.

With the exception of the enterobacteria and some of the vibrios, which have received detailed characterization because of their importance in human

TABLE 26.1. Key to some Gram-negative rods that grow on nutrient agar

	HL test	*Oxidase*	*Arginine hydrolysis*	*Gelatin liquefaction*	*Growth on MacConkey*	*Motility*
Achromobacter	Ox	+	−	+	−	−
Acinetobacter calcoaceticus	Ox	−	−	−	+	−
A. lwoffii	None	−	−	−	+	−
Pseudomonas mallei	Ox	v	+	v	−	−
P. pseudomallei	Ox	+	+	+	+	+
Pseudomonas spp.	Ox or none	+	v	v	+	+
Bordetella parapertussis	None	−	−	−	+	−
B. bronchiseptica	None	+	−	−	+	+
Aeromonas	F	+	+	+	+	+
Alcaligenes faecalis	None	+	−	+	+	+
A. odorans	None	+	−	+	+	+
Chromobacterium lividum	Ox	+	−	+	v	+
C. violaceum	F	+	+	+	v	+
Enterobacteria	F	−	v	v	+	v
Flavobacterium	Ox	+	−	+	v	−
Moraxella	None	+	−	v	v	−
Pasteurella	F	+	−	−	−	−
Yersinia	F	−	−	−	+	v[a]
Vibrio	F	+	−	+	+	+

Ox, oxidative; F, fermentative; v, variable;
[a] Some species motile at 22 °C

disease, the taxonomic positions and the specific names of some of these organisms are uncertain. We therefore recognize the inadequacy of the list of genera and groups in *Table 26.1*. In the text and in other tables we give cultural and biochemical properties that may permit the assignment of some of the organisms to species and others to a genus or group only.

Pseudomonas

The organisms in this genus are Gram-negative non-sporing rods about 3 μm × 0.5μm, which are motile by polar flagella, may produce a fluorescent pigment, are oxidase positive, utilize glucose oxidatively and do not produce gas.

They commonly occur in soil and water. Some species are recognized human and animal pathogens but some others, formerly regarded as saprophytes and commensals, have been incriminated as opportunist pathogens in hospital-acquired infections and have colonized distilled water supplies, soaps, disinfectants, intravenous infusions and other pharmaceuticals. Species colonizing swimming pools have been associated with ear infections[1,2].

Isolation and identification

Inoculate nutrient agar, MacConkey agar, nutrient agar containing 0.1% cetrimide or one of the pseudomonas media. Incubate at 20–25 °C (food, etc.) or 35–37 °C (animal material).

Colonies on nutrient agar are usually large (2–4 mm), flat spreading and pigmented. A greenish yellow or bluish yellow fluorescent pigment may diffuse into the medium. Occasionally, melanogenic strains are encountered. A brown pigment is formed around the colonies. Mucoid (encapsulated) strains are sometimes found in clinical material and may be confused with klebsiellas.

Do oxidase test, reaction in Hugh and Leifson medium, inoculate Thornley's arginine broth (Moeller's method is too anaerobic), MacConkey agar and test for gelatin liquefaction (five days' incubation). Inoculate nutrient broths and incubate at 4 and 42 °C. Use a subculture from a smooth suspension for this test.

Pseudomonads are motile (occasionally non-motile strains may be met), oxidative, non-reactive or produce alkali in Hugh and Leifson medium. The oxidase test is positive except possibly with *P. maltophilia* (*see below*). Thornley's arginine test is positive except for *P. maltophilia*, *P. cepacia*, *P. stutzeri* and *P. alcaligenes*, which are not common. Those pseudomonads which give a positive arginine test do so more rapidly than other Gram-negative rods and this test is very useful for recognizing non-pigmented strains. *Alcaligenes*, *Achromobacterium* and *Vibrio* spp. do not hydrolyse arginine. Pseudomonads vary in their ability to liquefy gelatin and grow at 4 and 42 °C (*see Table 26.2*). Penicillin disc sensitivity tests may be useful.

Pigment formation is best observed on a basal synthetic medium with 4% potassium gluconate, or on one of the commercial pseudomonas media.

Some fluorescent strains produce an opalescence (egg yolk reaction) on

TABLE 26.2. *Pseudomonas* spp. *Vibrio*, *Alkaligenes* and *Aeromonas*

	Fluorescent pigment	*HL test*	*Arginine hydrolysis*	*Gelatin liquefaction*	*Growth at 4 °C*	*Growth at 42 °C*
P. aeruginosa	+	Ox	+	+	–	+
P. fluorescens	+	Ox	+	+	+	–
P. putida	+	Ox	+	–	v	–
P. alkaligenes	–	Alk	–	–	–	+
P. pseudo-alkaligenes	–	Alk	+	–	–	v
P. maltophilia	–	Alk	–	+	–	–
P. cepacia	–	Ox	–	+	–	v
P. mallei	–	Ox	+	v	–	–
P. pseudomallei	–	Ox	+	+	–	+
P. stutzeri	–	Ox	v	–	v	v
Vibrio	–	F	–	+	v	–
Alkaligenes	–	Alk or none	–	–	–	–
Aeromonas	–	F	+	+	–	–

All of these organisms are oxidase positive. *Acinetobacter* spp. are oxidase negative.
Ox, oxidative; F, fermentative; Alk, alkaline reaction; v, variable

Willis and Hobbs' medium and also a lipase (pearly layer). To demonstrate the latter, flood the plate with saturated copper sulphate solution; the fatty acids released by lipolysis give a greenish blue precipitate of copper soaps.

Some varieties associated with food spoilage are psychrophiles and may be lipolytic. Brine-tolerant strains and phosphorescent strains are not uncommon.

Species of *Pseudomonas*

P. aeruginosa

Colonies on agar medium, 18–24 h at 25–30 °C, are large, flat spreading and irregular, greyish green in colour. The greenish pigment diffuses into the medium. Broth cultures are blue-green in colour. Two pigments are formed, pyocyanine and fluorescein; both are soluble in water but only pyocyanine is soluble in chloroform. Arginine is hydrolysed rapidly. Gelatin is liquefied. Growth takes place at 42 °C but not at 5 °C. The egg yolk reaction is negative. This organisms is very resistant to antibiotics except polymyxin, gentamicin and carbenicillin.

It is a saprophyte, found in soil and water, causes spoilage of foods, including 'blue milk', is pathogenic for man and animals ('blue pus'), often as a secondary infection, and is often found in clinical material[2,3].

It is an important agent in cross-infection in hospitals and strains may be typed by pyocin production. This is usually done at Reference Laboratories.

P. fluorescens

Colonies on agar and its biochemical properties are similar to those of *P. aeruginosa* but only one pigment (fluorescein) is produced. Growth occurs at 5 °C but not at 42 °C. The egg yolk reaction is positive. This organism is a saprophyte found in soil, water and sewage. It is a food spoilage organism. It may gel UHT milk if this is stored above 5 °C.

P. putida

Colonies resemble those of *P. aeruginosa* but only fluorescein is formed. It is a psychotroph that may not grow at 37 °C but grows at 22 °C. There are two biotypes; one grows at 4 °C, the other does not; both hydrolyse arginine. The egg yolk reaction is negative, gelatin is not liquefied and old cultures have a marked odour of trimethylamine (bad fish). It is an important fish pathogen and fish spoilage organism and has been isolated from human material.

P. maltophilia

Colonies resemble those of *P. aeruginosa* but a yellow or brown diffusible pigment may be produced. This species was reported as oxidase negative by Gilardi[4], who used Difco oxidase discs, but Snell[1] found oxidase positive strains when testing with tetramethyl-1-phenylenediamine dihydrochloride. The method of performing the tests seems to be important. This species fails to hydrolyse arginine and does not grow on cetrimide agar. It is the only pseudomonad that gives a positive lysine decarboxylase reaction. It has been isolated from a variety of sources and appears to be an opportunist pathogen[1,3].

P. cepacia

Colonies may produce a yellow, water-soluble, non-fluorescent pigment, occasionally may be purple, or may be non-pigmented. Growth at 42 °C has been reported. Widely distributed in nature, it occurs as a 'hospital infection'.

P. stutzeri

Colonies are rough, dry and wrinkled and can be removed entire from the medium. Older colonies turn brown. It may not hydrolyse arginine and may not grow at 4 or 42 °C or on cetrimide medium. It has been found in clinical material.

P. alcaligenes and P. pseudoalcaligenes

These organisms are reported to grow at 42 °C but not at 4 °C. They do not liquefy gelatin and may or may not grow on cetrimide agar. *P. alcaligenes* does not hydrolyse arginine.

P. mallei

In exudates this organism may appear granular or beaded and may show bipolar staining. It gives 1-mm, shining, smooth, convex, greenish yellow, buttery or slimy colonies which may be tenacious on nutrient agar, no haemolysis on blood agar and does not grow on MacConkey agar in primary culture. Growth may be poor on primary isolation. It grows at room temperature, does not change Hugh and Leifson medium, produces no acid in carbohydrates, is nitrate positive, gives a variable urease reaction and may or may not grow in KCN medium. The oxidase test gives variable results.

This is the causative organism of glanders (*see* caution *below*).

P. pseudomallei

Cultures on blood agar and nutrient agar at 37 °C give mucoid or corrugated, wrinkled, dry colonies in 1–2 days, and an orange pigment may develop. There is no growth on cetrimide agar. This organism is not easy to identify. It must be distinguished from non-pigmented strains of *P. aeruginosa*, *P. stutzeri* (*Table 26.2*) and from *Pseudomonas mallei* (*Table 26.3*).

TABLE 26.3. *Acinetobacter, Alkaligenes, Pseudomonas mallei, Pseudomallei* and *Bordetella*

	Growth on MacConkey	*Oxidase*	*HL test*	*Motility*	*Acid from Glucose*	*Nitrate reduction*	*Urease*	*KCN*
A. calcoaceticus	+	−	Ox	−	+	−	v	+
A. lwoffii	+	−	none	−	−	−	+	−
Alcaligenes spp.	+	+	none	+	−	v	−	+
P. mallei	−	v	none	−	(+)	+	v	v
P. pseudomallei	+	+	Ox	+	+	+	v	+
B. parapertussis	+	−	none	−	−	−	+	−
B. bronchiseptica	+	+	none	+	−	+	+	+

Ox, oxidative; (+) —acid in ammonium salt glucose but not in peptone water glucose; v, variable

This is an important pathogen of man (melioidosis) and farm animals in SE Asia, where it is endemic in rodents and is found in moist soil, on vegetables and on fruit. Cultures should be sent to a Reference Laboratory (*see* caution *below*).

Caution

Pseudomonas mallei and Pseudomonas pseudomallei are Risk Group III pathogens and should be handled only in Containment laboratories.

For further information about the pseudomonads *see* Hendrie and Shewan[5] and also the papers by Hugh and Gilardi[3]; Ballard *et al.*[6] and Stanier *et al.*[7].

Alteromonas

Alteromonas spp. are found in sea water, are difficult to distinguish from *Pseudomonas* except by specialized methods (*see* Hendrie and Shewan[5]).

Acinetobacter

Organisms have been moved in and out of this genus for several years. At present it seems to contain only two species.

Isolation and identification

Plate material on nutrient, blood and MacConkey agar (and Bordetella or similar media if indicated). Incubate at 22 °C and at 37 °C for 24–48 h. Do oxidase and catalase tests. Inoculate nutrient broth for motility, Hugh and Leifson medium, glucose peptone water, nitrate broth, urea medium, KCN medium.

Acinetobacter are non-motile, obligate aerobes usually oxidase negative, catalase positive and are oxidative or give no reaction in Hugh and Leifson medium (*see below* and *Table 26.3*).

Species of *Acinetobacter*

A. calcoaceticus (*A. anitratus*)

This gives large non-lactose fermenting colonies on MacConkey and DC agars and grows at room temperature. On Hugh and Leifson medium it is oxidative, producing acid but not gas from glucose. No other sugars are normally attacked. It is nitrate negative, gives a variable urease reaction and grows in KCN medium.

Intestinal and other infections in man have been reported and the organisms are found in the genitourinary tract. Sometimes in direct films this organism may resemble *Neisseria*.

A. lwoffii (*Moraxella lwoffii*)

Colonies on blood agar are small, haemolytic and may be sticky, but this organism also grows on nutrient and MacConkey agars. There is no change in Hugh and Leifson medium; it fails to attack carbohydrates, is nitrate negative, urease negative and does not grow in KCN medium. It can cause conjunctivitis.

For more information about infections with *Acinetobacter see* Vivian *et al.*[2].

Alcaligenes and *Achromobacter*

These organisms are widely distributed in soil, fresh and salt water and are economically important in food spoilage (especially of fish and meat). Many strains are psychrophilic. Some species have been isolated from human material and suspected of causing disease.

Isolation and identification

Plate on blood, skim milk and MacConkey agars. Incubate at 20–22 °C (foodstuffs) or at 35–37 °C (pathological material) for 24–48 h. Subculture white colonies of Gram-negative rods on agar slopes for oxidase test, in peptone water for motility, in Hugh and Leifson medium, dextrose peptone water, bromocresol purple milk, arginine broth and gelatin (*see Table 26.4*).

TABLE 26.4. ***Alcaligenes* and *Achromobacter***

	HL Test	*Acid from glucose*	*Bromocresol milk*
Alcaligenes	Alk or none	—	Alk
Achromobacter	Ox	A	Acid or no change

Alk = alkaline; Ox = oxidative; A = acid.

Species of *Alcaligenes*

The genus *Alcaligenes* is restricted here to motile oxidase positive, catalase positive Gram-negative rods which give an alkaline or no reaction in Hugh and Leifson medium and an alkaline reaction in bromocresol purple milk, grow on MacConkey agar and in KCN medium. Arginine is not hydrolysed, gelatin is not usually liquefied. Most strains are resistant to penicillin (10 IU disc) (*Tables 26.1* and *26.3*).

A. faecalis

Culturally resembles *B. bronchiseptica* but is urease negative. It is a widely distributed saprophyte and commensal.

A. odorans

This may be the same as *A. faecalis* (Snell[1]) but cultures are said to smell of apples and to show a greenish discoloration around colonies on blood agar.

A. viscolactis

This dubious species is one of several that are responsible for ropy milk.

Achromobacter

This is no longer officially recognized, but food bacteriologists find it a convenient 'dustbin' for non-pigmented Gram-negative rods which are non-motile, oxidative in Hugh and Leifson medium, oxidase positive, produce acid but no gas from glucose, acid or no change in purple milk, do not hydrolyse arginine and usually liquefy gelatin.

Flavobacterium

The genus *Flavobacterium* contains many species that are difficult to identify. Gram-negative bacilli forming yellow colonies are frequently isolated from food and water samples and although some undoubtedly belong to other genera (e.g. *Aeromonas*, *Cellvibrio*, *Cytophaga*, *Erwinia*, *Klebsiella*, *Myxobacteria*, *Vibrio*) it suits the convenience of many food bacteriologists to call them 'flavobacteria'. Some of these organisms are proteolytic or pectinolytic and are associated with spoilage of fish, fruit and vegetables.

For elucidation *see* McMeekin and Shewan[8], and Hayes *et al.*[9].

There is one species of medical importance.

Flavobacterium meningosepticum

This has been isolated from CSF in meningitis, from other human material and from hospital intravenous and irrigation fluids.

It grows on ordinary media; colonies at 28 h are 1–2 mm, smooth, entire, grey or yellowish and butyrous. There is no haemolysis on blood agar, no growth on DCA and none on primary MacConkey cultures, but growth on this medium may occur after several subcultures on other media. A yellowish green pigment may diffuse into nutrient agar media.

F. meningosepticum is oxidative in Hugh and Leifson medium, non-motile, oxidase and catalase positive, liquefies gelatin in five days, does not reduce nitrates and does not grow on Simmons' citrate medium. Acid is produced from (10%) glucose and lactose in ammonium salt medium but not from sucrose or salicin. Indole production seems to be positive by Ehrlich's but negative by Kovac's methods (Snell[1]), *See Table 26.5* for differentiation from pseudomonads and *Acinetobacter*.

TABLE 26.5. *Flavobacterium meningosepticum, Pseudomonas* and *Acinetobacter*

Species	*Oxidase*	*Nitrate reduction*	*Citrate*	*Acid from 10% solution in ammonium salt medium of*		
				Glucose	*Mannitol*	*Lactose*
F. meningosepticum	+	−	−	+	+	+
Pseudomonas	+	+	+	+	+	+
Acinetobacter	−	−	+/−	−	−	−

Chromobacterium

The organisms in this genus are characterized by a violet pigment, but this may not be apparent until the colonies are several days old. They grow well on ordinary media, giving cream or yellowish colonies that turn purple at the edges. The best medium to demonstrate the pigment is potato. The pigment is soluble in ethanol but not in chloroform or water and can be enhanced by adding mannitol or meat extract to the medium. Citrate but not malonate is utilized in basal synthetic medium, the catalase test is positive and the urease test negative. Ammonia is formed. Gelatin is liquefied. The oxidase test is positive but difficult to do if the pigment is well formed. Both species are motile (*see Table 26.6*).

TABLE 26.6. *Chromobacterium violaceum* and *Chromobacterium lividum*

Species	*Growth at*		*HL test*	*Hydrolysis of*			*KCN*
	5 °C	*37 °C*		*Arginine*	*Casein*	*Aesculin*	
C. violaceum	−	+	F	+	+	−	+
C. lividum	+	−	Ox	−	−	+	−

F = fermentative; Ox = oxidative.

C. violaceum

This species is mesophilic, growing at 37 °C but not at 5 °C. It is fermentative in Hugh and Leifson medium, hydrolyses arginine and casein but not aesculin and grows in KCN broth. It is a facultative anaerobe.

Although usually a saprophyte, cases of human and animal infection have been described in Europe, the USA and the Far East.

C. lividum

This organism is psychrophilic, growing at 5 °C but not at 37 °C. It is oxidative in Hugh and Leifson medium, does not hydrolyse arginine or casein, hydrolyses aesulin but does not grow in KCN broth. It is an obligate aerobe.

Other species, *C. fluviatile* and marine strains, have been described (*see* Sneath[10]).

Acetic acid bacteria

These bacteria are widely distributed in vegetation. They are of economic importance in the fermentation and pickling industries, e.g. in cider manufacture[11], as a cause of ropy beer and sour wine, and of off-odours and spoilage of materials preserved in vinegar. Ethanol is oxidized to acetic acid by *Gluconobacter* spp. but *Acetobacter* spp. continue the oxidation to produce carbon dioxide and water.

Isolation and identification to genus

Identification to species is rarely necessary. It is usually sufficient to recognize acetic acid bacteria and to determine whether the organisms produce acetic acid and/or destroy it.

Inoculate wort agar or unhopped beer solidified with gelatin and yeast extract broth containing 10% glucose and 3% calcium carbonate at pH 4.5. Incubate at 25 °C for 24–48 h. Colonies are large and slimy.

Subculture into yeast broth containing 2% ethanol and congo red indicator at pH 4.5. Acid is produced.

Test the ability to produce carbon dioxide from acetic acid in 2% acetic acid broth. Use either a Durham tube or the method described on p. 333 for hetero-fermentative lactic acid bacilli. Inoculate lactose and starch peptone water 'sugars' at pH 4.5 with congo red indicator.

Plant bacterial masses (heavy inoculum, do not spread) on yeast extract agar at pH 4.5 containing 5% glucose and at least 3% finely divided calcium carbonate in suspension. Clear zones occur around masses in three weeks but if less chalk is used some pseudomonads can do this.

Culture on yeast extract agar containing 2% calcium lactate. *Acetobacter* grow well and precipitate calcium carbonate. *Gluconobacter* grow poorly and give no precipitate.

Carr and Passmore[11] describe a medium to differentiate the two genera. It is yeast extract agar containing 2% ethanol and bromocresol green (1 ml of 2.2% solution/litre) made up in slopes. Both *Gluconobacter and Acetobacter* produce acid from ethanol and the indicator changes from blue-green to yellow. *Acetobacter* then utilize the acid and the colour changes back to yellow (*see Table 26.7*).

TABLE 26.7. Acetic acid bacteria

	Acid from ethanol	*CO_2 from acetic acid*	*Carbonate from lactate*
Gluconobacter	+	−	−
Acetobacter	+	+	+

For more information *see* Carr and Passmore[11].

References

1. SNELL, J.J.S. (1973) *The Distribution and Identification of Non-fermenting Bacteria*, Public Health Laboratory Service, Monograph No. 4. London: HMSO
2. VIVIAN, A., HINCHCLIFFE, E.J. and FEWSON, C.A. (1981) *Journal of Hospital Infection*, **2**, 199
3. HUGH, R. and GILARDI, G.L. (1980) Pseudomonas. In *Manual of Clinical Microbiology*, 3rd edn. Edited by E.H. Lennett, E.H. Spaulding and J.P. Truant, pp. 283–317. Washington: American Society for Microbiology
4. GILARDI, G.L. (1971) *Journal of Applied Bacteriology*, **34**, 623
5. HENDRIE, M.S. and SHEWAN, J.M. (1979) Identification of pseudomonads. In *Identification Methods for Microbiologists*, 2nd edn. Edited by F.A. Skinner and D.W. Lovelock, pp. 1–12. London: Academic Press
6. BALLARD, R.W., PALLERONI, M.J., DOUDOROFF, M., STANIER, R.Y. and MANDEL, M. (1970) *Journal of General Microbiology*, **60**, 199
7. STANIER, R.Y., PALLERONI, N.J. and DOUDOROFF, M. (1966) *Journal of General Microbiology*, **43**, 159
8. McMEEKIN, T.A. and SHEWAN, J.M. (1978) *Journal of Applied Bacteriology*, **45**, 321
9. HAYES, P.R., McMEEKIN, T.A. and SHEWAN, J.M. (1979) The identification of Gram-negative yellow pigmented rods. In *Identification Methods for Microbiologists*. 2nd edn. Edited by F.A. Skinner and D.W. Lovelock, pp. 177–185. London: Academic Press
10. SNEATH, P.H.A. (1979) Identification methods applied to Chromobacterium. In *Identification Methods for Microbiologists*, 2nd edn. Edited by F.A. Skinner and D.W. Lovelock, pp. 167–174. London: Academic Press
11. CARR, J.G. and PASSMORE, S.M. (1979) Methods for identifying acetic acid bacteria. In *Identification Methods for Microbiologists*, 2nd edn. Edited by F.A. Skinner and D.W. Lovelock, pp. 33–45. London: Academic Press

Chapter 27

Vibrios, aeromonas, plesiomonas and photobacterium

T.J. Donovan and A.L. Furniss

Vibrios

Vibrios are Gram-negative, non-sporing rods, motile by polar flagella enclosed within a sheath. Some have lateral flagella and may swarm on solid media. They are catalase positive, utilize carbohydrates fermentatively and rarely produce gas. All but one species (*Vibrio metschnikovii*) are oxidase and

TABLE 27.1 ***Vibrio, Plesiomonas, Aeromonas, Pseudomonas*** **and enterobacteria**

	0/129[a]	*Oxidase*	*Hugh and Leifson*	*Gas*	*Salt enhancement*
Vibrio	S	+[b]	F	–[c]	+
Plesiomonas	S	+	F	–	–
Aeromonas	R	+	F	v	–
Pseudomonas	R	+	O	–	–
Enterobacteria	R	–	F	v	–

[a]150 μg: S—sensitive; R—resistant
[b]except *Vibrio metschnikovii*
[c]except some strains of *V. fluvialis*
v, different strains show different reactions

nitratase positive. They are all sensitive to 150 μg discs of the vibriostatic agent 2,4-diamino 6,7-di-isopropyl pteridine (0/129). Most strains liquefy gelatin and hydrolyse deoxyribonucleic acid. Vibrios occur naturally in fresh and salt water. Species include some human and fish pathogens. *See Table 27.1* for differentiation between *Vibrios, Plesiomonas, Aeromonas, Pseudomonas* and Enterobacteria.

Isolation

Vibrios grow readily on most ordinary media but enrichment and selective media are necessary for faeces and other material containing mixed flora.

Faeces

Add 2 g of faeces to 20-ml amounts of alkaline peptone water (APW). Inoculate thiosulphate citrate bile salt sucrose agar (TCBS) medium.

Incubate APW at 37 °C for 5–8 h or at 20–25 °C for 18 h and subculture to TCBS.

Incubate TCBS cultures at 37 °C overnight.

Other pathological material

Vibrios may occur in wounds and other material particularly if there is a history of sea bathing. Examine primary blood agar plates (*see* below for colony appearance).

Water and foods

Halophilic vibrios may overgrow non-halophilic vibrios such as *V. cholerae* in cultures of sea water and sea food. Therefore test with APW with and without added sodium chloride. Add 20 ml of water to 100 ml of APW. Add 10 ml of food to 100 ml of APW and emulsify in a Stomacher Lab-Blender. Incubate at 20–30 °C for 18 h and subculture on TCBS as for faeces.

Colonial morphology

Table 27.2 shows the colony appearances of vibrios and related organisms on TCBS medium.

TABLE 27.2 Colony appearance of some vibrios, *Aeromonas* and *Plesiomonas* on TCBS after 18 h incubation at 37 °C

	Colony size	*Appearance*
V. cholerae	Medium (2–3 mm)	Yellow
V. mimicus	Medium (2–3 mm)	Green
V. metschnikovii	Medium (2–4 mm)	Yellow
V. parahaemolyticus	Large (2–5 mm)	Green
V. alginolyticus	Large (2–5 mm)	Yellow
V. fluvialis	Medium (2–3 mm)	Yellow
V. anguillarum	No growth or <2 mm	Yellow
Aeromonas	Variable	Yellow
Plesiomonas	No growth or <1 mm	Green

On non-selective media colonial morphology is variable. The colonies may be opaque or translucent, flat or domed, haemolytic or non-haemolytic, smooth or rough. One variant is rugose and adheres closely to the medium.

Electrolyte concentration

Increase the electrolyte concentration of conventional identification media by 1%. (This will not interfere with the identification of enterobacteria.) Some strains grow poorly, however, and need an electrolyte supplement: NaCl 10%, $MgCl_2$, $6H_2O$, 4%; KCl, 4% in distilled water. Add 0.1 ml to each 1.0 ml of medium.

Growth on CLED medium

Some vibrios, however, will grow without the addition of sodium chloride to the medium. Culture on an electrolyte deficient medium, e.g. CLED. This permits two groups, halophilic and non-halophilic vibrios to be distinguished. Inoculate CLED lightly with the culture and incubate at 30 °C overnight.

Salt tolerance

This varies with species. Inoculate peptone water containing 0, 3, 6, 8 and 10% NaCl. This technique must be standardized to obtain consistent results.

Sensitivity to 0/129

This was originally used as a disc method to distinguish between *Vibrio* (sensitive) and *Aeromonas* (resistant)[1], but the use of the two discs (150 μg and 10 μg) enables two groups of vibrios to be recognized[2]. Dissolve the phosphate derivative of 2,4-diamino-6,7-di-isopropyl pteridine (BDH) to give concentrations of 7500 and 500 μg/ml and place 0.02 ml (50 dropper) of the solution on blank antibiotic discs, and dry in air. Make a lawn of the organisms on nutrient agar, and place one disc of each concentration of 0/129 on it and incubate overnight. Do not use any special antibiotic sensitivity testing medium because the growth of vibrios and the diffusion characteristics of 0/129 differ from those on nutrient agar.

Oxidase test

Use Kovac's method (p. 111) and test colonies from nonselective medium. Do not use colonies from TCBS or other media containing fermentable carbohydrate because changes in pH may interfere with the reaction.

Decarboxylase tests

Use Moeller's medium containing 1% additional NaCl, but do not read the results too early as there is an initial acid reaction before the medium becomes alkaline. The blank should give an acid reaction; failure to do so may suggest poor growth and the electrolyte supplement should be added. Thornley's arginine medium is particularly useful provided that the salt concentration is adequate.

VP test

Use a semisolid medium[2] under controlled conditions. If incubation is prolonged and a sensitive method is used almost any vibrio may give positive results.

Swarming

Use Marine agar (*D*) or a medium with constant characteristics, at 37 °C for reliable results.

Luminescence

Some vibrios may show luminescence in the dark. It is most marked in young cultures and may be lost on continued incubation. A positive control must always be included. Growth is best examined after overnight incubation at 25 °C on special nutrient agar[2]. It is very important to allow at least 5 min for one's eyes to become adapted to the dark before examining cultures for luminescence.

Identification kits

Not all of these are at present ideal for identifying vibrios. All appropriate tests are not included and there may not be enough data in the matrices. The API 20 system gives satisfactory results if the organisms are suspended in 1% NaCl at pH 6.5 and a heavier than usual suspension is employed. Additional tests, e.g. 0/129 sensitivity and growth on CLED should be used.

Other tests

Test for fermentation with Hugh and Leifson medium, acid production from sucrose and arabinose in peptone water containing 1% NaCl.

Properties of vibrios, etc.

See Table 27.3.

Species of vibrios

V. cholerae

Is sensitive to 0/129, oxidase positive, decarboxylates lysine and ornithine but does not hydrolyse arginine. It produces acid but no gas from glucose (fermentative in Hugh and Leifson's medium) and sucrose, but not from arabinose or lactose. It is non-halophilic in that it grows on CLED medium.

All strains possess the same heat-labile H antigen but may be separated into serovars by their O antigens.

Serovar 0:1

Is the causative organism of epidemic or Asiatic cholera[3,4]. It is agglutinated by specific 0:1 cholera antiserum. It is possible, by using carefully absorbed sera, to distinguish two subtypes of *V. cholerae* 0:1. These are known as Ogawa and Inaba but as they are not completely stable and variation may occur *in vitro* and *in vivo* subtyping is of no great epidemiological value, unlike phage typing[5] which is of epidemiological value. Non-toxigenic strains of *V. cholerae* 0:1 have been isolated and have shown distinctive phage patterns.

Although there are two biotypes of *V. cholerae* 0:1, the 'classical' (non-haemolytic) and the 'eltor' (haemolytic) the former are now virtually non-existent. Biotyping is not, therefore, a useful epidemiological tool (*see Table 27.4*).

Table 27.3 Properties of *Vibrio* spp, *Aeromonas* and *Plesiomonas*

	Decarboxylase			*Gas from glucose*	*Acid from*				*VP*	*ONPG*	*0/129*		*Growth in % NaCl*				*Oxidase*	*Growth on CLED*
	Arginine	*Lysine*	*Ornithine*		*Arabinose*	*Inositol*	*Salicin*	*Sucrose*			*10*	*150*	*0*	*6*	*8*	*10*		
V. cholerae	−	+	+	−	−	−	−	+	+	+	S	S	+	−	−	−	+	+
V. mimicus	−	+	+	−	−	−	−	−	−	+	S	S	+	v	−	−	+	+
V. metschnikovii	+	v	−	−	−	v	−	+	+	v	S	S	v	+	v	−	−	v
V. parahaemolyticus	−	+	+	−	v	−	−	−	−	−	R	S	−	+	+	−	+	−
V. alginolyticus	−	+	+	−	−	−	v	+	+	−	R	S	−	+	+	+	+	−
V. vulnificus	−	+	+	−	−	−	+	−	−	+	S	S	−	+	−	−	+	−
V. fluvialis I	+	−	−	−	+	−	+	+	−	+	R	S	v	+	v	v	+	v
V. fluvialis II	+	−	−	+	+	−	+	+	−	+	R	S	v	+	v	−	+	v
V. anguillarum	+	−	−	−	v	−	−	+	+	+	S	S	v	v	−	−	+	v
Aeromonas	+	v	−	v	v	−	v	+	v	+	R	R	+	−	−	−	+	+
Plesiomonas	+	+	+	−	−	+	−	−	−	+	v	S	+	−	−	−	+	+

0/129 μg/ml; S, sensitive; R, resistant

0/129 μg/ml; S, sensitive; R, resistant

TABLE 27.4 Recognition of classical and eltor strains of *Vibrio cholerae*

	Classical	*Eltor*
Haemolysis	–	+
VP	–	+
Chick cell haemagglutination	–	+
Polymyxin (50 IU)	S	R
Classical phage IV	S	R
Eltor phage 5	R	S

S, sensitive; R, resistant

Serovars other than 0:1

Have identical biochemical characteristics as *V. cholerae* 0:1 and the same H antigen but possess different O antigens[6,7] and are not agglutinated by the 0:1 serum. They have been called 'non-agglutinating' (NAG) vibrios. This is an obvious misnomer; they are agglutinated both by cholera H antiserum and by antisera prepared against the particular O antigen they possess. Another term which has been used is 'Non-cholera vibrio' (NCV) and as both terms have been used in different ways this has resulted in considerable confusion. It is best if all these vibrios are referred to as non-0:1 *V. cholerae*, or, if a strain has been serotyped, by that designation.

Some of these strains are undoubtedly potential pathogens and produce a toxin similar to, if not identical with, that of the cholera vibrio. Some outbreaks have occurred but most isolates have been from sporadic cases. These vibrios are widespread in fresh and brackish water in many parts of the world, including the UK. They do not, however, cause true epidemic cholera.

V. mimicus

Resembles non-0:1 *V. cholerae* but is VP negative and does not ferment sucrose[8]. Colonies on TCBS media are therefore green. Antigenically it appears to be *V. cholerae* but it was given specific rank because of its low degree of DNA homology with *V. cholerae*. On the other hand it could be retained as a subspecies or biovar of *V. cholerae*. It has been isolated from the environment and is associated with sea foods. It has also been isolated from human faeces. Some strains produce a cholera-like toxin.

V. parahaemolyticus

Is a halophilic vibrio and will not grow on CLED. It does not ferment sucrose and therefore gives a (large) green colony on TCBS agar. O and K antigens may be used to serotype strains. It is recognized as the commonest cause of food poisoning in Japan[9]. It is usually present in coastal waters, although only in the warmer months in the UK. The Kanagawa haemolysis[10] test is said to correlate with pathogenicity. Controlled conditions for testing the haemolysis of the human red blood cells are essential. Only laboratories with sufficient experience should do this test. Most environmental isolates are negative.

V. vulnificus

Resembles *V. parahaemolyticus* but ferments lactose[11,12]. It has been isolated from blood cultures from patients (mostly in the USA) who are immunologically compromised or suffering from liver disease.

V. fluvialis

First reported as Group F vibrios[13], this is common in rivers, particularly in the brackish water of estuaries. Two biotypes can be recognized[14]. Biotype 1 is anaerogenic and may cause gastroenteritis in man. Biotype 2, which is aerogenic, does not appear to be pathogenic. Biotype 1 needs to be distinguished from *V. anguillarum.*

V. anguillarum

Is phenotypically far from being a uniform species. It is not uncommon in rivers. Some strains may be pathogenic for fish, although not especially for the eel, as its name might suggest. Many strains will not grow at 37 °C and would be missed in laboratories which confine their incubation temperatures to about 37 °C. There is no evidence that *V. anguillarum* has ever been pathogenic for man.

Plesiomonas shigelloides

There is no specific enrichment method or selective medium for *Plesiomonas.* It grows poorly on TCBS but well on deoxycholate and MacConkey agar. Colonies appear to be *Shigella*-like—hence the specific name—and some strains are agglutinated strongly by *Shigella sonnei* antiserum.

First described as C27 group of organisms[16] this organism has been variously classified and included within the genus *Vibrio*, because of its sensitivity to 0/129. It is now considered to be distinct enough to be placed in a separate genus containing one species.

It is now thought to be a cause of gastroenteritis, because most clinical isolates are from patients with diarrhoea; it is rarely isolated otherwise. Many of the isolates in the UK are from people returning from abroad.

Aeromonas

These are small Gram-negative rods which are motile by polar flagella. They are oxidase and catalase positive and resistant to 0/129. They attack carbohydrates fermentatively (gas may be produced from glucose) liquefy gelatin and reduce nitrates. The arginine dihydrolase test using arginine broth is positive. Indole and VP reaction vary with species. Some strains are psychrophilic and most grow at 10 °C. *See Table 27.3* for differentiation from other groups.

Isolation

Aeromonas species will grow on non-selective media, but for recovery from mixed flora culture on xylose deoxycholate agar (replace lactose by xylose at *10 g/l)*[17]. *Xylose is not fermented and colonies appear pale after overnight incubation.*

Identification

Test for growth at 20 and 37 °C. Do oxidase and nitrate reduction tests, OF test, incubate glucose, arabinose, salicin and aesculin media. Do lysine decarboxylase, arginine dihydrolase and VP tests, look for brown pigment on agar medium containing 1% tyrosine (e.g. Difco Furunculosis agar) and incubated at 20 °C (*see Tables 27.5* and *27.6*).

TABLE 27.5 ***Aeromonas salmonicida***

	Motility	*Growth at 37 °C*	*Brown pigment*
A. salmonicida	−	−	+
Other aeromonads	+	+	v

TABLE 27.6 ***Aeromonas* species**

	Aesculin hydrolysis	*Acid from*		*Gas from glucose*	*Lysine decarboxylase*	*VP reaction*
		Arabinose	*Salicin*			
A. hydrophila[a]	+	+	+	+	+	+
A. caviae[b]	+	+	+	−	−	−
A. sobria[a]	−	−	−	+	+	+

[a] Enterotoxigenic
[b] Not enterotoxigenic

Species of *Aeromonas*

There is still uncertainty about the status of species, but we favour the recognition of the four mentioned below.

A. salmonicida

Produces a brown pigment on media containing tyrosine but does not grow at 37 °C and is non-motile. It causes furunculosis of fish, a disease of economic importance in fish farming.

Other species

A. hydrophila, *A. caviae* and *A. sobria* are motile and usually grow at 37 °C. They are more commonly isolated from the stools of patients with diarrhoea than from normal faeces in the UK but in countries where aeromonads are common in drinking water they are as frequently isolated from normal as abnormal stools. It appears that enteropathogenicity, if it does occur, is confined to certain strains of *A. hydrophila* and *A. sobria*, as judged by animal models. Anaerogenic strains do not seem to be enteropathogenic and most environmental strains appear to nonpathogenic in the animal model.

Strains identifiable as *A. hydrophila* are found in food. They may be associated with spoilage, e.g. of eggs and some strains are suspected of being pathogenic for amphibians and fish.

Photobacterium

This genus was created to accommodate luminescent bacteria that have been isolated from sea water and marine fauna but which cannot be identified as vibrios. *Photobacterium* species are difficult to distinguish from luminescent vibrios. They do not grow well on TCBS medium and some strains grow poorly at 37 °C. They are less active biochemically than vibrios and utilize only a narrow range of carbon compounds. The following species are recognized: *P. phosphoreum*, *P. leiognathi* and *P. angustum*[18].

References

1. SHEWAN, J.M., HODGKISS, W. and LISTON, J. (1954) *Nature, London*, **173,** 208
2. FURNISS, A.L., LEE, J.V. and DONOVAN, T.J. (1978) *The Vibrios,* Public Health Laboratory Service Monograph No. 11. London: HMSO
3. GARDNER, A.D. and VENKATRAMAN, K.V. (1935) *Journal of Hygiene*, **35,** 262
4. DONOVAN, T.J. and FURNISS, A.L. (1982) *Lancet*, **ii,** 866
5. LEE, J.V. and FURNISS, A.L. (1981) In *Acute Enteric Infections in Children. New Prospects for Treatment and Prevention.* Edited by Holme, T., Holmgren, J., Merson, M. and Molby, R., pp. 119–122. Amsterdam: Elsevier/North Holland Biomedical Press.
6. SAKAZAKI, R., TAMURA, K., GOMEZ, C.Z. and SEN, R. (1970) *Japanese Journal of Medical Science & Biology*, **23,** 13
7. SHIMADA, T. and SAKAZAKI, R. (1977) *Japanese Journal of Medical Sciences & Biology*, **30,** 275
8. DAVIS, BETTY R., FANNING, G.R., MADDEN, J.M., STEIGERWALT, A.G., BRADFORD, H.B., SMITH, H.L. and BRENNER, D.J. (1981) *Journal of Clinical Microbiology*, **14,** 631
9. SAKAZAKI, R., IWANAMI, S. and FUKUMI, H. (1963) *Japanese Journal of Medical Sciences*, **16,** 161
10. MIYAMOTO, Y., KATO, T., OBARA, Y., AKIYAMA, S., TAKIZAWA, K. and YAMAI, S. (1969) *Journal of Bacteriology*, **100,** 1147
11. HOLLIS, D.G., WEAVER, R.E., BAKER, C.N. and THORNSBERRY, C. (1976) *Journal of Clinical Microbiology*, **3,** 425
12. FARMER, J.J. (1979) *Lancet*, **ii,** 631
13. FURNISS, A.L., LEE, J.V. and DONOVAN, T.J. (1977) *Lancet*, **ii,** 565
14. LEE, J.V., SHREAD, P., FURNISS, A.L. and BRYANT, T.N. (1981) *Journal of Applied Bacteriology*, **50,** 73
15. LEE, J.V., DONOVAN, T.J. and FURNISS, A.L. (1978) *International Journal of Systematic Bacteriology*, **28,** 99
16. FERGUSON, W.W. and HENDERSON, N.D. (1947) *Journal of Bacteriology*, **54,** 179
17. SHREAD, P., DONOVAN, T.J. and LEE, J.V. (1981) *Society for General Microbiological Quarterly*, **8,** 184
18. LEE, J.V., HENDRIE, MARGARET S. and SHEWAN, J.M. (1979) In *Identification Methods for Microbiologists.* Society for Applied Bacteriology Technical Series 14. Edited by Skinner, F.A. and Lovelock, D.W. London: Academic Press

Chapter 28

Escherichia, citrobacter, klebsiella, enterobacter, salmonella, shigella and proteus

These genera, collectively known as the enterobacteria, contain many species of small Gram-negative rods that ferment glucose to produce acid or acid and gas. They are oxidase negative; some are motile. Most are commensals or parasites in the human and animal intestine. *Table 28.1* shows their general properties.

Table 28.1 Biochemical differentiation within the enterobacteria

	MR	*VP*	*Urease*	H_2S	*Citrate*	*KCN*	*PA*
Escherichia	+	−	−	−	−	−	−
Shigella	+	−	−	−	−	−	−
Salmonella	+	−	−	+	+	−	−
Arizona	+	−	−	+	+	−	−
Citrobacter[a]	+	−	v	+	+	+	−
Klebsiella	v	v	v	−	v	v	−
Enterobacter	−	+	v	−	+	+	−
Serratia	v	+	−	−	+	+	−
Proteus	+	−	+	v	v	+	+
Providence	+	−	−	−	+	+	+

PA, ability to deaminate phenylalanine.
v, reactions variable but constant for the strain.
[a], Includes Bethesda–Ballerup group.

For medical and public health laboratory purposes, it is convenient to divide the enterobacteria into two groups according to the fermentation of lactose. This is an historical division, dating from the time when bacteriology was almost exclusively a medical science and the lactose fermenters were considered to include mostly saprophytic and commensal organisms while the non-lactose fermenters included the pathogens.

The lactose fermenters

Produce acid or acid and gas rapidly from this sugar. The genera *Escherichia, Klebsiella, Citrobacter* and *Enterobacter* are collectively known as coliform bacilli.

The non-lactose fermenters

Either completely fail to ferment lactose or ferment it late or irregularly. Included are *Salmonella*, the Arizona group, *Shigella*, the Betheseda–Ballerup and the Alkalescens–Dispar groups; some *Citrobacter*, *Proteus* and *Serratia*.

The taxonomy, biochemical properties, antigenic structure and identification procedures within the enterobacteria are complex. We have retained those names and descriptions that are still in common use in medicine and industry notwithstanding the many changes recently made by taxonomists.

The lactose fermenting enterobacteria

These organisms are nutritionally non-exacting and will grow on simple culture media. Media containing bile, which inhibit most cocci and Gram-positive bacilli, are most useful. Lactose and an indicator in the medium allow coliform bacilli to be recognized easily.

Isolation

Pathological material

Plate stools from children under three years old, intestinal contents of animals, urine deposits, pus, etc., on MacConkey, EMB or Endo agar and on blood agar. Cystine lactose electrolyte deficient medium (CLED) is useful in urinary bacteriology. Proteus does not spread on this medium. Incubate at 37 °C overnight.

Foodstuffs

The organisms may be damaged and may not grow from direct plating. Make 10% suspensions of the food in tryptone soya broth. Incubate at 25 °C for 2 h and then subculture into MacConkey broth or plate on MacConkey, Endo, EMB or CLED media.

Identification

Many laboratories use the kit methods described on p. 103. Other methods are described here.

For direct serological idenfication of enteropathogenic, urinary and animal pathogenic strains, *see below*.

Coliform bacilli show pink or red colonies, 2–3 mm in diameter, on MacConkey agar. Klebsiella colonies may be large and mucoid. On Levine EMB agar, colonies of *E. coli* are blue-black by transmitted light and have a metallic sheen by incident light. Colonies of klebsiellas are larger, brownish, convex and mucoid and tend to coalesce. On Endo medium, the colonies are deep red and colour the surrounding medium. They may have a golden yellow sheen.

Examine Gram-stained films of each kind of colony and pick Gram-negative bacilli into two tubes of peptone broth. Incubate one tube at 44 ± 0.2 °C

TABLE 28.2 The lactose fermenting enterobacteria and serratia

	Gas at 44 °C	*Indole*	*MR*	*VP*	*Citrate*	*Gelatin liquefaction*	*Motility*	*KCN*	*Urease*	*Malo-nate*	*Gluco-nate*	H_2S	*Lysine decarboxylase*
E. coli	+	+	+	−	−	−	+	−	−	−	−	−	+
C. freundii	−	v	+	−	+	−	+	+	v	v	−	+	−
K. aerogenes	−	−	−	+	+	v	−	+	+	+	+	−	+
K. pneumoniae	−	−	+	−	+	−	−	−	+	+	v	−	+
K. edwardsii	−	−	v	+	+	−	−	+	+	v	+	−	+
K. ozoenae	−	−	+	−	v	−	−	+	v	−	−	−	v
K. rhinoscleromatis	−	−	+	−	−	−	−	+	−	+	−	−	−
E. aerogenes	−	−	−	+	+	v	−	+	−	+	+	−	+
E. cloacae	−	−	−	+	+	+	+	+	v	v	+	−	−
S. liquefaciens	−	−	v	+	+	+	+	+	v	−	+	−	v
S. marcescens	−	−	v	+	+	+	+	+	v	v	+	−	+

v, reactions variable.
It is difficult to distinguish biochemically between *K. edwardsii* var. *edwardsii* and *K. edwardsii* var. *atlantae* or between *E. cloacea* and *E. liquefaciens*, *K. aerogenes* var. *oxytoca* is indole positive and may liquefy gelatin (*see* text). *S. marcescens*, which may not ferment lactose, is included because it may be confused with *Klebsiella* and *Enterobacter*.

for 24 h for the indole test. Incubate the other tube at 37 °C for 4–6 h and use to inoculate the following media:

Two tubes of MRVP (glucose phosphate) medium: incubate at 30 °C for five days for MR and VP tests.

Simmons citrate medium: inoculate with a straight wire and incubate at 30–35 °C for five days.

Lactose broth plus indicator and Durham's tube: incubate at 37 °C for 24 h.

Lactose broth plus indicator and Durham's tube: incubate at 44 ± 0.2 °C for 24 h (Eijkman test). Include control strains that:

(1) will yield gas, and
(2) will not yield gas at this temperature (*E. coli* and *K. aerogenes*).

KCN broth, malonate and gluconate broths, urease medium, H_2S medium and lysine decarboxylase medium.

Test the original peptone broth for motility and continue to incubate at 35–37 °C for the indole test (*see Table 28.2* for identification).

In food examinations, the tests for indole at 44 °C and gas at 44 °C may be done by inoculating peptone broth and BGB broth directly from the positive primary tubes. Cultures may not be pure, but *E. coli* is the only one likely to be present that produces both indole and gas at 44 °C or the membrane technique (p. 255) may be used.

It is usually necessary to know only of the presence of:

(1) coliforms, and
(2) *E. coli*.

The identity of the coliforms other than *E. coli* is unimportant. MacConkey broth should not be used for the direct 44 °C test: some *C. perfringens* strains grow in this medium and produce acid and gas.

For clinical purposes, we have found Donovan's system[1] useful. Stab Donovan's medium with a straight wire and inoculate peptone broth and Simmons citrate. Incubate overnight at 37 °C. Donovan's medium (p. 74) is blackened by hydrogen sulphide; a blue to yellow colour change indicates inositol fermentation; motile organisms give a diffuse pink cloud or red outgrowths from the stab (*see Table 28.3*).

TABLE 28.3 Donovan's short system for lactose fermenters[1]

	Donovan's medium			*Indole*	*Citrate*
	H_2S	*Motility*	*Acid from inositol*		
E. coli	−	+	−	+	−
Citrobacter spp.	+	+	−	−	+
E. cloacae	−	+	+	−	+
Klebsiella spp.	−	−	+	−	+

Antigens of enterobacteria

There are three kinds of antigens. The O or somatic antigens of the cell body are polysaccharides and are heat stable, resisting 100 °C. The H or flagellar antigens are protein and destroyed at 60 °C. The K and Vi are envelope, sheath or capsular antigens and heat labile. There are three kinds of K antigens: L, which is destroyed at 100 °C and is an envelope, occasionally capsular; A, which is destroyed at 121 °C and is capsular; and B, which is destroyed at 100 °C and is an envelope. These Vi and K antigens mask the O, and agglutination with O sera will not occur unless the bacterial suspension are heated to inactivate them.

Species of coliform bacilli

Escherichia coli

This species is motile, produces acid and gas from lactose at 44 °C and at lower temperatures, is indole positive at 44 and 37 °C, MR positive, VP negative, fails to grow in citrate and KCN media and is malonate and gluconate negative. It is H_2S negative and usually decarboxylates lysine.

These are the so-called 'faecal coli' that occur normally in the human and animal intestine and it is natural to assume that their presence in food indicates recent contamination with faeces. *E. coli* is, however, widespread in nature and although most strains probably had their origin in faeces, its presence, particularly in small numbers, does not necessarily mean that the food contains faecal matter. It does suggest a low standard of hygiene. It seems advisable to avoid using the term 'faecal coli' and to report the organisms as *E. coli*.

Some serotypes are pathogenic for man and animals, causing gastroenteritis in babies, urinary tract infections, travellers' diarrhoea, suppurative lesions, white scours in calves, mastitis, pyometria in bitches, coli granulomata in fowls, etc. Agglutination sera for the identification of some of these serotypes by their O and K(B) antigens are available commercially (Difco; Wellcome).

Do slide agglutination tests on at least five colonies from the MacConkey and blood agar plates with the three polyvalent O sera; if these are negative, test the confluent growth. Test doubtful positives with 1:500 acriflavine solution, as no enteropathogenic *E. coli* strains are agglutinated by this. If a polyvalent serum gives good slide agglutination, do slide tests with the individual sera that it contains.

Positive slide agglutination is usually due to the K(B) antigen on the surface of the organism. Confirm by tube O agglutination using a culture of a saline suspension that has been boiled for 1 h.

Polyvalent and monospecific sera are available from Difco and Wellcome Reagents. These companies have different combinations of serotypes in their polyvalent antisera.

Citrobacter (Escherichia) freundii

Is motile, indole variable, MR positive, VP negative and grows in KCN and citrate media. Hydrogen sulphide is produced but the lysine decarboxylase test is negative. Malonate utilization is variable and gluconate is not oxidized. Non- or late-lactose fermenting strains occur (*see below*).

This organism occurs naturally in soil and is therefore a useful indicator of pollution. It can cause urinary tract and other infections in man and animals.

Klebsiella aerogenes

Is non-motile, indole negative, MR negative, VP positive, grows in citrate and KCN medium, does not produce hydrogen sulphide but the lysine decarboxylase test is positive. It is malonate and gluconate positive. It is capsulated and most strains give large, mucoid, slimy colonies on media that contain carbohydrates. Normally present on grain and plants, it is also found in the human and animal intestine and can cause urinary infections. It is often isolated from the sputum of patients treated with antibiotics. Indole positive strains occur that sometimes also liquefy gelatin. These are known as *K. aerogenes* var. *oxytoca* or as *K. oxytoca*.

K. pneumoniae (*Friedlander's pneumobacillus*)

Is non-motile, indole negative, MR positive, VP negative, grows in citrate medium but not in KCN medium, does not produce hydrogen sulphide but gives a positive lysine decarboxylase test. It is malonate positive but the gluconate test is variable. Colonies resemble those of *K. aerogenes*. It is thought to be associated with respiratory infections and may be cultured from sputum.

K. edwardsii

Is difficult to distinguish from other *Klebsiella* spp. There are two varieties, *K. edwardsii* var. *atlantae* and *K. edwardsii* var. *edwardsii*. This species resembles *K. pneumoniae* in its pathogenicity and incidence.

K. ozoenae

Is also difficult to identify, but is not uncommonly found in the respiratory tract associated with chronic destruction of the bronchi.

K. rhinoscleromatis

Also difficult to identify, is rare in clinical material and may be found in granulomatous lesions in the upper respiratory tract.

Klebsiellas may be classified serologically by their O and K antigens. For more information on this genus and associated organisms, *see* Bascombe *et al.*[2].

Enterobacter cloacae

Is motile, indole negative, MR negative, VP positive, grows in citrate and KCN media and liquefies gelatin (but this property may be latent, delayed or lost). It does not produce hydrogen sulphide and the lysine decarboxylase test is negative. Growth in malonate is variable and the gluconate test is positive. It is found in sewage and in polluted water.

Enterobacter hafniae

This gives reliable biochemical reactions only at 20–22 °C. It gives acid and gas in dextrose and mannitol, does not hydrolyse urea, but grows in KCN medium. It is MR negative, VP variable, indole negative, grows in citrate but does not liquefy gelatin (*Tables 28.1* and *28.11*). It is a soil and water organism that is sometimes found, in equivocal circumstances, in human material.

E. aerogenes

Resembles *E. cloacae* but the lysine decarboxylase test is positive. Gelatin liquefaction is late. It is often confused with *K. aerogenes*. *E. aerogenes* is non-motile and urease negative; *K. aerogenes* is non-motile and urease positive.

E. liquefaciens

Is difficult to distinguish biochemically from *E. aerogenes* but is a psychrophile. Now included in *Serratia*.

E. agglomerans

Formerly *Erwinia herbicola*, produces acid and occasionally gas from glucose, acid from mannitol and sometimes (late) from lactose. It liquefies gelatin but is indole, hydrogen sulphide, urease and lysine decarboxylase negative. Other reactions are variable (*Table 28.11*).

The non-lactose fermenting enterobacteria

This group of organisms is, in general, no more exacting in its nutritional requirements than are the lactose fermenters and can be grown on the same media, although a few (*Salmonella typhi* and some strains of *Shigella*) are exacting for certain amino acids. The non-lactose fermenters rarely occur in pure culture but as minority populations among other enterobacteria. More selective media than MacConkey, EMB or Endo media are recommended to inhibit as many lactose fermenters as possible.

There are two fundamentally different approaches to the identification of intestinal pathogens; biochemical, followed by serological, and serological confirmed by biochemical. As the final identification of *Salmonella* and *Shigella* must be serological, the second method gives the answer sooner than the first (24–48 h after receiving the sample), but it requires expert knowledge of colonial appearance, skilled slide agglutination technique and media that do not impair slide agglutinability of the organisms. On the other hand, subculturing into a set of sugar media and awaiting the results delays identification unduly. Composite media such as the Kohn or Gillies two tube tests, triple sugar iron agar (TSI) are convenient but the 'kit' systems (p. 103) are very useful.

Biochemical reactions are variable. Serological identification should not be delayed until biochemical tests are completed. The full biochemical reactions will, however, also be described.

Before describing methods for identifying the non-lactose fermenting enterobacteria, the antigenic structure of some of the organisms will be considered. *Salmonella* and *Shigella* are identified serologically but although the antigens of most of the other members of the family have been investigated in detail, serology is not used in routine investigations.

Salmonella antigens

The system used was initiated by Kauffmann and White and the tables which are used bear their names.

There are more than 60 somatic or O antigens and these occur characteristically in groups. Antigens 1–50 are distributed among Groups A–Z. Subsequent groups are labelled 51–61. This enables more than 1700 *Salmonella* 'species' or serotypes to be divided into about 40 groups with the commoner organisms in the first six groups.

For example, in *Table 28.4*, each of the organisms in Group B possesses the antigens 4 and 12 but, although 12 occurs elsewhere, 4 does not. Similarly, in Group C all the organisms possess antigen 6; some possess in addition antigen 7 and other antigens 8. In Group D, 9 is the common antigen. Unwanted antigens may be absorbed from the sera produced against these organisms and single-factor O sera are available that enable almost any salmonella to be placed by slide agglutination into one of the groups. Various polyvalent sera are also available commercially (*see* note on p. 293).

To identify the individual organisms in each group, however, it is necessary to determine the H or flagellar antigen. Most salmonellas have two kinds of H antigen and an individual cell may possess one or the other. A culture may therefore be composed of organisms all of which have the same antigens or may be a mixture of both. The alternative sets of antigens are called *phases* and a culture may therefore be in phase I or in phase II or in both phases simultaneously. The antigens in phase I are identified by lower-case letters; thus, the H antigen of *S. paratyphi A* was called *a*, of *S. paratyphi B*, *b*, of *S. paratyphi C*, *c*, and of *S. typhi*, *d*, and so on. Unfortunately, after *z* was reached more antigens were found and so subsequent antigens were named z_1, z_2, z_3, etc. It is important to note that z_1 and z_2 are as different as are *a* and *b*; they are not merely subtypes of *z*.

Phase II antigens are given the Arabic numerals *1–7*.

Some of the lettered antigens in phase I also occur in phase II as alternatives to other lettered antigens. Whereas *S. paratyphi B* has antigens *b*, and *1* and *2*. *S. worthington* has *z*, and *l* and *w*, and *S. meleagridis* has *e* and *h*, and *l* and *w*. The lettered antigens in phase II do, however, occur mostly in groups, e.g. as *e*, *h*, *e*, *n*, z_{15}, *e*, *n*, *x*, *l*, *y*, etc., and lettered antigens in phase II, apart from these, are uncommon.

It is therefore necessary to identify both the phase I and the phase II antigens as well as the O antigens. It is usual to find the O group first. Various polyvalent and single-factor O antisera are available commercially for this (*see* note *below*). The H antigens are then found using the Spicer–Edwards system known as the Rapid Salmonella Diagnostic (RSD) sera. These are mixtures of antisera that enable the phase I antigens to be determined. In the

British set, there are three pools of sera and *S. typhimurium* is tested for separately. In the American set, there are four pools and *S. typhimurium* is included. Both sets are described below. The phase II antigens are found by using a polyvalent serum containing factors *1–7* and then individual sera.

Another antigen is the Vi, found in *S. typhi* and a few other species. This is a surface antigen which masks the O antigen. If it is present the organisms may not agglutinate with O sera unless the suspension is boiled.

TABLE 28.4 Antigenic structure of some of the common salmonellas (Kauffmann-White classification)

Absorbed O antisera available for identification	*Group*	*Name*	*Somatic (O) antigen*	*Flagellar (H) antigen*	
				Phase I	*Phase II*
Factor 2	A	*S. paratyphi A*	1, 2, 12	a	—
Factor 4	B	*S. paratyphi B*	1, 4, 5, 12	b	1, 2
		S. stanley	4, 5, 12	d	1, 2
		S. schwarzengrund	4, 12, 27	d	1, 7
		S. saintpaul	1, 4, 5, 12	e, h	1, 2
		S. reading	4, 5, 12	e, h	1, 5
		S. chester	4, 5, 12	e, h	e, n, x
		S. abortus equi	4, 12	—	e, n, x
		S. abortus bovis	1, 4, 12, 27	b	e, n, x
		S. typhimurium	1, 4, 5, 12	i	1, 2
		S. bredeney	1, 4, 12, 27	l, v	1, 7
		S. heidelberg	1, 4, 5, 12	r	1, 2
		S. brancaster	1, 4, 12, 27	z_{29}	—
Factor 7	C1	*S. paratyphi C*	6, 7, Vi	c	1, 5
		S. cholerae-suis	6, 7	c	1, 5
		S. typhi-suis	6, 7	c	1, 5
		S. braenderup	6, 7	e, h	e, n, z
		S. montevideo	6, 7	g, m, s	—
		S. oranienburg	6, 7	m, t	—
		S. thompson	6, 7	k	1, 5
		S. potsdam	6, 7	l, v	e, n, z
		S. bareilly	6, 7	y	1, 5
Factor 8	C2	*S. tennessee*	6, 7	z_{29}	—
		S. muenchen	6, 8	d	1, 2
		S. newport	6, 8	e, h	1, 2
		S. bovis morbificans	6, 8	r	1, 5
Factor 9	D	*S. typhi*	9, 12, Vi	d	—
		S. enteritidis	1, 9, 12	g, m	—
		S. dublin	1, 9, 12	g, p	—
Factors 3, 10	E1	*S. anatum*	3, 10	e, h	1, 6
		S. meleagridis	3, 10	e, h	1, w
		S. london	3, 10	l, v	1, 6
		S. give	3, 10	l, v	1, 7
Factor 19	E4	*S. senftenburg*	1, 3, 19	g, s, t	—
Factor 11	F	*S. aberdeen*	11	i	1, 2
Factors 13, 22	G	*S. poona*	13, 22	z	1, 6
		S. worthington	1, 13, 23	z	l, w

Reproduced by permission of Dr F. Kauffmann, formerly Director of the International Salmonella Centre.
The complete tables, revised regularly, are too large for inclusion here. They are obtainable from Salmonella Centres.

The Kauffmann–White Scheme (*Table 28.4*) is brought up to date every few years by the International Salmonella Centre.

Shigella antigens

The genus *Shigella* is divided into four antigenic and biochemical subgroups (*Table 28.5*).

TABLE 28.5 *Shigella* classification (*see also Table 28.6*)

Subgroup	*Species*	*Serotypes*
A: Mannitol not fermented	*S. dysenteriae*	1–10 all distinct
B: Mannitol fermented	*S. flexneri*	1–6 all related; 1–4 divided into subserotypes
C: Mannitol fermented	*S. boydii*	1–15 all distinct
D: Mannitol fermented, lactose fermented late	*S. sonnei*	None

Subgroup A

S. dysenteriae

Contains ten serotypes that have distinct antigens and do not ferment mannitol.

Subgroup B

S. flexneri

Contains six serotypes (I–VI) that can be divided into subserotypes according to their possession of some group factors designated 3,4; 4; 6; 7; and 7,8. (*Table 28.6*).

TABLE 28.6 Flexner subserotypes determined with Wellcome reagents sera

Agglutination with serotype	*Subserotype*	*Abbreviated antigenic formula*
1	1a	1:4
1 and 3	1b	1:6
2	2a	II:3, 4
2 and X (7, 8)	2b	II: 7, 8
3 and X (7, 8)	3a	III:6, 7, 8
3 and Y (3, 4)	3b	III:3, 4; 6
3	3c	III:6
4	4a	IV:3, 4
4 and 3	4b	IV:6
5	5	V:7, 8
6	6	VI:—

The X variant is an organism that has lost its type antigen and is left with the group factors 7,8.

The Y variant is an organism that has lost its type antigen and is left with the group factors 3,4.

Subgroup C

S. boydii

Contains 15 serotypes with distinct antigens; all ferment mannitol. There is some cross-agglutination with the Alcalescens–Dispar (AD) group.

S. sonnei

Contains only one, distinct serotype; this ferments mannitol. It may occur in phase I or in phase II, sometimes referred to as 'smooth' and 'rough'. The change from phase I to phase II is a loss variation and phase II organisms are often reluctant to agglutinate or give a very fine, slow agglutination. Phase II organisms are rarely encountered in clinical work but the sera supplied commercially agglutinate both phases.

Choice of media

Liquid media

These are generally inhibitory to coliforms but less so to salmonellas. Selenite F media are commonly used. If not overheated, they permit *S. sonnei* and *S. flexneri* 6 to grow but do not enrich the culture as they do salmonellas. Mannitol selenite and cystine selenite broths are preferred by some workers for the isolation of salmonellas from foods. Tetrathionate media enrich only salmonellas. Shigellas do not grow. *S. typhi* grows in 0.8% selenite.

Cultures from these media are plated on solid media.

Solid media

Many media exist for the isolation of salmonellas and shigellas but in our experience the Hynes modification of deoxycholate citrate agar (DCA) gives the best results for faeces. For the isolation of salmonellas from foods, the Leifson modification of DCA and brilliant green agar appear to be superior. Bismuth sulphite agar (Wilson and Blair medium) is an excellent medium for salmonellas but there are many versions and some are variable in their selectivity.

Colony appearance

DCA

On the ideal DCA medium, colonies of salmonellas and shigellas are not sticky. Salmonella colonies are creamy brown, 2–3 mm in diameter at 24 h and usually have black or brown centres. Unfortunately, colonies of proteus are similar. Shigella colonies are smaller, usually slightly pink and do not have black centres. Colonies of *S. typhi* are similar. *S. cholerae-suis* grows poorly on this medium. White, opaque colonies are not significant. Some klebsiella strains grow.

Bismuth sulphite agar

Salmonella colonies are black, 2–3 mm across and usually surrounded by a halo showing a metallic sheen due to the diffusion of sulphide. The black coloration of the colony is due to the release of hydrogen sulphide by the organism and the formation of bismuth sulphide. This is a very variable medium.

Brilliant green agar

Salmonella colonies are pale pink with a pink halo. Other organisms have yellow-green colonies with a yellow halo.

Isolation

Faeces, rectal swabs and urines

Plate out faeces and rectal swabs on DCA medium. Inoculate selenite F broth with a portion of stool about the size of a pea or with 0.5–1.0 ml of liquid faeces. Break off rectal swabs in the broth. Add about 10 ml of urine to an equal volume of double-strength selenite F broth.

Incubate plates and broths at 37 °C for 18–24h. Incubation of the selenite F cultures at 43 °C may yield more salmonellas. Plate broth cultures on DCA and incubate at 37 °C for 18 h (*see also* Harvey and Price)[3,4].

Drains and sewers

Fold three or four pieces of cotton gauze (approx. 15 × 20 cm) into pads of 10 × 4 cm. Tie with string and enclose in small-mesh chicken wire to prevent sabotage by rats. Autoclave in bulk and transfer to individual plastic bags or glass jars. Suspend in the drain by a length of wire and leave for several days. Squeeze out the fluid into a jar and add an equal volume of double-strength selenite F broth. Place the pad in another jar containing 200 ml of single-strength selenite F broth. Incubate both jars at 43 °C for 24 h and proceed as for faeces. This is a modification of the method originally described by Moore[3,4] (*see also* Harvey and Price[5] and Vassiliades *et al.*[6]).

Foods, feedingstuffs and fertilizers

The method of choice depends on the nature of the sample and on the total bacterial count. In general, raw foods such as comminuted meats have very high total counts and salmonellas are likely to be present in large numbers. On the other hand, heat-treated and deep-frozen foods should have lower counts and if any salmonellas are present they may suffer from heat shock or cold shock and require resuscitation and pre-enrichment.

Highly contaminated foods

Emulsify about 50 g in 100 ml of 0.1% peptone-water diluent, preferably in a Stomacher Lab-Blender. Add about 50-ml aliquots to equal volumes of double-strength selenite F broth or mannitol or cystine selenite broth in duplicate.

Incubate at 37 and 43°C for 24 h and plate on DCA and brilliant green agar. Examine DCA plates after 18–24 h and brilliant green plates after 24 and 48 h (*see also* Harvey and Price[3, 4]).

Heated or frozen foods

Emulsify as above but use nutrient broth as diluent. Incubate at 37 °C for 4 h (to avoid overgrowth by heat-resistant organisms). Add aliquots to double-strength selenite medium and incubate and plate as above (*see also* Harvey and Price[3, 4]).

Waters

See p. 253 and the monograph by Harvey and Price[3, 4].

Biochemical screening tests

Method A

Pick representative colonies on a urea slope as early as possible in the forenoon. Incubate in a water-bath at 35–37 °C. In the afternoon, reject any urea positive cultures and from the others inoculate the following: Triple Sugar Iron (TSI) medium by spreading on slope and stabbing butt; broth for indole test; lysine decarboxylase medium (paper strip method may be used, *see* p. 103). Incubate overnight at 37 °C (*see Table 28.7*) and proceed to serological identification if indicated.

TABLE 28.7 Biochemical reactions of *Salmonella, Shigella,* etc.: Method A

	TSI[a]		H_2S	*Indole*	*ONPG*	*Lysine decarboxylase*
	Butt	*Slope*				
S. typhi	A	–	+[b]	–	–	+
S. paratyphi A	AG	–	–	–	–	–
Other salmonellas	AG	–	+/–	–	–	+
Shigella	A	–	–	+/–	+/–	–
Arizona	AG	–	+	–	+	+
Citrobacter[c]	AG	–	+	–	+	–
Alcalescens-Dispar[d]	A	–	–	+	+/–	v
Serratia	A(G)	–	–	–	+	+

A = acid; AG = acid and gas; A(G) = some gas may be formed.
[a] In TSI medium acid in the butt indicates the fermentation of glucose in the slope that of lactose and/or sucrose.
[b] H_2S production by *S. typhi* may be minimal.
[c] *Citrobacter* includes the Bethesda-Ballerup group.
[d] Regarded as late lactose fermenting escherichias.

Method B

Pick representative colonies in broth and incubate at 37 °C in a water-bath for 4 h. Inoculate the following: Gillies (Kohn) medium 1 by spreading on slope and stabbing butt, and medium 2 by stabbing; insert test papers for indole and hydrogen sulphide in the top of medium 2 by wedging them under the cap; ONPG broth; lysine decarboxylase medium (or use a paper strip). Incubate overnight at 37 °C. Reject any culture that splits urea (cerise colour

TABLE 28.8 Biochemical reactions of *Salmonella, Shigella,* etc.: Method B

	Gillies 1[a]		Gillies 2[a]	Motility	H_2S	Indole	ONPG	Lysine decarboxylase
	Butt	Slope						
S. typhi	A	A	–	+	+[b]	–	–	+
S. paratyphi A	AG	A	–	+	–	–	–	–
Other salmonellas	AG	A	–	+	+/–	–	–	+
Shigella	A	A/–	–	–	–	+/–	+/–	–
Arizona	AG	A	–	+	+	–	+	+
Citrobacter[c]	AG	A	–	+	+	–	+	–
Alcalescens-Dispar[d]	A	A	+/–	–	–	+	+/–	v
Serratia	A(G)	A	+/–	+	–	–	+	+

A = acid production; AG = acid and gas; A(G) = some gas may be formed.
[a] In Gillies 1, acid in the butt indicates the fermentation of dextrose, in the slope that of mannitol. In Gillies 2, acid production indicates the fermentation of sucrose and/or salicin.
[b] H_2S production by *S. typhi* may be minimal.
[c] *Citrobacter* includes the Bethesda-Ballerup group.
[d] Late lactose fermenting escherichias.

in medium 1) and/or ferments sucrose and salicin (yellow colour in medium 2) (*see Table 28.8*), and proceed to serological identification if indicated.

Method C

Use one of the kit systems described on p. 103. Follow the manufacturer's directions and proceed to serological identification if indicated.

Full biochemical tests

For a more academic investigation, perform the following tests: fermentation of glucose, sucrose, salicin, dulcitol, lactose and mannitol; indole; hydrogen sulphide; growth in citrate and KCN media; gelatin liquefaction; malonate, gluconate; ONPG; lysine and ornithine decarboxylases. Check that the organism is oxidase negative and fermentative in Hugh and Leifson (OF) medium. If neither, then it is not in the enterobacteria group (*see Tables 23.9, 23.10* and *23.11*).

Serological identification of salmonellas

Note on commercial agglutination sera

These are obtainable from BBL, Difco and Wellcome Reagents. These firms vary in the range of sera they offer and in the packaging, e.g. of polyvalent sera.

Test the suspected organism by slide agglutination with Polyvalent O sera. If positive, do further slide agglutinations with the relevant single-factor O sera; only one should be positive. If in doubt, test against 1:500 acriflavine; if this is positive, the organisms are unlikely to be salmonellas.

Subculture into glucose broth and incubate in a water-bath at 37 °C for 3–4 h, when there will be sufficient growth for tube H agglutinations. Tube H

TABLE 28.9 Salmonellas and Arizona

Species	*Acid from*			*Citrate*	*Gelatin liquefaction*	*Malonate*	*ONPG*	*Lysine decarboxylase*	*Ornithine decarboxylase*
	Glucose	*Dulcitol*	*Mannitol*						
S. typhi	+	–/+	+	–	–	–	–	+	–
S. paratyphi A	+G	+	+	+	–	–	–	–	–
Other salmonellas	+G	+	+	+	–	–	–	+	+
Arizona	+G	–	+	+	+slow	+	+	+	+

+G, fermentation with gas; –/+, most strains negative.
None of these ferment lactose (but *see* text), sucrose or salicin; all are motile, and are indole, KCN and gluconate negative.

TABLE 28.10 Shigellas and Alcalescens-Dispar

Species		*Acid from*				*Indole*	*Lysine decarboxylase*	*Ornithine decarboxylase*
		Lactose	*Sucrose*	*Dulcitol*	*Mannitol*			
S. dysenteriae	+	–	–	–	–	–/+	–	–
S. flexneri 1–5	+	–	–	–	+	v	–	–
S. flexneri 6 (Boyd 88)	+(G)	–	–	–	+/–	–	–	–
S. sonnei	+	(+)	(+)	–	+	–	–	+
S. boydii 1–15	+	–	–	–	+	+/–	–	–
Alcalescens–Dispar	+	+/–	+/–	+/–	+	–	+/–	+/–

+(G), a small bubble of gas may be formed; –/+, most strains negative; +/–, most strains positive.
None of these organisms grow in citrate or ferment salicin; none produce H_2S, liquefy gelatin or are motile.

TABLE 28.11 Late or non-lactose fermenting *Citrobacter*, *Serratia* and *Enterobacter*

	Acid from				H_2S	*Gelatin liquefaction*	*Malonate*	*Gluconate*	*Lysine decarboxylase*
	Glucose	*Lactose*	*Dulcitol*	*Salicin*					
Citrobacter	+G	(+)	+/–	+/–	+	–	–	–	–
Serratia	+(G)	(+)/–	–	+	–	+	+/–	+	v
E. hafniae (at 22 °C)	+G	–	–	–	–	–	v	+	+
E. agglomerans	+	–(+)	–/(+)	+/–	–	+	+/–	–/+	–

+G, acid and gas; +/–, most strains positive; (+), late; (G), some strains may produce gas; –/+, most strains negative.
These organisms are all motile, usually ferment mannitol and sucrose, are indole negative, grow in citrate medium and in KCN and are usually ONPG positive.

agglutinations are more reliable than slide H agglutinations because cultures on solid media may not be motile. The best results are obtained when there are about 6–8 × 10^8 organisms/ml. Dilute, if necessary, with carbol saline. Add 0.5 ml of formalin to 5 ml of broth culture and allow to stand for at least 10 min to kill the organisms.

TABLE 28.12 Spicer–Edwards Rapid Salmonella Diagnostic (RSD) H agglutination sera (BBL and Difco)

H antigen	*Agglutination with serum*			
	A *1*	*B* *2*	*C* *3*	*D (BBL)* *4 (Difco)*
a	+	+	+	–
b	+	+	–	+
c	+	+	–	–
d	+	–	+	+
eh	+	–	+	–
G complex	+	–	–	+
i	+	–	–	–
k	–	+	+	+
r	–	+	–	+
y	–	+	–	–
z_4 *complex*	–	–	+	–
z_{10}	–	–	–	+
z_{29}	–	+	+	–

The G complex (Difco) and the g complex (BBL) contain all those antigenic groups which include *f, g, m, p, q, s, t,* and *u*.
The Z_4 complex (Difco) and the z_4 complex (BBL) contain all those antigenic groups which include z_4.

Place one drop of Polyvalent (phase 1 and 2) Salmonella H serum in a 75 × 9 mm tube. Add 0.5 ml of the formolized suspension and place in a water-bath at 52 °C for 30 min. If agglutination occurs, the organism is probably a salmonella.

Identify the H antigens as follows. Do tube agglutinations against the Rapid Salmonella Diagnosis (RSD) sera, *S. typhimurium* (*i*) if it is not included in the RSD set, phase 2 complex *1–7*, EN (en) complex and L (l) complex. If agglutination occurs (*see Tables 28.12* and *28.13*). Check the result, if possible by doing tube agglutinations with the indicated monospe-

TABLE 28.13 Spicer–Edwards Rapid Salmonella Diagnostic (RSD) agglutination sera (Wellcome Reagents)

H antigen	*Agglutination with serum*		
	1	*2*	*3*
b	+	+	–
d	+	–	+
E complex	+	+	+
G complex	–	–	+
k	–	+	+
L complex	–	+	–
r	+	–	–

The E complex contains *e, h; e, n, x;* and *e, n,* z_{15}.
The G complex contains all those antigenic groups which include *f, g, m, p, q, s, t* and *u*.
The L complex contains *l, v* and *l, w*.

cific serum. If the test indicates EN (en), G or L (l) complexes, send the culture to a Public Health or Communicable Diseases Laboratory.

If agglutination occurs in the phase 2 *1–7* serum, test each of the single factors *2*, *5*, *6* and *7*. Note that with some commercial sera, factor *1* is not absorbed and it may be necessary to dilute this out by using one drop of 1:10 or weaker serum.

Changing the H phase

If agglutination is obtained with only one phase, the organism must be induced to change to the other phase. The simplest way of doing this is our modification of Jameson's method[7].

Cut a 50 × 20 mm ditch in a well-dried nutrient agar plate. Soak a strip of previously sterilized filter-paper (36 × 7 mm) in the H serum by which the organism is agglutinated and place this strip across the ditch at right-angles. At one end, place one drop of 0.5% thioglycollic acid to neutralize any preservative in the serum. At the other end, place a filter-paper disc, about 7 mm in diameter, so that half of it is on the serum strip and the other half on the agar. Inoculate the opposite end of the strip with a young broth culture of the organism and incubate overnight. Remove the disc with sterile forceps, place it in glucose broth and incubate it in a water-bath for about 4 h, when there should be enough growth to repeat agglutination tests to find the alternative phase. Organisms in the original phase demonstrated will have been agglutinated on the strip. Organisms in the alternate phase will not be agglutinated and will travel across the strip.

The phase may be changed with a Craigie tube. Place 0.1 ml of serum and 0.1 ml of 0.5% thioglycollic acid in the inner tube of a Craigie tube and inoculate the inner tube with the culture. Incubate overnight and subculture from the outer tube into glucose broth.

Some organisms, e.g. *S. typhi* and *S. montevideo*, have only one phase. These should be sent to a reference expert.

Antigenic formulae

List the O, phase I and phase II antigens in that order and consult the Kauffmann–White tables (*Table 28.4*) for the identity of the organism. In the table, the antigenic formula is given in full, but as we use single factor sera for identification it is usually written thus:

S. typhimurium	4, *i*, *2*
S. typhi	9, *d*, –
S. newport	6, 8, *e,h*, 2

(The – sign indicates that the organism is monophasic.)

The Vi antigen

If this antigen is present and prevents the organisms agglutinating with O group sera, make a thick suspension in saline from an agar slope and boil it in a water-bath for 1 h. This destroys the *Vi* and permits the O agglutination.

Salmonella subgenera

The genus *Salmonella* has been divided into four subgenera, numbered I, II, III and IV. Subgenus I is the most important because it contains the majority of the human and animal pathogens. Serotypes ('species') are named. Subgenus II contains serotypes commonly found in reptiles but rarely found in man. Some are named; others known only by their antigenic formula. Subgenus III contains the organisms known as the Arizona group. Subgenus IV contains rare serotypes, only some of which are named.

Identification of subgenus is by biochemical tests but as this is of limited practical value it is not pursued here (*see* Edwards and Ewing[8]).

Salmonella species and serotypes

There seems to be some confusion about the status of species of salmonellas: some have been named as species (e.g. *S. typhi*) others as serotypes (e.g. *S. enteritidis* ser. *paratyphi* B). For practical purposes the binomials used in routine laboratory practice are to be preferred.

Salmonella typhi

Produces acid but no gas from glucose and mannitol. It may produce acid from dulcitol but fails to grow in citrate and KCN media and does not liquefy gelatin. It decarboxylates lysine but not ornithine. It is malonate and ONPG negative. Freshly isolated strains may not be motile at first. This organism causes enteric fever.

Other serotypes

Produce acid and usually gas from glucose, mannitol and dulcitol (anaerogenic strains occur) and rare strains ferment lactose. They grow in citrate but not in KCN and do not liquefy gelatin. They decarboxylate lysine and ornithine (except *S. paratyphi A*) and are malonate and ONPG negative. *S. paratyphi A*, *B* and *C* may also cause enteric fever. Strains with the antigenic formula 4; *b*; *1*, *2* may be *S. paratyphi B*, usually associated with enteric fever, OR *S. java*, which is usually associated with food poisoning. Many other serotypes cause food poisoning.

Arizona organisms

These may be confused with salmonellas. They have the antigenic formula 23; z_4, z_{23}, z_{26}. There are a number of subserotypes. They fail to ferment dulcitol, may ferment lactose late, slowly liquefy gelatin, decarboxylate lysine and ornithine and are malonate and ONPG positive (*see Table 28.9*).

Reference laboratories

It is not possible to identify all of the 1000 or more salmonella species or serotypes with the commercial sera available and in the smaller laboratory.

The following organisms should therefore be sent to a Reference or Communicable Diseases Laboratory.

Salmonellas placed by RSD sera into Groups E, G, or L.

Salmonellas agglutinated by Polyvalent H or monospecific H sera but not by phase II sera.

Salmonellas agglutinated by Polyvalent H but not by available phase I or phase II sera.

Salmonellas giving the antigenic formula 4; *b*; 1, 2, which may be either *S. paratyphi B* or *S. java*.

S. typhi, *S. paratyphi B*, *S. typhimurium*, *S. thompson* and *S. enteritidis* should be sent for phage typing.

Note that *Brevibacterium* spp., which are sometimes found in foods, possess Salmonella antigens. These organisms are Gram positive.

Identification of shigellas and the Alcalescens-Dispar (AD) group

Agglutinating sera for these organisms are obtainable from BBL, Difco and Wellcome Reagents.

The dysentery bacilli do not grow in citrate or KCN media, do not decarboxylate lysine, may or may not ferment mannitol and may or may not produce indole. The ONPG test is variable. The AD organisms have similar reactions. Other reactions, including TSI results are given in *Tables 28.7*, *28.8* and *28.10*.

Mannitol non-fermenters

Test by slide agglutination against polyvalent *S. dysenteriae* antiserum. If positive, test with individual type sera (1–10). If negative, test against *S. flexneri* 6 antiserum (some strains of *S. flexneri* 6 fail to ferment mannitol).

Mannitol fermenters

Test indole-negative strains against *S. sonnei* (phase I and II) serum. It is rarely necessary to use phase I and II sera separately. Test indole-negative strains that do not agglutinate, and indole-positive strains against Polyvalent Flexner and Polyvalent Boyd sera. If either is positive, test with individual type sera (Flexner, 1–6; Boyd, 1–15). Test any strain that is not agglutinated by *S. sonnei*, *S. flexneri* and *S. boydii* sera against Alcalescens-Dispar (AD) serum.

If the correct biochemical reactions are obtained but there is no slide agglutination, boil a suspension of the organism for 1 h to destroy possible masking surface (K) antigens and re-test.

Strains that cannot be identified locally, i.e. give the correct biochemical reactions but are not agglutinated by available sera, should be sent to a Reference Laboratory.

Note that the motile organism *Plesiomonas shigelloides* (p. 277) is agglutinated by *S. sonnei* serum.

Species of *Shigella* (*Table 28.10*)

S. dysenteriae (*S. shiga*)

Forms acid but no gas from glucose but does not ferment mannitol. Types 1, 3, 4, 5, 6, 9 and 10 are indole negative. Types 2, 7 and 8 are indole positive and are also known as *S. schmitzii* (*S. ambigua*). *S. dysenteriae* is responsible for the classical bacillary dysentery of the Far East.

S. flexneri

Serotypes 1–5 form acid but no gas from glucose and mannitol and are indole positive. Some type 6 strains (Newcastle strains) may not ferment mannitol and may be indole negative. A bubble of gas may be formed in the glucose tube. This serotype comes through selenite media. Serotypes 1–6 are widely distributed, especially in the Mediterranean area, and often cause outbreaks in mental and geriatric hospitals, nurseries and schools in temperate climes.

S. flexneri subserotypes

Wellcome Reagents do not absorb group 6 factor from their type 3 serum and they also provide X and Y variant sera containing group factors 7, 8 and 3, 4. These sera may therefore be used to determine the Flexner subserotypes for epidemiological purposes (*Table 28.6*).

S. sonnei

Forms acid but no gas from glucose and mannitol, is indole negative and decarboxylates ornithine. It causes the commonest and mildest form of dysentery, which mostly affects babies and young children and spreads rapidly through schools and nurseries. It may be typed for epidemiological purposes by colicine production (Reference Laboratory). It survives in some selenite media.

S. boydii

Forms acid but no gas from glucose and mannitol; indole production is variable. Growth occurs in KCN medium. The 15 serotypes of this species are widely distributed but not common and cause mild dysentery.

Alcalescens–Dispar Group

These organisms give reactions similar to those of dysentery bacilli but they are much more active biochemically. They are now considered to be anaerogenic, non-motile, late-lactose fermenting members of the genus *Escherichia*. Their pathogenicity is dubious.

Other non- or late-lactose fermenters

Citrobacter–Bethesda–Ballerup group (*C. ballerupensis*)

Some of these resemble *C. freundii* but ferment lactose late or not at all. They differ from salmonellas by growing in KCN medium and by giving a negative lysine decarboxylase reaction (*Tables 28.1*, *28.7*, *28.8* and *28.11*). They have some antigens that give cross-agglutination with salmonellas, including the *Vi* antigen. Their pathogenicity is questionable.

Serratia marcescens

This is motile, produces acid or acid plus a small amount of gas from glucose, ferments lactose late or not at all and liquefies gelatin. The VP reaction is positive, it is lysine decarboxylase variable and the DNAse test is positive. A red or pink pigment may be produced on agar at room temperature. Old cultures smell of trimethylamine (*Tables 28.1*, *28.2*, *28.7*, *28.8* and *23.11*). Originally thought to be a saprophyte and hence used in aerosol and filter pore-size tests, this species is now considered to be important in hospital-acquired infections, e.g. meningitis, endocarditis, septicaemia and urinary tract infections.

Paracolon bacilli

This was a dustbin or ragbag group formed by early bacteriologists to accommodate non- or late-lactose fermenters that could not otherwise be identified and whose pathogenicity was doubtful. These organisms are now in the Citrobacter, Bethesda-Ballerup, Arizona and Alcalescens-Dispar groups or have been identified as species.

Proteus and *Providencia*

These organisms are widely distributed. Some strains are easily recognized by their ability to 'swarm' or spread over the surface of agar media in a series of successive waves. This is inconvenient in diagnostic bacteriology as the swarming obscures the presence of other organisms. Swarming does not take place on salt-free agar, e.g. cystine lactose electrolyte deficient (CLED) medium. Swarming is also inhibited on agar media containing 1:500 chloral hydrate, 1:1000 sodium azide or 6% agar. Non-motile strains are not uncommon.

Isolation and identification

Plate on nutrient agar or blood agar to observe swarming, and on CLED and chloral hydrate blood agar to inhibit it and to reveal other organisms if present. Incubate overnight at 37 °C.

Subculture to test for phenylalanine deaminase (PPA test) and urease (PathoTec paper strip tests are convenient for both of these). Test PPA positives for fermentation of lactose, maltose, mannitol and dulcitol; for indole production, growth in citrate, gelatin liquefaction, hydrogen sulphide formation and ornithine decarboxylase. Proteus species are PPA and urease positive; Providence are PPA positive but urease negative (*see Table 28.14* for species differentiation).

TABLE 28.14 Biochemical properties of *Proteus* and *Providencia*

	Acid from		*Indole*	*Citrate*	*Gelatin liquefaction*	*Urease*	H_2S	*Ornithine decarboxylase*
	Mannitol	*Maltose*						
P. vulgaris	−	+	+	v	+	+	+	−
P. mirabilis	−	−	−	+[a]	+	+	+	+
P. morganii	−	−	+	−	−	+	−	+
P. rettgeri	+	−	+	+	−	+	−	−
Providencia	+	−	+	+	−	−	−	−

[a] May be delayed.

Proteus species

Proteus vulgaris

Common in soil and vegetation and in the animal intestine. It frequently contaminates food, leading to spoilage and decomposition, but is rarely found in pure culture. It is frequently found in hospital infections.

P. morganii

Is also a commensal but may be associated with infantile (summer) diarrhoea. Found in hospital infections.

P. mirabilis

Associated with putrid and decaying animal and vegetable matter. It is a doubtful agent of gastroenteritis but is found in hospital infections, e.g. in urinary tract infections and suppurative lesions.

P. rettgeri

This differs from other *Proteus* spp. in failing to produce hydrogen sulphide and in fermenting mannitol. It is associated with fowl typhoid and similar diseases of poultry and with human infections.

Providencia spp.

Are found in urinary tract infections and are possibly associated with diarrhoea in man.

References

1. DONOVAN, T.J. (1966) *Journal of Medical Laboratory Technology*, **23,** 194
2. BASCOMBE, S., LAPAGE, S.P., WILCOX, W.R. and CURTIS, M.A. (1971) *Journal of General Microbiology*, **66,** 279
3. HARVEY, R.W.S. and PRICE, T.H. (1975) *Isolation of Salmonellas*. Public Health Laboratory Service Monograph No. 8. London: HMSO
4. HARVEY, R.W.S. and PRICE, T.H. (1979) *Journal of Applied Bacteriology*, **46,** 27

5. MOORE, B. (1948) *Monthly Bulletin of the Ministry of Health and Public Health Laboratory Service*, **7,** 241
6. VASSILIADIS, P., TRICHOPOULOS, D., KALASDIDI, A. and XIROUCHAKI, E. (1978) *Journal of Applied Bacteriology*, **44,** 233
7. JAMESON, E.J. (1961) *Monthly Bulletin of the Ministry of Health and Public Health Laboratory Service*, **20,** 14
8. EDWARDS, P.R. and EWING, W.H. (1972) *Identification of the Enterobacteriaceae*. 3rd edn. Minneapolis: Burroughs

Chapter 29

Brucella, haemophilus, gardnerella, moraxella, bordetella, actinobacillus, yersinia, pasteurella, francisella, bacteroides, campylobacter and legionella

Brucella

The genus Brucella contains Risk Group III pathogens. All manipulations that might produce aerosols should be done in microbiological safety cabinets in Containment laboratories.

There are three important and several other species of small, regular Gram-negative bacilli in this genus. They are non-motile, reduce nitrates to nitrites, but carbohydrates are not attacked when normal cultural methods are used. There are several biotypes within each species.

Isolation

Human disease

Do blood cultures, using the Castenada double phase method (p. 147) in which the solid and liquid media are in the same bottle. This not only lessens the risk of contamination during repeated subculture but also minimizes the hazards of infection of laboratory workers. Use tryptone glucose media and to suppress contaminants add to each 100 ml of medium: bacitracin, 10 units; polymyxin B, 4 units; cycloheximide, 0.001 mg. Inoculate four bottles, each with 5 ml of blood. Replace the bottle caps with cotton wool plugs and incubate at 37 °C in an atmosphere of 10% carbon dioxide (candle jars are inadequate).

Examine weekly for six weeks for growth on the solid phase. If none is seen, tilt the bottle to flood the agar with the liquid, restore it to a vertical position and reincubate.

Treat bone marrow and liver biopsies in the same way. Subculture on serum glucose agar or chocolate blood agar.

Animal material

Plate uterine and cervical swabbings or homogenized fetal tissue on Farrell's selective medium[1]. This is serum glucose agar containing (per ml): bacitracin,

25 units; polymyxin, 5 units; nalidixic acid, 5 μg; vancomycin, 5 μg; cycloheximide, 100 mg; amphotericin B, 10 μg (Oxoid Brucella Supplement).

When guinea pigs have been inoculated with milk deposits or cream, plate homogenized spleen and marrow from the femur on blood agar and Farrell's medium.

Milk

Dip a throat swab in the gravity cream and inoculate Farrell's medium and the fluid enrichment medium of Bodie and Sinton[2]. This is tryptone soya broth containing 5% horse serum and these antibiotics (per ml): bacitracin, 20 units; polymyxin 5 units; nalidixic acid, 5 μg; vancomycin, 20 μg; nystatin, 100 units; cycloheximide 100 μg; amphotericin B, 4 μg; cycloserine, 312.5 μg.

Incubate under 10% carbon dioxide (loosen the caps) and subculture in duplicate on serum glucose agar and Farrell's medium. Incubate one culture in air and the other in 10% carbon dioxide for ten days.

Identification

Colonies on primary media are small, flat or slightly raised and translucent. Subculture on slopes of glucose tryptone agar with a moistened lead acetate paper in the upper part of the tube to test for hydrogen sulphide production.

Do dye inhibition tests. In Reference laboratories three concentrations of thionin (1:25 000; 1:50 000 and 1:100 000) and two of basic fuchsin (1:50 000 and 1:100 000) are used to identify biotypes.

For diagnostic purposes, inoculate tubes of prepared *Brucella* thionin and *Brucella* fuchsin media (Difco; BBL), or use glucose tryptone serum agar containing

(1) 1:50 000 thionin, and
(2) 1:50 000 basic fuchsin (National Aniline Division, Allied Chemical and Dye Corp.).

Inoculate tubes or plates with a small loopful of a 24-h culture.

It is advisable to control *Brucella* identification with reference strains, obtainable from Type Culture Collections or CDCs: *B. abortus* 554, *B. melitensis* 16M and *B. suis* 1330 (*see Table 29.1*).

TABLE 29.1 *Brucella* species

	Growth in		*Needs* CO_2	H_2S
	Fuchsin	*Thionin*		
B. melitensis	+	+	−	−
B. abortus	+	−	+	+ (most strains)
B. suis (American)	−	+	−	++
B. suis (Danish)	−	+	−	−

Do slide agglutination tests with available commercial sera (BBL, Difco, Wellcome). Absorbed monospecific sera are not yet available commercially. Reference laboratories make their own and also use bacteriophage identification and typing methods, and metabolic tests (*see* Robertson *et al.*[3]).

Species of *Brucella*

Brucella melitensis

Grows on blood agar and on glucose tryptone serum agar aerobically in 3–4 days. Does not need carbon dioxide to initiate growth. Produces hydrogen sulphide and is not inhibited by fuchsin or thionin in the concentrations used in the commercial media. This organism causes brucellosis in man, the Mediterranean or Malta fever or undulant fever. The reservoirs are sheep and goats and infection occurs by drinking goat milk.

B. abortus

Requires 5–10% carbon dioxide to initiate growth, produces hydrogen sulphide and is inhibited by thionin but not by fuchsin. It causes contagious abortion in cattle. Drinking infected milk can result in undulant fever in man, Veterinarians and stockmen are frequently infected from aerosols released during birth or abortion of infected animals. Laboratory infections are usually acquired from aerosols released by faulty techniques.

B. suis

This species does not require carbon dioxide for primary growth. American strains produce abundant hydrogen sulphide but Danish strains produce none. It is inhibited by fuchsin but not by thionin. It causes contagious abortion in pigs and may infect man, reindeer, hares and geese.

The Ring test

Purchase the stained antigen from a veterinary or commercial laboratory. It is a suspension of *Brucella abortus* cells stained with haematoxylin or another dye. Store the milk samples overnight at 4 °C before testing.

To 1 ml of well mixed raw milk in a narrow tube (75 × 9 mm), add one drop (0.03 ml) of the stained antigen. Mix immediately by inverting several times and allow to stand for 1 h at 37 °C.

If the milk contains antibodies, these will agglutinate the antigen and the stained aggregates of bacilli will rise with the cream, giving a blue cream line above a white column of milk. Weak positives give a blue cream line and a blue colour in the milk. Absence of antibodies is shown by a white cream line above blue milk. It may be necessary to add known negative cream to a low-fat milk.

False positives may be obtained with milk collected at the beginning and end of lactation, probably due to leakage of serum antibodies into the milk.

The Ring test does not give satisfactory results with pasteurized milk or with goat milk.

Whey agglutination test

This is a useful test when applied to milk from individual cows but is of questionable value for testing bulk milk.

Centrifuge quarter milk and remove the cream. Add a few drops of rennin to the skimmed milk and incubate at 37 °C for about 6 h. When coagulated, centrifuge and set up doubling dilutions of the whey from 1:10 to 1:2560, using 1-ml amounts in 75 × 9 mm tubes. Add one drop of standard concentrated *B. abortus* suspension and place in a water-bath at 37 °C for 24 h. Read the agglutination titre. More than 1:40 is evidence of udder infection unless the animal has been vaccinated recently.

Serological diagnosis

Apart from the standard agglutination test (above) there are others that are outside the scope of this book. For information on the mercaptoethanol test, antihuman globulin and complement fixation tests *see* Robertson *et al.*[3].

Useful information about the laboratory diagnosis of brucellosis is also given in the WHO Monograph No. 5[4] and the PHLS Monograph No. 14[3].

Haemophilus

For laboratory culture, these organisms require either or both of two factors that are present in blood: X factor, which is haematin, and V factor, which is diphosphopyridine nucleotide and can be replaced by co-enzymes I or II. The X factor is heat stable; the V factor is heat labile. V factor is synthesized by *Staphylococcus aureus*. The bacilli are small (1.5 μm × 0.3 μm), non-motile, usually regular, fail to grow on ordinary media and reduce nitrate to nitrite. There are several species.

Isolation and identification

Plate sputum, spinal fluid, eye swabs, etc., on chocolate agar, which is the best medium for primary isolation because V factor may be inactivated by enzymes present in fresh blood. Incubate at 37 °C for 18–24 h. Colonies may be haemolytic or non-haemolytic, are 1 mm in diameter on blood, larger on chocolate agar, grey and translucent. Bacilli are usually small and regular but filamentous forms may be seen (especially in spinal fluids and cultures from them).

'Satellitism,' i.e. large colonies around *S. aureus* colonies and around colonies of other organisms synthesizing V factor, may be observed on mixed, primary cultures on blood agar.

TABLE 29.2 *Haemophilus* species

	Haemolysis	*Factors required*			*Growth enhanced*
		XV	*V only*	*X only*	*by 10% CO_2*
H. influenzae[a]	−	+			−
H. parainfluenzae	−		+		−
H. haemolyticus	+	+			−
H. parahaemolyticus	+		+		−
H. haemoglobinophilus[b]	−			+	−
H. ducreyi	±			+	−
H. aphrophilus	−			+	+

[a] *H. aegyptius* [b] *H. canis*

Test for X and V factor requirements as follows.

Pick several colonies and spread on nutrient agar to make a lawn. Place X, V and X + V factor discs (BBL, Difco, Oxoid) on the medium and incubate at 37 °C overnight. Observe growth around the discs, i.e. if the organisms require both X and V or X or V or neither. To test sugar reactions, add Fildes' extract to glucose, maltose, lactose and mannitol fermentation broths. These reactions are not entirely reliable (*see Table 29.2*).

Agglutination and capsular swelling tests

Do slide agglutination tests using commercial sera (e.g. Wellcome) to confirm identity. These are not satisfactory with capsulated strains as agglutination can occur with more than one serum. To test for capsule swelling with homologous sera use a very light suspension of the organism in saline, coloured with filtered methylene blue solution. Mix a drop of this with a drop of serum on a slide and cover with a cover-slip. Examine with an oil immersion lens with reduced light. Swollen capsules should be obvious, compared with non-capsulated organisms. More information about these tests is given by Turk[5].

Species of *Haemophilus*

Haemophilus influenzae

Requires both X and V factors, produces acid from glucose but not lactose or maltose and varies in indole production. Although associated with upper respiratory infection, it occurs naturally in thc nasopharynx. In eye infections ('pink eye'), it is known as the Koch–Weekes bacillus. Is also found in the normal vagina and is one of the causative organisms of purulent meningitis. There are several serological types and growth 'phases.' *H. aegyptius* is probably this species.

H. parainfluenzae

This differs only in not requiring X factor. Normally present in the throat, but may be pathogenic.

H. haemolyticus and parahaemolyticus

Haemolytic organisms with the same properties as *H. influenzae* and *H. parainfluenzae*, respectively.

H. haemoglobinophilus (*H. canis*)

Requires X but not V factor. It is found in preputial infections in dogs and in the respiratory tract of man.

H. suis

This has properties similar to those of *H. influenzae* but fails to attack carbohydrates in ordinary sugar media. It occurs in swine and is associated with swine influenza.

H. ducreyi

Ducrey's bacillus, associated with chancroid or soft sore, is difficult to grow but may be obtained in pure culture by withdrawing pus from a bubo and inoculating inspissated whole rabbit blood slopes or 30% rabbit blood nutrient agar at 35 °C in a 10% carbon dioxide atmosphere. It requires X but not V factor.

H. piscium

Has been isolated in furunculosis-like diseases of salmonid fish. In addition to X and V factors it requires an aqueous extract of the tissues of these fish.

H. aphrophilus

Colonies on chocolate agar are small (0.5 mm) at 24 h, smooth and translucent. Better growth is obtained in a 10% carbon dioxide atmosphere. Some strains require X factor. There is no growth on MacConkey agar. Acid is produced from glucose, maltose, lactose and sucrose but not from mannitol. Fermentation media should be enriched with Fildes' extract. It is oxidase and catalase negative.

Human infections, including endocarditis have been reported[6, 7].

This organism closely resembles *Actinobacillus actinomycetemcomitans* (p. 314) but it does not produce acid from lactose and sucrose (*see Table 29.5*).

Gardnerella

The species *Gardnerella vaginalis* was previously known as *Haemophilus vaginalis* or *Corynebacterium vaginale*.

Plate vaginal swabs on blood agar and incubate overnight anaerobically or under carbon dioxide. Colonies of this medium are about 0.5 mm in diameter at 24 h and are colourless, transparent and usually β haemolytic. Neither X nor V factor is required. There is no growth on MacConkey agar and apart from acid from glucose (add Fildes' extract), biochemical tests seem to be all negative.

Subculture on blood agar and apply discs of metronidazole (50 μg) and sulphonamide (100 μg) (Oxoid). *G. vaginalis* is sensitive to metronidazole but resistant to sulphonamide (Bailey *et al.*)[8].

This organism is found in the human vagina and may be associated with nonspecific vaginitis.

Moraxella

The bacilli are plump (2 μg × 1 μg), often in pairs end to end, and non-motile. They are oxidase positive and do not attack sugars. They are indole negative, do not produce hydrogen sulphide and are sensitive to penicillin. Some species require enriched media.

Plate exudate, conjunctival fluid, etc., on blood agar and incubate at 37 °C overnight. Subculture colonies of plump, oxidase positive Gram-negative bacilli on blood agar and on nutrient agar (not enriched), on gelatin agar (use the plate method) and on Loeffler medium and incubate at 37 °C. Do the nitrate reduction test (*see Table 29.3*).

TABLE 29.3 Species of *Moraxella*

	Growth on nutrient agar	*Gelatin liquefaction*	*Nitrate reduction*	*Catalase*	*Urease*
M. lacunata[a]	–	+	+	+	–
M. nonliquefaciens	v	–	v	+	–
M. bovis	+	+	v	v	–
M. osloensis	+	–	v	+	–
M. kingii[b]	+	v	v	–	–
M. phenylpyruvica	+	–	+	+	+

[a] Also known as *Moraxella liquefasciens*
[b] *Kingella kingii*

Species of *Moraxella*

Moraxella lacunata

Colonies on blood agar are small and may be haemolytic. There is no growth on non-enriched media. Colonies on Loeffler medium are not visible but are indicated by pits of liquefaction ('lacunae'). Gelatin is liquefied slowly. Nitrates are reduced.

This organism is associated with angular conjunctivitis and is known as the Morax–Axenfeld bacillus.

M. liquefaciens

This is similar to *M. lacunata* and may be a biotype of that species. It liquefies gelatin rapidly.

M. nonliquefaciens

This is also similar to *M. lacunata* but fails to liquefy gelatin and to reduce nitrates.

M. bovis

Requires enriched medium, liquefies gelatin but does not reduce nitrates. It causes pink eye in cattle but has not been reported in human disease.

M. osloensis (*M. duplex*, *Mima polymorpha* var. *oxidans*)

Enriched medium is not required. Gelatin is not liquefied; nitrates may be reduced. It is found on the skin and in the eyes and the respiratory tract of man but its pathogenicity is uncertain.

M. kingii (now Kingella kingii)

Colonies are haemolytic and may be mistaken for haemolytic streptococci or haemolytic haemophilus. It does not require enriched media. This is the only member of the group which is catalase negative. It has been found in joint lesions and in the respiratory tract[9].

M. phenylpyruvica

Another 'new' species[10]. It does not require enriched media. It is strongly urea positive and reduces phenylalanine to phenylpyruvic acid (PPA positive)—the only species in this group to do this. It has been isolated from assorted human material but its pathogenicity is uncertain.

M. lwoffii

As this is oxidase negative, it is included in *Acinetobacter* (p. 265).

Bordetella

The bacilli in this genus are small, 1.5 μm × 0.3 μm and regular. Nitrates are not reduced and carbohydrates are not attacked.

Isolation and identification

Perinasal swabs are better than cough plates, but swabs are best conveyed in one of the commercial transport media (p. 60). Plate on Bordetella, Bordet-Gengou or Thayer Martin medium containing cephalexin 40 mg/l (Oxoid Bordetella supplement). Incubate under conditions of high humidity and at 37 °C for three days. Examine daily and identify by slide agglutination, using commercially available sera. FA reagents are also available.

Test for growth and pigment formation on nutrient agar, urease, nitratase and motility (*see Table 29.4*).

TABLE 29.4 *Bordetella* species

	Growth on nutrient agar	*Brown coloration*	*Urease*	*Nitrate reduction*	*Motility*
B. pertussis	−	−	−	−	−
B. parapertussis	+	+	+	−	−
B. bronchiseptica	+	−	+	+	+

Species of *Bordetella*

Bordetella pertussis

Growth on one of the above media has been described as looking like a 'streak of aluminium paint'. Colonies are small (about 1 mm in diameter) and pearly grey. This organism cannot grow in primary culture without heated blood, but can be accustomed to nutrient agar. *B. pertussis* causes whooping cough. For detailed information about the bacteriological diagnosis of this disease, *see* Lautrop and Lacey[11] and Brooksaler and Nelson[12].

B. parapertussis

Growth on blood agar may take 48 h and a brown pigment is formed under the colonies. It grows on nutrient and on MacConkey agar. On Bordetella or similar media, the pearly colonies ('aluminium paint') develop earlier than those of *B. pertussis*. It does not change Hugh and Leifson medium and does not attack carbohydrates. It is nitrate negative, urease positive and does not grow in KCN medium.

This organism is one of the causative organisms of whooping cough.

B. bronchiseptica

This organism has been in the genera *Brucella* and *Haemophilus*. It forms small, smooth colonies, occasionally haemolytic on blood agar, grows best as 37 °C and is motile, urease positive and grows in citrate medium.

It causes bronchopneumonia in dogs, often associated with distemper, bronchopneumonia in rodents and snuffles in rabbits. It has been associated with whooping cough.

Actinobacillus

This genus contains non-motile Gram-negative rods that are fermentative but do not produce gas and vary in their oxidase and catalase reactions.

Isolation and identification

Plate pus, which may contain small white granules, or macerated tissue on blood agar and incubate at 37 °C under a 10% carbon dioxide atmosphere. Subculture small flat colonies on to MacConkey agar, in Hugh and Leifson medium and test for fermentation of glucose, lactose and sucrose in enriched fermentation media (*see Table 29.5*).

TABLE 29.5 ***Actinobacillus, Haemophilus aphrophilus*** **and** ***Cardiobacterium***

	Growth on MacConkey agar	*Acid from*		*Oxidase*	*Catalase*
		Lactose	*Sucrose*		
A. lignieresii	+	late	+	+	+
A. equuli	+	early	+	+	+
A. actinomycetemcomitans	−	−	−	−	+
H. aphrophilus	−	+	+	−	−
Cardiobacterium hominis	−	−	+	+	−

Species of *Actinobacillus*

Actinobacillus ligniersii

This species grows on MacConkey agar, is fermentative and produces acid only from glucose, lactose (late) and sucrose.

It is oxidase and catalase positive. It is associated with woody tongue in cattle and human infections have been reported.

A. equuli

This species resembles *A. ligniersii* but ferments lactose rapidly. The oxidase and catalase tests give variable reactions. It is associated with joint ill and sleepy disease of foals, and is also known as *Shigella equirulis* or *S. equulis*.

A. actinomycetemcomitans

This organism requires carbon dioxide on primary isolation and some strains require the X factor. Use an enriched medium and place an X factor disc on the heavy part of the inoculum. There is no growth on MacConkey agar and the oxidase test is negative. Acid is produced from glucose (fermentatively) but not from lactose or sucrose. It is catalase negative.

This organism is difficult to distinguish from *H. aphrophilus* (*see Table 29.4*). It is sometimes associated with infections by *Actinomyces israelii* and it has been recovered from blood cultures of patients with endocarditis.

Cardiobacterium hominis

This species is included here for convenience.

It is a facultative anaerobe and requires an enriched medium and high humidity. There is no growth on MacConkey medium, the oxidase test is positive and glucose and sucrose, but not lactose, are fermented without gas production. The catalase, urea and nitrate reduction tests are negative. Hydrogen sulphide is produced but gelatin is not liquefied (*see Table 29.4*).

This organism has been isolated from blood cultures of patients with endocarditis but may also be found in the upper respiratory tract.

Yersinia, Pasteurella and *Francisella*

These genera were originally included in *Pasteurella*. Cowan and Steel[13] and Philip and Owen[14] have removed the plague and pseudotuberculosis organisms to *Yersinia*. The genus *Pasteurella* now contains the haemolytic septicaemia organisms. The tularaemia organism has been placed in *Francisella*. They are small, ovoid, often pleomorphic, Gram-negative, non-sporing bacilli, which may require blood agar or Fildes' extract to grow on primary culture but which thereafter grow well on ordinary medium (except *Francisella tularensis*). They are catalase positive, fermentative in Hugh and Leifson medium (slow reaction), reduce nitrates to nitrites and fail to liquefy gelatin.

The plague bacillus, Yersinia pestis, is in Risk Group III and all work should be done in a microbiological safety cabinet in a Containment laboratory. After isolation and presumptive identification cultures should be sent to a Reference laboratory for any further tests.

Isolation and identification

Plague

In suspected bubonic plague examine pus from buboes; in pneumonic plague, sputum; and in rats the heart blood, enlarged lymph nodes and spleen. Before examining rats immerse them in disinfectant for several hours to kill fleas which might be infected.

Make Gram- and methylene blue-stained films and look for small oval Gram-negative bacilli with capsules. Plate on blood agar containing 0.025% sodium sulphite to reduce the oxygen tension, on 3% salt agar and on MacConkey agar. Incubate for 24 h.

Colonies of plague bacilli are flat or convex, greyish white and about 1 mm in diameter at 24 h. Subculture into nutrient broth and cover with liquid paraffin. Test for motility at 22 °C, urease, indole, ornithine decarboxylase, growth in KCN and inoculate sucrose, cellobiose, amygdalin and rhamnose media (*see Table 29.6*).

Films of colonies on salt agar show pear-shaped and globular forms. *Yersinia pestis* grows on MacConkey agar and usually produces stalactite growth in broth covered with paraffin. It is motile at 22 °C, is urease, indole and ornithine decarboxylase negative, does not grow in KCN and does not produce acid from sucrose, cellobiose, amygdalin and or rhamnose.

It is the causative organism of plague in rats, transmitted from rat to rat and from rat to man by the rat flea. In man plague is bubonic, pnuemonic or septicaemic.

Materials and cultures should be sent to the appropriate Reference Laboratory or Communicable Diseases Centre. If plague investigations are seriously contemplated, the papers by Goldenberg[15] and Baltazard[16] should be consulted. Reference laboratories use FA methods, bacteriophage, specific agglutination tests, precipitin tests and animal inoculations.

Intestinal infections

Look for *Yersinia enterocolitica*, which is cold tolerant, grows well at 22–29 °C but poorly, if at all at 37 °C.

Inoculate selenite broth with faeces, selenite broth and/or phosphate buffered saline with 10% suspensions of food made in a Stomacher Lab-Blender. Keep at 4 °C for six weeks and subculture weekly on deoxycholate media containing the minimum amount of the bile salt (SS or the Oxoid Selective Yersinia medium). Direct plating may be successful. If DCA cultures have been incubated at 37 °C for salmonellas and shigellas reincubate them at 22–29 °C, but again, too much deoxycholate is inhibitory.

Lactose-Sucrose-Urea (LSU) medium is recommended[17]. *Y. enterocolitica* colonies have indistinct edges and granular centres at 24 h. On further incubation the colonies become clearly visible and the centres look like ground glass. Test for motility at 22 °C, urease (Christensen), indole, KCN tolerance, ornithine decarboxylase and inoculate sucrose, cellobiose, amygdalin and rhamnose media (*see Table 29.6*).

TABLE 29.6 Species of *Yersinia, Pasteurella* and *Francisella*

	Growth on		Indole	Motility at 22°C	Urease	Ornithine decarboxylase	Acid from				KCN	Growth at 4°C
	Nutrient agar	MacConkey					Sucrose	Cellobiose	Amygdalin	Rhamnose		
Y. pestis	+	+	−	−	−	−	−	−	−	−	−	−
Y. enterocolitica	+	+	v	+	+	+	+	+	+	+/−[a]	−	+
Y. pseudotuberculosis	+	+	−	+	+	−	−	−	−	+	−	−
Y. multocida	+	−	+	−	v	+					+	
F. tularensis	−	−	−	−	−	−	−	−	−	−	−	−

[a] Toxigenic strains usually positive. Environmental strains usually negative

Toxigenic strains are said not to ferment rhamnose. Those that do are the environmental, or YELO (Yersinia enterocolitica-like organisms).

Y. enterocolitica causes enteritis in man and animals. The reservoirs are pigs, cattle, poultry, rats, cats, dogs and chinchillas. It has been found in milk and milk products, oysters, mussels and in water. Person to person spread occurs, mainly in families.

For further information *see* Morris and Feeley[18] and Schieman[19].

Pseudotuberculosis

This is caused by *Yersinia pseudotuberculosis*. Make suspensions of stool and other material and store it at 4 °C for 48 h. Then plate on MacConkey and DCA.

Y. pseudotuberculosis

Ovoid and coccoid forms are common, and also bipolar staining. It is motile at 22 °C. Colonies on nutrient agar are raised, sometimes umbonate, granular, about 1 mm in diameter. Rough variants occur. Growth on MacConkey agar is obvious but poor. Indole is not formed; acid is produced in glucose and maltose; it is urease positive, KCN negative and ornithine decarboxylase negative. It causes pseudotuberculosis in rodents and is important in experimental animal houses.

Human infections have been reported in Europe.

Haemorrhagic disease in animals

Look for *Pasteurella multocida (P. septica)*. Examine Gram-stained films; culture on blood agar and MacConkey agar.

Pasteurella multocida

Bipolar staining and pleomorphism are less obvious than in other species. The bacilli are very small (about 1.5 μg × 0.3 μm) and non-motile. Colonies on nutrient agar are translucent, slightly raised, about 1 mm in diameter at 24 h; on blood agar they are slightly larger, more opaque and non-haemolytic. There is no growth on MacConkey agar. Acid is produced from glucose; this is the only species which may not produce acid from maltose and is indole positive and KCN positive. It is urease variable and ornithine decarboxylase positive (*see Table 29.6*).

This organism causes haemorrhagic septicaemia in domestic and wild animals and birds, e.g. fowl cholera, swine plague, transit fever. Man may be infected by animal bites or by inhaling droplets from animal sneezes. It has also been isolated from septic fingers.

It is often named after host species and given specific epithets such as *avicida, aviseptica, suilla, suiseptica, bovicida, boviseptica, ovicida, oviseptica, cuniculocida, lepiseptica, muricida, muriseptica*. There is some host interspecificity: Cattle strains are also pathogenic for mice but not for fowls; fowl cholera strains are pathogenic for cattle and mice; lamb septicaemia strains are not pathogenic for rodents.

Tularaemia

The causative organism is *Francisella tularensis.*

> *Francisella tularensis is in Risk Group III. It is highly infectious and has caused many laboratory infections. It should be handled with great care in microbiological safety cabinets in Containment laboratories.*

Culture blood, exudate, pus or homogenized tissue on several slopes of blood agar (enriched), cystine glucose agar and inspissated egg yolk medium. Also inoculate tubes of plain nutrient agar. Incubate for 3–6 days and examine for very small drop-like colonies. If the material is heavily contaminated, add 100 μg/ml of cycloheximide or 200 units/ml of nystatin and 2.5–5 μg/ml of neomycin.

Subculture colonies of small, swollen or pleomorphic Gram-negative bacilli on blood and nutrient agar, on MacConkey agar and in enriched nutrient broth. Test for acid production from glucose and maltose, motility at 22 °C, urease activity and growth in KCN broth (*see Table 29.6*).

Francisella tularensis

The bacilli are very small, show bipolar staining and are non-motile. There is no growth on nutrient or MacConkey agar, poor growth on blood agar with very small, grey colonies; on inspissated egg, at three to four days, colonies are minute and drop-like. It is a strict aerobe. Acid is formed in glucose and maltose. Indole and urease tests are negative; there is no growth in KCN broth and no motility at 22 °C.

The organism is responsible for a plague-like disease in ground squirrels and other rodents in western USA and is transmissible to man.

Cultures and material in cases of suspected tularaemia should be sent to a Centre for Disease Control. Fluorescent antibody procedures are used for rapid diagnosis.

Bacteroides, Fusobacterium, Eikenella

These Gram-negative anaerobic bacilli are found in the alimentary and genitourinary tracts of man and other animals and in necrotic lesions, often in association with other organisms. They are pleomorphic, often difficult to grow and die easily in cultures. Identification is often difficult. Reference experts use gas–liquid chromatography. It seems desirable, therefore, for diagnostic laboratories to report their isolates as belonging to, e.g. the *Bacteroides fragilis* group, the *Bacteroides melanogenicus* group, the *Fusobacterium necrophorus* group and as *Eikenella*.

In pathological material they are not uncommonly mixed with anaerobic or capnophilic cocci. In general, primary infections above the diaphragm are due to *Fusobacterium*, but *Bacteroides* may occur anywhere and are frequently recovered from blood cultures.

Isolation and identification

Examine Gram films of pathological material. Diagnosis of Vincent's angina can be made from films counterstained with dilute fuchsin, if fusiform bacilli can be seen in large numbers, associated with spirochaetes (*Borrelia vincenti*). In pus from abdominal, brain, lung or genital tract lesions, large numbers of Gram-negative bacilli of various shapes and sizes, with pointed or rounded ends suggest *Bacteroides* infection. Such material often has a foul odour.

Culture pus and other material anaerobically, aerobically and under 5–10% carbon dioxide. For anaerobic cultures use an enriched blood agar containing (per ml): neomycin, 75 μg; menadione, 0.5 μg. Most culture media manufacturers will supply ready-to-use supplements containing these and other materials containing haemin. Inoculate Robertson's cooked meat broth and thioglycollate broth containing 75 μg/ml neomycin. For aerobic and carbon dioxide culture use neomycin blood agar but do not add menadione.

Incubate at 37 °C for 24 h and compare the cultures. If there is no or poor growth reincubate and examine again after 3–5 days. Look for brown or black pigmentation and greening around colonies.

Examine Gram-stained films of colonies on the anaerobic and capnophilic cultures which do not grow aerobically. Subculture the liquid medium at 24-h intervals for 5–6 days on appropriate blood agar and incubate in the same way. Test suspected organisms for antibiotic sensitity using these discs: erythromycin, 60 μg; rifampicin, 15 μg; colistin sulphate, 10 μg; penicillin G, 2 units; kanomycin, 1000 μg; and vancomycin, 5 μg (Leigh and Simmons)[20]. Sets of these discs are supplied by Mast (MID 8) and Oxoid (An-ident). Test also for bile resistance with discs dipped in ox-gall (Difco, Oxoid), and for β-lactamase (p. 103) (*see Table 29.7*). The API 20E methods is also useful. For more information about these organisms *see* Holdeman and Johnson[21], Finegold and Barnes[22] and Hamman[23].

Campylobacter

These small, curved, actively motile rods are microaerophilic, reduce nitrates to nitrites but do not attack carbohydrates. They have become very important in recent years as a cause of food poisoning (Chapter 14).

Isolation

Plate emulsions of faeces on blood agar containing (commercially available) campylobacter growth and antibiotic selective supplements (p. 59). Incubate duplicate cultures at 37 and 42 °C for 24–48 h in an atmosphere of approximately 5% oxygen, 10% carbon dioxide and 85% nitrogen, preferably using a BBL or Oxoid gas generating kit, otherwise a candle jar.

Identification

Campylobacter colonies are about 1 mm in diameter at 24 h, grey, watery and flat. Older colonies become opaque. Examine Gram-stained films and inoculate chocolate agar for sensitivity to nalidixic acid disc (30 μg). Test for hydrogen sulphide in cysteine medium with lead acetate strips and for growth in the presence of 1% glycine and 8% glucose in broth media (*see Table 29.8*).

TABLE 29.7 Groups of *Bacteroides*, *Fusobacterium* and *Eikenella*

	Resistance to							β-lactamase
	Penicillin	*Erythromycin*	*Rifampicin*	*Vancomycin*	*Kanamycin*	*Colistin*	*Bile*	
B. fragilis group	+	–	–	+	+	+	+	+
B. melanogenicus group	–	–	–	+	+	v	–	+
F. necrophorus group	–	+	+	+	–	–	–	–
Eikenella	–	–	–	+	–	–	+	–

Adapted from Leigh and Simmons[20], Mast and Oxoid

TABLE 29.8 Species of *Campylobacter*

European name	*N. American name*	Growth at		Resistance to		Growth in	
		25°C	*42°C*	*nalidixic acid*	H_2S	*1% glycine*	*8% glucose*
C. fetus venerealis	*C. fetus fetus*	+	–	+	–[a]	–	+
C. fetus fetus	*C. fetus intestinalis*	+	–	+	+	+	+
C. jejuni	*C. fetus jejuni*	–	+	–	+	+	–
C. coli	*C. fetus jejuni*	–	+	–	+	+	+
NARTC[b]		v	+	+			

[a] a sub-type produces limited H_2S
[b] NARTC—nalidixic acid resistant thermophilic campylobacters

Species of *Campylobacter*

The names in current use vary with the geographical location of their authors and do not always agree with the Approved Lists of Skerman *et al.*[23]. The text and tables therefore give alternative names.

Campylobacter venerealis (*C. fetus* subsp. *fetus*)

Is found in the genitalia of bulls, who act as carriers, and causes abortion and sterility in cows. It cannot multiply in the intestine and therefore is not known to cause human disease.

C. fetus fetus (*C. fetus intestinalis*)

Is found in sheep, when it may cause abortion. It can multiply in the intestine and is the usual cause of systemic campylobacteriosis, but rarely enteritis in humans.

C. jejuni (*C. fetus subsp. jejuni*)

Forms part of the normal flora of many domestic and wild animals and is the cause of most cases of campylobacter enteritis.

C. coli (*C. fetus subsp. jejuni*)

Is also found in animals. It multiplies in the small bowel and causes infectious abortion in sheep.

Other species

There are un-named nalidixic acid resistant thermophilic campylobacters (NARTC) and also a species, *C. sputorum*, which is not regarded as significant.

For more information about campylobacters *see* Skirrow[25], Raz *et al.*[26] and Newell[27].

Legionella

Legionella haemophila and related organisms are small Gram-negative bacilli which will not grow on ordinary, unenriched media. They are responsible for severe often fatal respiratory tract infections, sometimes in explosive outbreaks. They have been found in water supplies, in association with, as well as independent of human disease.

Isolation and identification

Examine bronchial secretions and lung tissue by FA methods. Homogenize the material in buffered saline or peptone water diluent and culture on Greaves[28] medium, which is blood agar containing (per ml): L-cysteine HCl,

400 μg; ferric pyrophosphate, 250 μg; sodium selenate, 10 μg; colistin, 15 units; vancomycin, 5 μg; trimethoprim, 2.5 μg; amphotericin B 2.5 μg. These are conveniently supplied as Legionella Growth and Legionella Selective Supplements by Oxoid.

Incubate under 5% carbon dioxide and humid conditions at 37 °C and examine daily for five days and then at seven and 14 days. Colonies of *L. pneumophilia* are up to 3 mm in diameter, circular, low convex, with moist, glistening surface and are grey- or greenish brown in colour. They are weakly catalase and oxidase positive. Confirm their identity with FA techniques.

The review by Fallon gives more information[29].

References

1. FARRELL, I.D. and ROBERTSON, L. (1972) *Journal of Applied Bacteriology*, **35,** 625
2. BRODIE, J. and SINTON, G.P. (1975) *Journal of Hygiene, Cambridge*, **74,** 359
3. ROBERTSON, L., FARRELL, I.D., HINCHCLIFFE, P.M. and QUAIFE, R.A. (1980) *Benchbook on Brucella.* Public Health Laboratory Service Monograph No. 14. London: HMSO
4. WORLD HEALTH ORGANIZATION (1981) *A Guide to the Diagnosis, Treatment and Prevention of Human Brucellosis.* Edited by S.S. Elberg. Geneva: WHO
5. TURK, D.C. (1982) *Haemophilus influenzae.* Public Health Laboratory Service Monograph No. 17. London: HMSO
6. SUTTER, V.L. and FINEGOLD, S.M. (1970) *Annals of the New York Academy of Science*, **174,** 468
7. PAGE, M.I. and KING, E.O. (1966) *New England Journal of Medicine*, **275,** 181
8. BAILEY, R.K., VOSS, J.L. and SMITH, R.F. (1979) *Journal of Clinical Microbiology*, **9,** 65
9. BØVRE, K. and HENRIKSEN, S.D. (1967) *International Journal of Systematic Bacteriology*, **17,** 343
10. HENRIKSEN, S.D. and BØVRE, K. (1968) *Journal of General Microbiology*, **51,** 377
11. LANTROP, H. and LACEY, B.W. (1961) *Bulletin of the World Health Organization*, **23,** 15
12. BROOKSALLER, F. and NELSON, J.D. (1967) *American Journal of Diseases of Children*, **114,** 389
13. COWAN, S.T. (1974) *Cowan and Steel's Manual for the Identification of Medical Bacteria.* 2nd edn, Cambridge: Cambridge University Press
14. PHILIP, C.B. and OWEN, C.R. (1961) *International Bulletin of Bacterial Nomenclature and Taxonomy*, **11,** 67
15. GOLDENBERG, M.I. (1968) *Health Laboratory Science*, **5,** 38
16. BALTAZARD, M. (1960) *Bulletin of the World Health Organization*, **14,** 313
17. JUHLIN, I. and ERICSON, A. (1961) *Acta Pathologia et Microbiologia, Scandinavica*, **52,** 185
18. MORRIS, G.K. and FEELEY, J.C. (1976) *Bulletin of the World Health Organization*, **54,** 79
19. SCHIEMAN, D.A. (1982) *Culture* (Oxoid), March
20. LEIGH, D.A. and SIMMONS, K. (1988) *Journal of Clinical Pathology*, **30,** 991
21. HOLDEMAN, L.V. and JOHNSON, J.L. (1977) *International Journal of Systematic Bacteriology*, **27,** 337
22. FINEGOLD, S.M. and BARNES, E.M. (1977) *International Journal of Systematic Bacteriology*, **27,** 388
23. HAMMAN, R. (1982) *Journal of Medical Microbiology*, **15,** 293
24. SKERMAN, V.D.B., MCGOWAN, V. and SNEATH, P.H.A. (1980) *International Journal of Systematic Bacteriology*, **30,** 225
25. SKIRROW, M.B. (1977) *British Medical Journal*, **2,** 9
26. RAZ, M.H.H., PARK, R.W.A. and SKIRROW, M.B. (1981) *Journal of Applied Bacteriology*, **50,** 55
27. NEWELL, D.G. (1982) (Editor) *Campylobacter: Epidemiology, Pathogenesis and Biochemistry.* Lancaster: MTP Press
28. GREAVES, P.W. (1980) *Journal of Clinical Pathology*, **33,** 581
29. FALLON, J. (1979) *Culture*, (Oxoid), September, pp. 3–4

Chapter 30

Neisseria, branhamella, gemella and veillonella

The genera *Neisseria*, *Branhamella*, *Gemella* and *Veillonella* are included in this chapter. *Neisseria* and *Branhamella* are oxidase and catalase negative; both are aerobes; *Veillonella* are obligate anaerobes.

Neisseria

These occur mostly as oval or kidney-shaped cocci arranged in pairs with their long axes parallel. They are non-motile, non-sporing, oxidase positive and reduce nitrates to nitrites. Most species are aerobic, facultatively anaerobic but obligate anaerobes are known. The two important pathogens are *N. gonorrhoeae* (the gonococcus) causative organism of gonorrhoea, and *N. meningitidis* (the meningococcus), one of the organisms causing cerebrospinal meningitis.

Isolation and identification

Gonococcus

If material cannot be examined immediately use Transgrow media for transport as gonococci die rapidly. In the male, examine Gram-stained films of urethral discharge for intracellular Gram-negative diplococci. Culture as below. In the female, culture material from the cervix or urethra (not vagina; the vaginal secretion is too acidic for these organisms to survive).

Culture pus from 'sticky eyes'. Inoculate New York City (NYC) or Thayer Martin (TM) media supplemented with enriched additives and antibiotics. Incubate all cultures at 35–36 °C (better than 37 °C) in 5–10% CO_2 and 70% humidity. Examine at 24 and 48 h. Colonies of *N. gonorrhoea* are transparent discs about 1 mm in diameter, later increasing in size and opacity, when the edge becomes irregular. Test suspicious colonies by the Gram film and oxidase test. Typical morphology and a positive oxidase test are presumptive evidence of the gonococcus in material from the male urethra, but not from other sites.

For rapid results test colonies by FA method or use coagglutination (Phadebact).

Subculture oxidase positive colonies on chocolate agar and incubate overnight in a 10% carbon dioxide atmosphere. Immediate subculture into carbohydrate test media may not be satisfactory as the primary medium contains antibiotics; other organisms may grow. Repeat the oxidase test on the subculture and emulsify positive colonies in about 1 ml of serum broth. Use this to test for acid production from glucose, maltose, sucrose and lactose. Use rabbit serum media because horse serum contains a maltase that may give a false-positive reaction with maltose. The best results are obtained with a semisolid medium, as gonococci do not like liquid media.

Disc methods are satisfactory. Place BBL or Difco carbohydrate discs on cystine tryptose agar heavily inoculated from the primary culture. Incubate overnight in 5–10% CO_2. Place Oxoid discs in small test-tubes containing a heavy emulsion of colonies in saline. Read results after 4–5 h incubation at 37 °C (*Table 30.1*). Confirm fermentation results by agglutination or coagglutination tests.

TABLE 30.1 ***Neisseria*** **species**

	Acid from				*Growth on*	
	Glucose	*Maltose*	*Sucrose*	*Lactose*	*Nutrient agar*	*Thayer Martin medium*
N. gonorrhoeae	+	−	−	−	−	+
N. meningitidis	+	+	−	−	−	+
N. lactamicus	+	+	−	+	+	+
Other *Neisseria*	v	v	v	v	v	−
Branhamella catarrhalis	−	−	−	−	+	−
Gemella	+	+	+	−	+	−

v, variable

Exercise caution in reporting these organisms without adequate experience, especially if they are from children, eye swabs, anal swabs or other sites.

They may be meningococci or other *Neisseria* which are genital commensals of no venereal significance. Sugar reactions and colonial morphology are not entirely reliable. Certain other organisms, *Gemella* (*see below*), *Moraxella*, (p. 308) and *Acinobacter calcoaceticus* (p. 266), may resemble gonococci in direct films and on primary isolation. *Moraxella*, like *Neisseria*, are oxidase positive but *Gemella* and *Acinetobacter* are oxidase negative.

For further information on gonococci: *see* the WHO Report[1], Skinner[2] and Oxoid[3].

Meningococcus

Examine films of spinal fluid for intracellular Gram-negative diplococci. Inoculate blood agar and chocolate agar with spinal fluid, or its centrifuged deposit. Incubate, preferably in a 5% carbon dioxide atmosphere, at 37 °C overnight.

Incubate the remaining CSF overnight and repeat culture on chocolate agar.

Thayer Martin medium may also be used but there is always the possibility of a mixed infection with organisms (e.g. *Haemophilus* spp. or streptococci) that do not grow on TM medium.

Meningococcal colonies are transparent, raised discs about 2 mm in diameter at 18–24 h. In spinal fluid culture, the growth will be pure, while in eye discharges and vaginal swabbings of young children, etc., other organisms will be present. Subculture oxidase positive colonies of Gram-negative diplococci in semisolid serum sugar media (glucose, maltose, sucrose). Meningococci give acid in glucose and maltose (*see Table 30.1*).

Agglutinating sera are available but should be used with caution. There are at least four serological groups. Slide agglutinations may be unreliable. It is best to prepare suspensions in formol saline for tube agglutination tests.

N. lactamicus (*N. lactamica*)

This is important in that it might be confused with pathogenic species because it is oxidase positive and grows on Thayer Martin medium, but it produces acid from lactose and is fermentative. The pathogens do not change lactose and are oxidative. *N. lactamicus* is also ONPG positive.

Other species and similar organisms

These grow on nutrient agar. Colonies are variable: 1–3 mm, opaque, glossy or sticky, fragile or coherent, very rough and granular and often showing a lemon or deep yellow pigment.

Branhamella

The single species in the genus, *B. catarrhalis,* was known for many years as *N. catarrhalis*. It is a commensal of the upper respiratory tract, grows on ordinary agar and produces large colonies that may be 'smooth' or 'rough' but fail to emulsify in saline. Carbohydrates are not fermented.

Gemella

The single species *G. haemolysans* resembles *Neisseria* morphologically but is oxidase and catalase negative. It grows at 22 °C on ordinary agar, produces acid from glucose, maltose and sucrose and is haemolytic on blood agar. It is found in the respiratory tract of mammals (*Table 30.1*).

Veillonella

These are obligate anaerobes and occur as irregular masses of very small Gram-negative cocci, cultured from various parts of the respiratory and alimentary tracts and genitalia. They do not appear to be significant. Colonies on blood agar are minute and may be haemolytic. Gas bubbles are formed in stab cultures of peptone media containing agar. *V. alcalescens*, a possible pathogen of rabbits, does not produce hydrogen sulphide or attack carbohydrates. It is catalase positive. *V. parvula*, a commensal, does both and is catalase negative. Both are oxidase positive. These organisms die very rapidly in air.

Use the API anaerobic system for identification.

For information about these and other anaerobic cocci, *see* Rosenblatt[4].

References

1. WORLD HEALTH ORGANIZATION (1978) *Neisseria gonorrhoea and gonococcus infection.* Technical Report Series 616. Geneva: WHO
2. SKINNER, F.A., WALKER, P.D. and SMITH, H. (Editor) (1977) *Gonorrhoea: Epidemiology and Pathogenesis.* FEMS Symposium. London: Academic Press
3. OXOID (1982) *Pathogenic Neisseria Isolation Media and Techniques.* Oxoid Manual, pp. 230–237
4. ROSENBLATT, J.E. (1980) Anaerobic cocci. In *Manual of Clinical Microbiology.* 3rd edn. Edited by E.H. Lennett and E.H. Spaulding and J.P. Truant, pp. 426–430. Washington: American Society for Microbiology

Chapter 31

Staphylococcus, micrococcus, peptococcus and sarcina

Staphylococci and micrococci are frequently isolated from pathological material and foods. Distinguishing between the two groups is important. Some staphylococci are known to be pathogens; some are doubtful or opportunist pathogens; others, and micrococci, appear to be harmless but are useful indicators of pollution. Staphylococci are fermentative, capable of producing acid from glucose anaerobically; micrococci are oxidative and produce acid from glucose only in the presence of oxygen.

As with other groups of micro-organisms recent taxonomic and nomenclatural changes have not exactly facilitated the identification of the Gram-positive cocci which are isolated in routine laboratories from clinical material and foods. The term 'micrococci' is still loosely but widely used to include many that cannot be easily identified (*see also* under *Aerococcus* and *Pediococcus* (pp. 340; 342). The genera *Peptococcus* and *Sarcina* are anaerobes.

Isolation

Pathological material

Plate pus, centrifuged urine deposits, swabs, etc., on blood agar and chloral hydrate blood agar to prevent the spread of proteus and inoculate salt meat broth. Incubate overnight at 37 °C. Plate the salt meat broth on blood agar.

Foodstuffs

Prepare 10% suspensions in 0.1% peptone water in a Stomacher Lab-Blender. If heat stress is suspected add 0.1-ml amounts to several tubes containing 10 ml of brain heart infusion broth. Incubate for 3–4 h and then subculture. Use one or more of the following media. Milk salt agar (MSA), phenolphthalein phosphate polymyxin agar (PPP), Baird-Parker medium (BP) or tellurite polymyxin egg yolk agar (TPEY). Incubate for 24–48 h. Test the PPP medium with ammonia for phosphatase-positive colonies (*see below*).

MSA agar relies on the high salt content to select staphylococci. In PPP medium, the polymyxin suppresses many other organisms. BP medium is very selective but may be overgrown by proteus. Add 50 μg/ml of sulphamethazine.

For enrichment or to assess the load of staphylococci, use salt meat broth or mannitol salt broth. Add 0.1 and 1.0 ml of the emulsion to 10 ml of broth and 10 ml to 50 ml. Incubate overnight and plate on one of the solid media. If staphylococci are present in large numbers in the food, there will be a heavy growth from the 0.1-ml inoculum; very small numbers may give growth only from the 10-ml sample.

If counts are required, use the Miles and Misra method (p. 133) and one of the solid media.

Anaerobic cocci

Plate on blood agar and brain heart infusion agar. Incubate replicate cultures at 25 and 27 °C.

Identification

Colonies of staphylococci and micrococci on ordinary media are golden brown, white, yellow or pink, opaque, domed, 1–3 mm in diameter after 24 h on blood agar and are usually easily emulsified. There may be β-haemolysis on blood agar. Aerococci show α-haemolysis.

On Baird-Parker medium after 24 h, *Staphylococcus aureus* gives black, shiny, convex colonies, 1–1.5 mm in diameter; there is a narrow white margin and the colonies are surrounded by a zone of clearing 2–5 mm in diameter. This clearing may be evident only at 36 h.

Other staphylococci, micrococci, some enterococci, coryneforms and enterobacteria may grow and may produce black colonies but do not produce the clear zone. Some strains of *S. epidermidis* have a wide opaque zone surrounded by a narrow clear zone. Any grey or white colonies can be ignored. Most other organisms are inhibited (but not proteus: *see above*).

On TPEY medium, colonies of *S. aureus* are black or grey and give a zone of precipitation around and/or beneath the colonies.

Examine Gram-stained films. Do coagulase and DNAse tests on Gram-positive cocci growing in clusters grown from clinical material. This is a short cut: strains positive by both tests are probably *S. aureus*.

Coagulase test

Possession of the enzyme coagulase which coagulates plasma is an almost exclusive property of *S. aureus*. There are two ways of performing this test:

(1) *Slide coagulase test* Emulsify one or two colonies in a drop of water on a slide. If no clumping occurs in 10–20 s dip a straight wire into human or rabbit plasma (EDTA) and stir the bacterial suspension with it. *S. aureus* agglutinates, causing visible clumping in 10 s.

Use water instead of saline because some staphylococci are salt sensitive, particularly if they have been cultured in salt media. Avoid excess (e.g. a loopful) of plasma as this may give false positives. Check the plasma with a known coagulase positive staphylococcus.

(2) *Tube test* Do this (a) to confirm the slide test, (b) if the slide test is negative.

Add 0.2 ml of plasma to 0.8 ml of nutrient (not glucose) broth in a small tube. Inoculate with the suspected staphylococcus and incubate at 37 °C in a water-bath for 6 h. Examine at 1 h and thereafter half-hourly. Include known positive and negative controls. In the tube test, citrated plasma may be clotted by any organism that can utilize citrate, e.g. by faecal streptococci (but these are catalase negative), *Pseudomonas* and *Serratia*. It is advisable therefore to use EDTA plasma (available commercially) or oxalate or heparin plasma. Check Gram films of all tube coagulase positive organisms.

S. aureus produces a clot, gelling either the whole contents of the tube or forming a loose web of fibrin. Longer incubation may result in disappearance of the clot due to digestion (fibrinolysis).

The slide test detects 'bound' coagulase ('clumping factor'), which acts on fibrinogen directly; the tube test detects 'free' coagulase, which acts on fibrinogen in conjunction with other factors in the plasma. Either or both coagulases may be present.

DNAse test

Inoculate DNAse agar plates with a loop so that the growth is in plaques about 1 cm in diameter. Incubate at 37 °C overnight. Flood the plate with 1 N hydrochloric acid. Clearing around the colonies indicates DNAse activity. The hydrochloric acid reacts with unchanged deoxyribonucleic acid to give a cloudy precipitate. A few other bacteria, e.g. *Serratia*, may give a positive reaction.

If the organisms are coagulase and DNAse negative or have been grown on a selective medium distinguish between staphylococci and micrococci with at least two of these tests:

(1) OF test using Baird-Parker's method (p. 76);
(2) Evans and Kloos' test[1],
(3) Schleifer and Kloos' test[2],
(4) Lysostaphin sensitivity[2].

Subculture on blood agar to repeat coagulase and DNAse tests and do phosphatase test.

Evans and Kloos' test

Prepare shake cultures in Brewer's thioglycollate medium containing 0.3% agar. Incubate at 37 °C and examine for submerged, i.e. anaerobic growth (staphylococci) or surface growth (micrococci).

Schleifer and Kloos' test

Use commercial purple agar base containing 10 ml glycerol and 0.4 mg erythromycin/litre. Staphylococci change the colour of the indicator (bromocresol purple) to yellow: micrococci do not grow.

Lysostaphin test

Prepare a solution of Lysostaphin (Sigma Chemicals) by dissolving 10 mg in

40 ml phosphate buffer (0.02 M). Adjust to pH 7.4 and add NaCl to give 1% (w/v). Dispense in 1-ml amounts and store at −60 °C. To use, add 1 ml to 9 ml phosphate buffer and mix five drops of this with five drops of an overnight broth culture in a small test tube. Incubate at 35 °C and examine for lysis at 30, 60 and 120 min. Include a control using phosphate buffer instead of Lysostaphin. Most staphylococci are lysed.

Phosphatase test

Inoculate phenolphthalein phosphate agar and incubate overnight. Expose to ammonia vapour. Colonies of phosphatase positive staphylococci will turn pink.

S. aureus gives a positive test (but negative strains have been reported). Coagulase-negative staphylococci and micrococci are usually phosphatase negative.

See Table 31.1.

TABLE 31.1 ***Staphylococcus, Micrococcus*** **and** ***Aerococcus***

	OF	*EK*	*SK*	*Lys*	*Coagulase*	*DNAse*	*Phosphatase*	*Morphology*
S. aureus	F	AN	+	+	+	+	+	Clusters
Staphylococcus spp.[a]	F	AN	+	+/−	−	−	−	Clusters
Micrococcus spp.	O	A	−	−	−	−	−	Clusters, tetrads or packets
Aerococcus viridans[b]	F	AN	−	−	−	−		Clusters or pairs

OF, Hugh and Leifson; EK, Evans and Kloos; SK, Schleifer and Kloos; Lys, lysostaphin sensitivity; F, fermentative; O, oxidative; AN, anaerobic; A, aerobic.
[a] Most clinical isolates are *S. epidermidis*
[b] Includes *Gaffkya*

If identification to species is required (other than *S. aureus*) use the API Staph system or inoculate Baird-Parker's carbohydrate media (glucose, lactose, maltose, sucrose, mannitol), test for nitratase and gelatin liquefaction. See under *Species* below and for further information Baird-Parker[3] and Kloos and Schleifer[4].

Staphylococci

These are Gram-positive, oxidase negative, catalase positive, fermentative cocci arranged in clusters.

Staphylococcus aureus

This species is coagulase and DNAse positive, forms acid from lactose, maltose and mannitol, reduces nitrate, hydrolyses urea and reduces methylene blue. It is usually phosphatase positive but does not grow on ammonium phosphate agar.

Most strains are haemolytic on horse blood agar but the zone of haemolysis is relatively small compared with the diameter of the colony (differing from the haemolytic streptococcus).

Production of the golden yellow pigment is probably the most variable characteristic. Young cultures may show no pigment at all. The colour may develop if cultures are left for one or two days on the bench at room temperature.

Pigment production is enhanced by the presence in the medium of lactose or other carbohydrates and their breakdown products. It is best demonstrated on glycerol monoacetate agar.

In clinical laboratories, *S. aureus* is usually identified by *either* the coagulase *or* the DNAse test. False-positive coagulase tests are posssible with enterococci, *Pseudomonas* and *Serratia* if citrated plasma is used. *Serratia* may give a positive DNAse test. Coagulase negative, DNAse positive strains do occur. We believe, therefore, that *both* coagulase and DNAse tests should be done.

The fermentation of mannitol is not reliable. Mannitol-fermenting strains of *S. epidermidis* occur.

S. aureus is a common cause of pyogenic infections and food poisoning (Chapter 14). Staphylococci are disseminated by common domestic and ward activities such as bedmaking, dressing or undressing. They are present in the nose, on the skin and in the hair of a large proportion of the population.

Air sampling, with scatter plates or slit samplers, is an interesting exercise.

In food poisoning and epidemiological investigations, *S. aureus* strains should be sent to a reference expert for bacteriophage typing. Phage typing techniques are described by Blair and Williams[5] and Asheshov[6].

Staphylococcus epidermidis (*S. albus*)

Is coagulase, DNAse and phosphatase negative and may liquefy gelatin. Occasional strains ferment mannitol. It produces no pigment; the colonies are 'china white'. Antibiotic-resistant strains are not uncommon and are often isolated from clinical material, including urine, where they may be opportunist pathogens (*see* Oeding and Digrainis[7]).

Several other species have been described including *S. saprophyticus*, *S. hominis*, *S. warneri* and *S. cohni* (*see* Schleifer and Kloos[8]).

Peptococcus and *Sarcina*

Peptococcus

Cocci are single, in pairs, tetrads or irregular clusters. Obligate anaerobes, growing well on blood agar at 37 °C[3]. Not haemolytic. Normally found in alimentary and respiratory tract and vagina but may be opportunists (secondary invaders).

Sarcina

Cocci in packets of eight. Obligate anaerobes, requiring 20–30 °C and carbohydrate media. Found in air, water and soil.

Micrococci

These are Gram-positive, oxidase negative, catalase positive cocci that differ from the staphylococci in that they utilize glucose oxidatively or do not produce enough acid to change the colour of the indicator in the medium. They are common saprophytes of air, water and soil and are often found in foods.

The classification is at present confused: the genus contains the organisms formerly called *Gaffkya* and *Sarcina*, tetrad- and packet-forming cocci. These morphological characteristics vary with cultural conditions and are not considered constant enough for taxonomic purposes. The genus *Sarcina* now includes only anaerobic cocci.

At present, it does not seem advisable to describe newly isolated strains by any of the specific names that abound in earlier textbooks, but two 'old' organisms are:

(1) *M. luteus* Forms yellow colonies, is biochemically inactive and differs from other micrococci in being sensitive to novobiocin.
(2) *M. roseus* Forms pink colonies.

References

1. KLOOS, W.E. and SMITH, P.B. (1980) Staphylococci. In *Manual of Clinical Microbiology*, 3rd edn. Edited by E.H. Lennette, A. Ballows, W.J. Housler and J.P. Truant. Washington DC: American Society for Microbiology
2. SCHLEIFER, K.H. and KLOOS, W.E. (1975a) *Journal of Clinical Microbiology*, **1**, 337
3. BAIRD-PARKER, A.C. (1979) Methods for identifying staphylococci and micrococci. In *Identification Methods for Microbiologists*. 2nd edn. Edited by F.A. Skinner and D.W. Lovelock, pp. 201–209. London: Academic Press
4. KLOOS, W.E. and SCHLEIFER, K.H. (1975) *Journal of Clinical Microbiology*, **1**, 82
5. BLAIR, J.E. and WILLIAMS, R.E.O. (1961) *Bulletin of the World Health Organization*, **24**, 771
6. ASHESHOV, E. (1967) Bacteriophage typing of staphylococci. In *Progress in Microbiological Techniques*. Edited by C.H. Collins, pp. 173–186. London: Butterworths
7. OEDING, P. and DIGRAINIS, A. (1977) *Acta Pathologicia et Microbiologica Scandinavica*, **85**, 136
8. SCHLEIFER, K.H. and KLOOS, W.E. (1975b) *International Journal of Systematic Bacteriology*, **25**, 50

Chapter 32

Streptococcus, aerococcus, leuconostoc and pediococcus

This group includes organisms of medical, dental and veterinary importance as well as starters used in the food and dairy industries, spoilage agents and saprophytes. The cells divide in one plane or in two planes at right angles to one another. Their action on glucose is either homofermentative (producing dextrolactic acid) or heterofermentative (producing laevolactic acid, ethanol or acetic acid and carbon dioxide).

	Cell division	*Action on glucose*
Streptococcus	one plane	homofermentative
Aerococcus	two planes	homofermentative
Leuconostoc	one plane	heterofermentative
Pediococcus	two planes	homofermentative

Streptococcus

Gram-positive cocci that always divide in the same plane, forming pairs or chains; the individual cells may be oval or lanceolate. They are Grampositive, non-sporing, non-motile and some are capsulated. Most strains are aerobic or microaerophilic but there are anaerobic species. The catalase test is negative.

Isolation

From clinical material

Blood agar is the usual primary medium. For recovery of streptococci from material that contains many other organisms, place a 30-μg neomycin disc on the heavy part of the inoculum, when many other organisms will be inhibited. Or use blood agar containing 1:500 000 crystal violet. Incubate at 37 °C.

MacConkey agar should be inoculated to allow early differentiation of human pathogens, most of which are inhibited.

Dental plaque

Plate on blood agar and trypticase yeast extract cystine agar or mitis salivarius agar.

Bovine mastitis

Use Edwards' medium. The crystal violet and thallous sulphate inhibit most saprophytic organisms; aesculin-fermenting saprophytic streptococci give black colonies and mastitis streptococci pale grey colonies.

Foods

Prepare 10% suspensions in 0.1% peptone water using a Colworth Lab-Stomacher. For cold-stressed organisms inoculate tryptone yeast glucose broth. For heat-stressed cocci inoculate glucose broth or tryptone soy broth. Incubate for 3–4 h and plate on one of the azide or thallous acetate media (Mead's or Slanetz and Bartley's). On these most other organisms are inhibited and 2,3,5-triphenyltetrazolium chloride (TTC) is reduced by enterococci, giving red colonies. Other streptococci give white or pink colonies. On Mead's medium, human enterococci (*Streptococcus faecalis sensu strictu*) ferment the sorbitol and decompose tyrosine. Typical colonies are maroon in colour and are surrounded by clear zones. *See* Chapter 25 for isolation from water.

Dairy products

Use yeast glucose agar for mesophiles and yeast lactose agar for thermophiles.

Water, mineral waters and sewage

Membrane filtration gives best results with one of the membrane enterococcus agars (*see also* p. 257).

Air

For β-haemolytic streptococci, in hospital cross-infection investigations use crystal violet agar containing 1:500 000 crystal violet with slit samplers. For evidence of vitiation use mitis salivarius agar.

Identification of streptococci

Colonies on blood agar are usually small, 1–2 mm in diameter and convex with an entire edge. The whole colony can sometimes be pushed along the surface of the medium. Colonies may be 'glossy', 'matt' or 'mucoid'. Growth in broth is often granular, with a deposit at the bottom of the tube.

The primary classification is made on the basis of alteration or haemolysis on horse blood agar.

α-Haemolytic or 'viridans' streptococci produce a small, greenish zone around the colonies. This is best observed on chocolate blood agar.

α'(Alpha prime)-haemolytic streptococci are surrounded by an area of haemolysis which superficially resembles that of β-haemolytic streptococci (below) but with a hazy outline and unaltered red blood cells within the haemolysed area.

β-Haemolytic streptococci give small colonies surrounded by a much larger, clear haemolysed zone in which all the red cells have been destroyed.

Some streptococci show no haemolysis.

Haemolysis on blood agar is only a rough guide to pathogenicity. The β-haemolytic streptococci include those strains which are pathogenic for man and animals but the type of haemolysis may depend on conditions of incubation.

Some saprophytic streptococci are β-haemolytic; so are organisms in other genera having similar colonial morphology. Haemolytic *Haemophilus* spp. are often reported as haemolytic streptococci in throat swabs because films were not made. *C. pyogenes* is also haemolytic.

Clinical isolates

It is usually sufficient to identify haemolytic streptococci from clinical material by one of the rapid coagglutination methods (*see below*) or simply as Group A or not Group A by the bacitracin disc method. Streptococci showing α-haemolysis may be identified by their sensitivity to optochin and bile solubility. Other streptococci require more detailed examination.

Streptococcal antigens

Species and strains of streptococci are usually identified by their serological group and type. There are 15 groups (A–P, excluding I) characterized by a series of carbohydrate (hapten) antigens contained in the cell wall. Rabbits immunized with known strains of each group produce serum which will react specifically *in vitro* by a precipitin reaction with an extract of the homologous organism. The carbohydrate is known as the C substance. The streptococci within *Group A* and a few other groups can be divided into serological types (1–30) by means of two protein antigens, M and T.

M is a type-specific antigen near the surface of the organism which can be removed by trypsinization and is present in matt and mucoid types of colonies but not in glossy types. It is demonstrated by a precipitin test.

T is not type-specific, may be present with or independent of the M antigen, and is demonstrated by agglutination tests with appropriate antisera.

Grouping is usually carried out in the laboratory where the organisms are isolated, using sera and control extracts obtained commercially. Only sera for Groups A, B, C, D and G need be used for routine purposes.

Typing of haemolytic (Group A, etc.) strains is necessary for epidemiological purposes. This is best done by Reference laboratories.

Serological grouping of streptococci

There are two approaches. Coagglutination, which requires commercially available kits, is rapid and gives results within 1 h of obtaining a satisfactory growth on the primary culture. Precipitin tests take longer but allow a larger number of groups to be identified.

Coagglutination: latex agglutination

At least three kits are in common use. The streptococcal antibody is attached to staphylococci or latex particles. These antibody coated particles are agglutinated when mixed on a slide with suspensions or extracts of streptococci of the same group. In the Phadebact (Pharmacia) method colonies from the primary plates are mixed with the reagent to identify streptococci in Groups A,B,C and G. The Streptosec (Organon) reagents are freeze-dried on the slides provided and they also allow streptococci in Groups A,B,C and G to be identified. The Streptex (Wellcome) method is slightly different. The organisms are incubated in an enzyme solution for 1 h to extract the antigen before mixing with the latex reagent on a slide. More groups may be identified—A,B,C,D,F and G. It is recommended that Group D streptococci are further identified by biochemical tests as enterococci, other Group D or 'viridans' streptococci.

Preparation of extract for grouping streptococci

Centrifuge 50 ml of an overnight culture of the streptococcus in 0.1% glucose broth or a suspension prepared by scraping the overnight growth from a heavily inoculated blood agar plate.

For Lancefield's method[1], suspend the deposit in 0.4 ml of 0.2 N hydrochloric acid and place in boiling water-bath for 10 min. Cool, add 1 drop of 0.02% phenol red and loopsful of 0.5 N sodium hydroxide solution until the colour changes to faint pink. Centrifuge; the supernatant is the extract.

For Fuller's formamide method[2] suspend the deposit in 0.1 ml of formamide and place in an oil-bath at 160 °C for 15 min. Cool and add 0.25 ml of acid alcohol (95:5 ethanol:2 N hydrochloric acid). Mix and centrifuge. Remove the supernatant and add to it 0.5 ml of acetone. Mix, centrifuge and discard the supernatant. To the deposit add 0.4 ml of saline, 1 drop of 0.02% phenol red and loopsful of 0.2 N sodium hydroxide solution until neutral. This is the extract. This is not a very good method for Group D streptococci.

For Rantz' method[3] suspend the organisms in 0.12 ml of saline and autoclave at 121 °C for 15 min. Centrifuge and use the supernatant.

For Hamilton's method[4] centrifuge the broth culture, remove all fluid and heat the deposit at 180 °C for 1 min in a boiling oil bath. Then resuspend the dried deposit in 0.4 ml of saline. Centrifuge again and use the supernatant.

In Maxted's method[5] the C substance can also be made from a streptococcal suspension by incubation with an extract prepared from a strain of *Streptomyces*.

To prepare this extract, obtain *Streptomyces* sp. No. 787 from the National Collection of Type Cultures and grow it for several days at 37 °C on buffered yeast extract agar. The medium is best sloped in flat bottles (120-ml 'medical flats'). When there is good growth, place the cultures in a bowl containing broken pieces of solid carbon dioxide. Allow to thaw and remove the fluid; this contains the enzyme. Bottle it in small amounts and store in a refrigerator.

To prepare the streptococcal extract, scrape the growth from a heavily inoculated 24-h blood agar culture of the organisms into 0.5 ml of enzyme solution in a small tube and place in a water-bath at 37 °C for 2 h. Centrifuge and use the extract to do precipitin tests as described below.

Grouping methods

Prepare capillary tubes as in *Figure 32.1* from pasteur pipettes. Dip the narrow end in the grouping serum so that a column a few millimetres long enters the tube. Place it in a block of Plasticine and, with a very fine pasteur pipette, layer extract on the serum so that the two do not mix but a clear interface is preserved. Some practice is necessary in controlling the pipette. If an air bubble develops between the two liquids, introduce a very fine wire, when an interface is usually produced.

Figure 32.1 Precipitin tube

A positive result is indicated by a white precipitate that develops at the interface. It is sometimes necessary to dilute the extract 1:2 or 1:5 with saline to obtain a good precipitate.

The bacitracin disc method

Inoculate a blood agar plate heavily and place a bacitracin disc (Difco, Mast, Oxoid, Pasteur) on the surface and incubate overnight. A zone of inhibition appears around the disc if the streptococci are Group A[6]. This test is not wholly reliable and is declining in popularity except as a screening method.

Biochemical and other tests

These are rarely used in clinical laboratories except for distinguishing between enterococci and other Group D streptococci. They are useful in oral (dental) bacteriology and in food bacteriology. A simple scheme for Groups A,B,D and viridans streptococci is shown in *Table 32.1*. A full range of tests is shown

TABLE 32.1 Groups A, B, D and 'viridans' streptococci

	Haemolysis	*Hippurate hydrolysis*	*Growth in 6% NaCl medium*	*Bile-aesculin*
Group A	β	–	–	–
Group B	β	+	+/–	–
Group D enterococci	α/β/none	–	+	+
Group D not enterococci	α	–	–	–

in *Table 32.2* but there is no need to do all the tests to identify all the species. The most useful tests may be selected according to the known source of the material. Nevertheless, identification of streptococci is far from easy. The API 20 system is generally useful and the AP ZYM enables *S. faecalis*, *S. bovis*, *S. mutans*, *S. sanguis*, *S. mitior* and *S. milleri* to be identified. The Minitek (p. 104) system identifies *S. mutans*, *S. sanguis*, *S. salivarius* and *S. mitior*.

Fluorescent antibody tests

These are useful for Group A streptococci but not so reliable for other groups.

Groups and species of streptococci

The general properties are shown in *Table 32.2*.

Group A

These are β-haemolytic, are the so-called haemolytic streptococci of scarlet fever, tonsillitis, puerperal sepsis and other infections of man, and are known as *S. pyogenes*. Some strains are capsulated and form large (3 mm) colonies like water drops on the surface of the medium. '*S. mucosus*' and '*S. epidemicus*' were names once used to describe these mucoid strains, which have been associated with milk-borne outbreaks. Capsule formation is, however, not uncommon when streptococci are grown in milk.

Group B

The β-haemolytic streptococci in this group correspond to *S. agalactiae*, the causative organism of chronic bovine mastitis. This streptococcus is rarely pathogenic for man but may be commensal in the human vagina. It is usually less strongly β-haemolytic than is *S. pyogenes* and grows on bile salt media.

Group C

The most important members of this group of β-haemolytic streptococci are *S. equi*, which causes strangles in horses, and *S. dysgalactiae*, which is associated with acute bovine mastitis (but is less common than *S. agalactiae*). *S. equisimilis* is responsible for some human infections and *S. zooepidemicus* for outbreaks among animals.

Group D

Some of these strains are β-, some α- and some nonhaemolytic. *S. faecalis* and *S. faecium* are the enterococci, commensals in human and animal intestines and are used as an indicator of 'faecal' pollution in sanitary bacteriology. Most of the organisms in this group are able to resist 60 °C for 30 min, all grow at 45 °C and in the presence of 10% bile and hydrolyse aesculin, but these properties are not exclusive to the group.

S. faecalis is common in the human and poultry but rare in other animal intestines. This is the only enterococcus which reduces 2,3,5-triphenyltetrazolium chloride (TTC), produces acid from sorbitol, decomposes tyrosine and grows in the presence of 0.03% potassium tellurite and at 10 °C.

S. faecium, present in the intestines of pigs and other animals, has caused spoilage of canned ham, which is pasteurized ('commercially sterilized') (*see* p. 233). TTC is not reduced; there is no growth on medium containing 0.03% potassium tellurite. *S. faecium* ferments arabinose; *S. durans* does not. Most

TABLE 32.2 Properties of some streptococci

			Growth at or in						*Hippurate hydrolysis*	*Arginine hydrolysis*	*Aesculin hydrolysis*	*Acid from*						*Inulin*	*Levan*
	Group	*Haemolysis*	*10°*	*45°*	*60°*	*40% bile*	*6.5% NaCl*	*pH 9.6*				*Mannitol*	*Lactose*	*Sorbitol*	*Trehalose*	*Salicin*	*Rafinose*		
S. pyogenes	A	β	−	−	−	−	−	−	−	+	v	−	+	−	+	+	−	−	−
S. agalactiae	B	α/β	−	−	−	+	−	−	+	+	−	−	v	−	+	v	−	−	−
S. dysgalactiae	C	α	−	−	−	−	−	−	v	+	v	−	v	v	+	v	−	−	−
S. equisimilis	C	β	−	−	−	−	−	−	−	+	v	−	v	−	+	+	−	−	−
S. equi	C	β	−	−	−	−	−	−	−	+	−	−	−	−	−	+	−	−	−
S. zooepidemicus	C	β	−	−	−	−	−	−	−	+	v	−	v	+	−	+	−	−	−
S. faecalis	D	$\beta/-$	+	+	+	+	+	+	+	+	+	+	+	+	+	+	−	+	−
S. faecium	D	α/β	+	+	+	+	+	+	+	+	+	v	v	v	v	+	v	−	v
S. bovis	D	$\alpha/-$	−	+	−	+	−	v	−	−	+	v	+	v	v	−	+	+	v
S. equinus	D	$\alpha/-$	−	+	−	+	−	−	−	−	+	−	−	−	v	+	−	v	−
S. milleri	ACFG	$-/\beta$	−	−	−	−	−	−	−	+	+	−	+	−	+	+	−	−	−
S. sanguis	HK	$\alpha(\beta)$	−	v	−	v	−	−	−	+	+	−	+	−	+	−	v	+	+
S. salivarius	K	$-/(\alpha)$	−	−	−	−	−	−	−	−	+	−	+	−	+	+	+	+	+
S. mitior	OKM	α	−	v	−	−	−	−	−	v	−	−	+	−	−	−	v	−	v
S. mutans	−	−	−	−	−	−	−	−	−	−	+	+	+	+	+	+	+	+	−
S. uberis	ECDP	−	−	−	−	+	v	−	+	+	+	+	+	+	+	+	−	+	
S. lactis	N	$(\alpha)-$	+	−	−	−	−	+	v	+	v	v	+	−	v	v	−	−	−
S. cremoris	N	$(\alpha)-$	+	−	−	−	−	−	−	−	v	−	+		v	v	−	−	−
S. thermophilus	−	−	−	+	+	−	−	−	−	−	−	−	+	−	−	−	−	−	−

10°,45°, temperature at which growth occurs; 60°, resistant

enterococci reduce litmus milk. Other Group D streptococci do not (use a heavy inoculum).

S. bovis and *S. equinus* are also found in the animal intestine and in milk and are difficult to tell apart. The Sims test allows *S. bovis* to be distinguished from other streptococci: inoculate Rogosa medium stabs in small screw-capped bottles in which little air space is left and cap tightly. *S. bovis* will grow in about four days if carbon dioxide is not allowed to escape. Other streptococci do not grow.

These organisms do not reduce TTC, are inhibited by 0.03% potassium tellurite. *S. bovis* ferments raffinose; *S. equinus* does not.

The D antigen is deeper in the cell than are the antigens of other streptococci. For this reason, extracts prepared by acid hydrolysates of the streptococci and the antisera prepared with them do not give good precipitation. The commercial antisera are satisfactory but Fuller's and Maxted's methods should be used for grouping.

Group D streptococci can be confused with aerococci but the latter do not grow at 45 °C and do not hydrolyse arginine (*Table 32.2* and p. 340). Both cause greening of bacon.

Group N

These are non-haemolytic and non-pathogenic and are of importance in the dairy industry. *S. lactis* and *S. cremoris* are the most common. They are used in the ripening and curing of Cheddar-type cheese and in the preparation of cultured buttermilk and the manufacture of sauerkraut. Both organisms are susceptible to a phage which rapidly destroys them, interfering with the commercial processes.

Other groups and species

Group G is known to contain some human pathogens and some strains in Groups L and M are pathogenic for animals. Groups H and K contain oral streptococci. Some species react with several group specific sera and others have not yet been grouped.

S. salivarius

These are commensals in the human upper respiratory tract and are therefore useful indicators in air hygiene and ventilation investigations. They are readily identified by their ability to produce a levan when grown on media containing 5% sucrose. The colonies are large and mucoid.

'Viridans' streptococci

This name is often given in error to any streptococcus that shows α-haemolysis. It should be restricted to *S. mitior*, which is found in the mouth. It may react with sera of Groups O, K or M. *S. uberis*, which also gives α-haemolysis, is a saprophyte found in soil, often gaining access to milk. This may react with sera of Groups E,C,D or P.

S. mutans

Appears to play a major role in dental caries[7], but does not appear to belong to any group.

S. milleri

This is also found in the mouth, in dental abscesses and other deeper abscesses. It may require 5% carbon dioxide for growth and some cultures smell of caramel. It is difficult to group but some strains may react with sera of Groups A,C,G or F.

S. sanguis

Is found in the mouth and in dental plaque and has been reported in heart valve disease. It usually reacts with Groups H or K sera.

S. thermophilus

Resembles Group D streptococci biochemically but has no D antigen. The optimum temperature for growth is 50 °C. It is used (along with *Lactobacillus bulgaricus*) in the manufacture of yoghurt.

Peptostreptococcus spp.

These obligate anaerobes may be pathogenic for man and animals, associated with gangrenous lesions. They are commensals in the intestine and have been isolated from the vagina in health and in puerpural fever. They produce large amounts of hydrogen sulphide from high-protein media, e.g. blood broth, and a putrid odour in ordinary media. The API system is useful for identifying these organisms.

S. pneumoniae

The pneumococcus is a causative organism of lobar pneumonia in man and of various other infections in man and animals, including (rarely) mastitis. In pus and sputum, the organism appears as a capsulated diplococcus but usually grows on laboratory media in chains and then shows no capsule. There are several serological types, but serological identification is rarely attempted nowadays. One type produces highly mucoid colonies but most strains give flat 'draughtsman'-type colonies 1–2 mm in diameter with a greenish haemolysis.

Pneumococci often resemble 'viridans' streptococci on culture. The following tests allow rapid differentiation.

(1) Pneumococci are sensitive to optochin (ethylhydrocupreine hydrochloride). Optochin discs are supplied by BBL, Difco, Mast, Oxoid and Pasteur Institute.

A disc placed on a plate inoculated with pneumococci will give a zone of inhibition of at least 10 mm when incubated aerobically (not in carbon dioxide).

(2) Pneumococci are bile soluble; other streptococci are not. Add 0.2 ml of 10% sodium deoxycholate in saline to 5 ml of an overnight broth culture (do not use glucose broth). Incubate at 37 °C. Clearing should be complete in 30 min.

Pneumococci are alone among the streptococci in fermenting inulin and are not heat resistant as are many 'viridans' streptococci. These two criteria should not be used alone for differentiation.

For a full review of the genus *Streptococcus see* Skinner and Quesnel[8].

Aerococcus

These are Gram-positive, oxidase negative, fermentative cocci that are usually in clusters, pairs, tetrads or short chains.

Isolation

Use blood agar for food investigations and in slit samplers for aerobiology. Incubate at 30–35 °C.

Identification

Aerococci give green (α) haemolysis on blood agar and a weak or negative catalase test. It is not easy to distinguish them from 'viridans' streptococci. They differ from *S. mitior* in growing at pH 9.6, in 6.6% sodium chloride broth and in being resistant to 60 °C for 30 min. They grow on bile salt media but differ from Group D streptococci in not possessing the D antigen, in failing to grow at 45 °C and in not hydrolysing arginine[9] (*Tables 32.1* and *32.2*).

Aerococcus viridans

Is the only named species. It is a common airborne contaminant and is also found in curing brines. It appears to be the same organism as *Gaffkya homari* which causes an infection in lobsters; they develop a pink discoloration on the ventral surface and the blood loses its characteristic bluish green colour, becoming pink.

Leuconostoc

This genus contains several species of microaerophilic streptococci of economic importance. Some produce a dextran slime on frozen vegetables.

Isolation

Culture material on yeast glucose agar at pH 6.7–7.0. Incubate at 20–25 °C under 5% carbon dioxide.

Identification

Inoculate MRS sugars: glucose, lactose, sucrose, mannose, arabinose and xylose and look for dextran slime on solid medium containing glucose (*see Table 32.3*).

TABLE 32.3 ***Leuconostoc*** **species**

	Acid from						*Slime*
	Arabinose	*Xylose*	*Glucose*	*Mannose*	*Lactose*	*Sucrose*	
L. cremoris	−	−	+	−	+	−	−
L. dextranicum	−	v	+	v	+	+	++
L. lactis	−	−	+	v	+	+	−
L. mesenteroides	+	v	+	+	v	+	+
L. paramesentericus	v	v	+	+	v	+	−

Species of *Leuconostoc*

Leuconostoc mesenteroides

This is the most important species. It produces a large amount of dextran slime on media and vegetable matter containing glucose. Colonies on agar media without the sugar are small and grey. Growth is very poor on media without yeast extract. It takes part in sauerkraut fermentation and silage production and is responsible for slime disease of pickles (unless *Lactobacillus plantarum* is encouraged by high salinity). It is also responsible for 'slimy sugar' or 'sugar sickness'. It spoils frozen peas and fruit juices.

L. paramesenteroides

Is similar to *L. mesenteroides* but does not produce a dextran slime. Widely distributed on vegetation and in milk and dairy products.

L. dextranicum

Produces rather less slime than *L. mesenteroides* and is less active biochemically. Widely distributed in fermenting vegetables and in milk and dairy products.

L. cremoris

Is rare in nature but extensively used as a starter in dairy products.

L. lactis

Not common. Found in milk and dairy products.

Pediococcus

Pediococci are non-capsulated microaerophilic cocci that occur singly and in pairs. They are nutritionally exacting and are of economic importance in the brewing, fermentation and food processing industries.

Isolation

Use enriched media and wort agar for beer spoilage pediococci. Tomato juice media at pH 5.5 is best. Incubate at temperatures according to species sought (*see Table 32.4*) under 5–10% carbon dioxide.

TABLE 32.4 ***Pediococcus* species**

	Growth at		*Acid from*						
	37°C	*45°C*	*Lactose*	*Sucrose*	*Maltose*	*Mannitol*	*Galactose*	*Salicin*	*Arabinose*
P. damnosus	−	−	−	−	+	−	−	−	−
P. acidi-lactici	+	+	v	v	−	−	+	−	−
P. pentosaceus	+	+	+	v	+	−	+	+	+
P. urinae-equi	+	−	+	+	+	−	+	+	v
P. halophilus	+	−	v	+	+	+	+	+	v

Identification

If it is necessary to proceed to species identification inoculate MRS sugar media: lactose, sucrose, galactose, salicin, maltose, arabinose and xylose. Incubate at appropriate temperature. Gas is not produced by any species (*see Table 32.4*).

Species of *Pediococcus*

Pediococcus damnosus (*P. cerevisiae*)

Is found in yeasts and wort. Contamination of beer results in a cloudy, sour product with a peculiar odour—'sarcina sickness'. *P. damnosus* is resistant to the antibacterial agents in hops.

P. acidi-lactici

Is found in sauerkraut, wort and fermented cereal mashes. Is sensitive to the antibacterial activity of hops.

P. pentosaceus

Is also found in sauerkraut as well as in pickles, silage and cereal mashes but not in hopped beer.

P. urinae-equi

Is a contaminant of brewers' yeasts. (The name indicates that it was originally isolated from urine of a horse.)

P. halophilus

As the name suggests, this is salt-tolerant (up to 15% NaCl). It is found in soy mash.

References

1. LANCEFIELD, R.C. (1933) *Journal of Experimental Medicine*, **57,** 571
2. FULLER, A.T. (1938) *British Journal of Experimental Pathology*, **19,** 130
3. RANTZ, L.A. (1942) *Journal of Infectious Disease*, **71,** 61
4. HAMILTON, W.J. (1972) *Medical Laboratory Technology*, **29,** 30
5. MAXTED, W.R. (1948) *Lancet*, **ii,** 255
6. MAXTED, W.R. (1953) *Journal of Clinical Pathology*, **6,** 224
7. SCULLY, C. (1981) *Journal of Infection*, **3,** 107
8. SKINNER, F.A. and QUESNEL, L.B. (Editors) (1978) *Streptococci.* Society for Applied Bacteriology Symposium, Series 7. London: Academic Press
9. SKERMAN, V.B.D. (1967) *The Genera of Bacteria.* 2nd edn, p. 161. Baltimore: The Williams and Wilkins Co.

Chapter 33

Corynebacterium, microbacterium, brochothrix, propionibacterium, brevibacterium, erysipelothrix, listeria, lactobacillus and bifidobacterium

Corynebacterium

These organisms are frequently club shaped, thin in the middle with swollen ends that contain metachromatic granules. These are best seen when Albert's or Neisser's stain is used. The stained bacilli may also be barred, with a 'palisade' or 'Chinese letter' arrangement, due to a snapping action in cell division.

This section is restricted to corynebacteria of medical and veterinary importance. Other 'corynebacteria' or 'coryneforms' are discussed under their generic headings. For a full review of these organisms *see* Bousfield and Calley[1].

The diphtheria group

Diphtheria is now a comparatively rare disease and may be clinically atypical. If an organism is suspected of being a diphtheria bacillus because of its colonial and microscopical appearance, inform the physician at once and before proceeding to identify the bacillus. It is in the best possible interests of the patient that this is done. No harm will result if the final result is negative, but a delay of 24 h or more to confirm a positive case may have serious consequences.

A culture should be sent immediately to the nearest reference expert. A case of diphtheria will usually generate large numbers of swabs from contacts. For the logistics of mass swabbing and the examination of large numbers of swabs in the shortest possible time *see* the paper by Collins and Dulake[2].

Isolation and identification of the diphtheria organism and related species

Plate throat, nasal swabs, etc., on Hoyle's or other tellurite media and on blood agar. Incubate at 37 °C and examine at 24 and 48 h.

For colonial characteristics, *see* species descriptions.

Make Gram-stained films of suspicious colonies. *C. diphtheriae* tends to decolorize easily and appears Gram-negative compared with diphtheroids, which stain solidly Gram-positive. When staining, make a control film of staphylococci on the same slide. This apparent Gram-negative reaction is a useful characteristic.

Subculture a colony from the tellurite to Loeffler serum as soon as possible. Incubate at 37 °C for 4–6 h. The best results are obtained with Loeffler medium containing an appreciable amount of water of condensation and when the tube has a cotton wool plug or the cap is left loose. Examine Gram- and Albert-stained films for typical beaded bacilli which may show metachromic granules at their extremities. Inoculate Robinson's serum water sugars (glucose, sucrose and starch) but do not wait for the result before doing the plate toxigenicity test.

Note: Robinson's sugar medium is buffered and contains horse serum. Unbuffered horse serum medium may give false-positive results because of fermentation of the small amount of glycogen it contains, which is attacked by many corynebacteria. Similarly, media that contain unheated rabbit serum may give false-positive starch fermentation reactions. Natural amylases hydrolyse starch, forming glucose, which is fermented by the organism. Misidentification of *C. diphtheriae* by unskilled workers using various other fermentation tests is not uncommon. We maintain that Robinson's sugars are the most reliable media for identification and differentiation.

The biochemical reactions are given in *Table 33.1*.

TABLE 33.1 Corynebacteria from human sources

	Acid from			*β-Haemolysis*	*Urea*
	Glucose	*Sucrose*	*Starch*		
C. diphtheriae gravis	+	−	+	−	−
C. diphtheriae mitis	+	−	−	+	−
C. diphtheriae intermedius	+	−	−	−	−
C. ulcerans	+	−	+	−	+
C. pseudodiphtheriticum	−	−		−	−
C. xerosis	+	+	+	−	−

Plate toxigenicity test

Elek's method, modified[3]: inoculate a moist slope of Loeffler's medium with a colony from the primary plate as early as possible in the morning. Also inoculate Loeffler slopes with the stock control cultures maintained on Dorset egg medium. NCTC No. 10648 is toxigenic and NCTC 10356 is non-toxigenic. Incubate at 37 °C for 4–6 h.

Melt two 15-ml tubes of Elek's medium (p. 74) and cool to 50 °C. Add 3 ml of sterile horse serum to each and pour into plastic or very clear glass petri dishes. Place immediately in the still liquid medium in each plate a strip of filter-paper (60 × 15 mm) which has been soaked in diphtheria antitoxin. With the antitoxin available until 1982 we found that 750 units/ml gave satisfactory results. This material is no longer available and newer products vary. Laboratories should titrate each new batch against stock cultures, using concentrations between 500 and 1000 units/ml. This should be done at leisure, not when a diphtheria investigation is imminent. Place the strip along a diameter of the plate. Dry the plates. These may be stored in a refrigerator for several days.

On each plate, streak the 4–6 h Loeffler cultures at right-angles to the paper strip so that the unknown strain is between the toxigenic and non-toxigenic controls as in *Figure 33.1*, approximately 10 mm apart. Incubate at 37 °C for 18–24 h. If no lines are visible re-incubate a further 12 h. Examine by transmitted light against a black background. If the unknown strain is toxigenic its precipitation lines should join those of the toxigenic control, i.e. a reaction of identity. Strains with attenuated toxigenicity may not show lines but turn the lines of adjacent toxigenic strains.

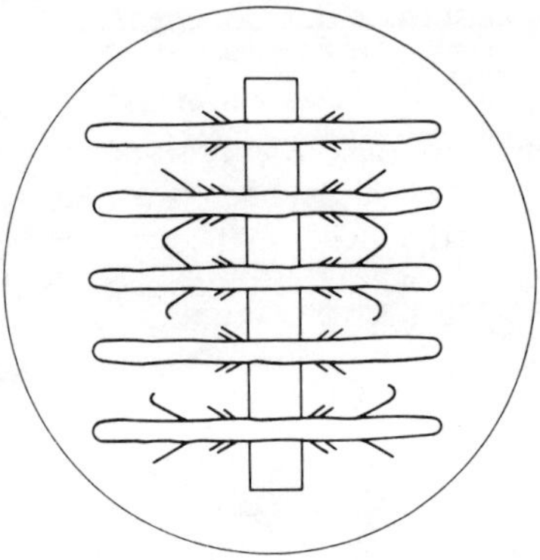

Figure 33.1 Elek plate toxigenicity test

An alternative method is that of Jameson[4]. Prepare the Loeffler cultures as above. Pour the plates but do not add the antitoxin strip. Dry the plates and place along the diameter of each a filter-paper strip (75 × 5 mm) which has been soaked in diphtheria antitoxin (for concentration *see above*). Place the plates over a cardboard template marked out as in *Figure 33.2* and heavily inoculate them from the Loeffler cultures in the shape of the arrowheads. Place the unknown strain between the two controls. Incubate at 37 °C for 18–24 h and examine for reactions of identity.

To be of value to the clinician, the plate toxigenicity test must give clear results in 18–24 h. If the toxigenic control does not show lines at 24 h, the

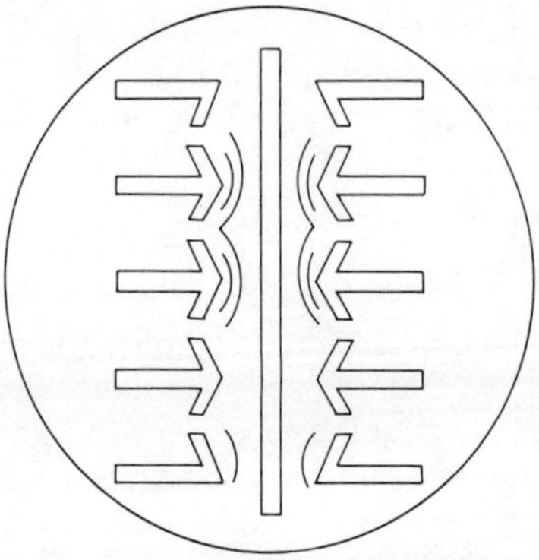

Figure 32.2 Jameson's method for plate toxigenicity test

medium or serum is unsatisfactory. The test requires a certain amount of experience and the medium and serum used must be tested frequently. For reasons not yet appreciated, certain batches of medium and serum give poor results. To test the medium and serum, use the control strains noted above and also NCTC 3894, which is weakly toxigenic. The strength of the antitoxin used is critical.

Report the organism as toxigenic or non-toxigenic *C. diphtheriae* cultural type *gravis*, *mitis* or *intermedius*.

Note that in any outbreak of diphtheria due to a toxigenic strain, a proportion of contacts may be carrying non-toxigenic *C. diphtheriae*.

Species of corynebacteria from human sources

The following general descriptions of colony appearance and microscopic morphology of diphtheria bacilli and diphtheroids is a guide only. The appearance of the colony varies considerably with the medium and it is advisable to check this periodically with stock (NCTC and ATCC) organisms and, if possible, strains from recent cases of the disease. Even then, strains may be encountered that present a new or unusual colony appearance.

C. diphtheriae

Three biotypes, *gravis*, *mitis* and *intermedius*, were originally given these names because of their respective association with the severe, mild and intermediate clinical manifestations of the disease. These names have become attached to the colonial forms and related to starch fermentation. These properties do not always agree and it is more important to carry out the toxigenicity test than to attempt to interpret the colonial and biochemical properties, which should be used only to establish that the organism belongs to the species. The following colony appearances relate to tellurite medium of the Hoyle type.

C. diphtheriae, gravis type

At 18–24 h, the colony is 1–2 mm in diameter, pearl grey with a darker centre; the edge is slightly crenated. It fractures easily when touched with a wire but is not buttery. At 48 h, it has enlarged to 3–4 mm and is much darker grey with the edge markedly crenated and with radial striations. This is the 'daisy head' type of the colony. The organism does not emulsify in saline. Films stained with methylene blue show short, often barred or beaded, bacilli arranged in irregular 'palisade' form or 'Chinese letters'. Well decolorized Gram films compared on the same slide with staphylococci or diphtheroids appear relatively Gram-negative. Glucose and starch are fermented, but not sucrose. There is no haemolysis on blood agar.

C. diphtheriae, mitis type.

At 18–24 h, colonies are 1–2 mm across and are dark grey, smooth and shining. The colonies fracture when touched but are much more buttery than the gravis type. At 48 h, they are larger (2–3 mm diameter), less shiny and darker in the centre, which may be raised, giving the 'poached egg' appearance. The organisms emulsify easily in saline. Films show long, thin bacilli, usually beaded and with granules, which may not be very Gram-positive (*see above*). Glucose is fermented but not starch or sucrose. Colonies on blood agar are haemolytic.

C. diphtheriae, intermedius type

Colonies at 18–24 h are small (1 mm diameter), flat with sharp edges, black and not shining or glossy. At 48 h, there is little change. They are easily emulsified and films show beaded bacilli, intermediate in size between *gravis*- and *mitis* types. Glucose is fermented but not starch or sucrose. Colonies on blood agar are non-haemolytic.

C. ulcerans

This organism resembles very closely the *gravis* type of diphtheria bacillus and is often mistaken for it. The colony tends to be more granular in the centre. Glucose and starch are fermented but not sucrose. *C. ulcerans* splits urea; *C. diphtheriae* does not. *C. ulcerans* also grows quite well at 25–30°C. It gives a reaction of incomplete identity with the plate toxigenicity test and *C. diphtheriae* antitoxin. This organism does not cause diphtheria and is not readily communicable, but can cause ulcerated tonsils.

C. pseudodiphtheriticum (*C. hofmanii*)

Hoffman's bacillus at 18–24 h gives colonies 2–3 mm in diameter that are domed, shiny and dark grey or black. They are buttery (occasionally sticky), easily emulsified and films show intensely Gram-positive bacilli that are smaller and much more regular in shape, size and arrangement than are diphtheria bacilli. Carbohydrates are not attacked. These organisms are common and harmless commensals.

C. xerosis

At 18–24 h, colonies are 1–2 mm in diameter, flat, grey and rough, with serrated edges. At 48 h, they are 2–3 mm in diameter. Films show beaded bacilli, very Gram-positive. This is a dubious species, a commensal of the conjunctivae and often found in the female genitalia. It is not a pathogen but resembles very closely the diphtheria bacillus in microscopical (but not colonial) morphology. It is, however, much more strongly Gram-positive than *C. diphtheriae* and produces acid from glucose, usually sucrose but not starch.

Corynebacteria from animal sources

Isolation and identification

Culture pus, etc., on blood agar and tellurite medium, incubate at 37 °C and examine at 24 and 48 h for colonies of beaded Gram-positive bacilli. Some corynebacteria are haemolytic on blood agar. Small grey-black colonies are usually found on tellurite medium.

Subculture on Loeffler medium and examine Albert-stained films at 6–18 h for clubbed or beaded bacilli which may have metachromatic granules. From the Loeffler inoculate gelatin medium, nutrient agar, Robinson's serum water glucose, sucrose, mannitol, maltose, lactose and urea medium. Incubate at 37 °C for 24–48 h (*see Table 33.2*).

TABLE 33.2 Corynebacteria from animal sources

	Liquefaction of gelatin and Loeffler	*Acid from*					*β-Haemo lysis*	*Urea*
		Glucose	*Sucrose*	*Mannitol*	*Maltose*	*Lactose*		
C. pyogenes	+	+	+/−	−	+	+	+	−
C. kutscheri	−	+	+	+	+	−	−	+/−
C. pseudotuberculosis	v	+	v	−	+	v	+	+
C. equi	−	−	−	−	−	−	−	+
C. renale	−	+	−	−	v	v	+/−	+

Species of corynebacteria infecting animals

C. pyogenes

The colonies on blood agar are pin-point and haemolytic and may be mistaken for haemolytic streptococci except for their slow rate of growth. They are often not apparent for 36–48 h. Films show very small bacilli that may not easily be recognized as corynebacteria. The catalase test is negative (other corynebacteria mentioned here are positive). It does not grow on nutrient agar but Loeffler's serum cultures are liquefied. This organism causes suppurative lesions including mastitis in domestic and wild animals.

C. kutscheri (*C. muris*)

Colonies on blood agar are larger than *C. pyogenes*, flat, grey and non-haemolytic. The bacilli are slender, often filamentous. Loeffler medium is not liquefied. It is responsible for fatal septicaemia in mice and rats.

C. pseudotuberculosis

Colonies are small, usually grey but may be yellowish and are haemolytic, especially in anaerobic cultures. Loeffler serum may be liquefied. The bacilli are slender and clubbed, very like *C. diphtheriae*. It is responsible for epizootics in rodents and is a nuisance in laboratory animal houses.

This is the Preisz–Nocard bacillus which causes pseudotuberculosis in sheep, cattle and swine and lymphangitis in horses.

C. equi

This is the only one of these species which grows well on nutrient agar, forming large, moist, pink colonies. The bacilli are very pleomorphic, varying from coccoid to large clubbed forms with granules. It is non-haemolytic, does not liquefy Loeffler medium and does not form acid from carbohydrates. It is responsible for serious pyaemia in young horses.

C. renale

Causes pyelitis and cystitis in cattle.

Microbacterium

The coryneforms in this genus are found mostly in dairy products and in association with animals. They are relatively heat resistant (70 °C for 15 min).

Microbacterium lacticum

This is thermoduric and a strict aerobe. It is non-motile, grows slowly at 30 °C on milk agar, forms acid from glucose and lactose, is catalase positive, hydrolyses starch, is lipolytic and does not liquefy gelatin in seven days. A pale yellow pigment may be formed. These organisms may cause spoilage in milk and indicate unsatisfactory dairy hygiene. They are also found in deep litter in hen houses.

Brochothrix

Brochothrix thermosphacta (*formerly Microbacterium thermosphactum*)

Is found in souring sausages of the British fresh type and in meat. It grows aerobically and anaerobically on Gardner's STAA medium (p. 75) and is highly pleomorphic; coccoid or filamentous forms may be seen. It is Gram-variable and grows best at 22 °C. It will grow at 1 °C but not at 37 °C and does not survive heating at 63 °C for 3 min. It is non-motile, catalase positive and produces acid from glucose, maltose and cellobiose and it grows in the presence of 6.5% sodium chloride (*see* Gardner[5] and Roberts *et al.*[6]).

Propionibacterium

These are pleomorphic coryneforms varying from coccoid to branched forms. They are aerobic but may be aerotolerant, non-motile and grow poorly in the absence of carbohydrates. They are catalase positive and do not usually liquefy gelatin (but *see Propionibacterium acnes, below*). The optimum temperature is 30 °C. There are several species of interest to food microbiologists. They occur naturally in the bovine stomach and hence in rennet and are responsible for flavours and 'eyes' in Swiss cheeses. A method for presumptive isolation is given on p. 218. It is difficult to identify species.

There is one species of medical interest: *P. acnes*.

Propionibacterium acnes

The bacilli are small, almost coccoid, or about 2.0×0.5 μm, and may show unstained bands. The best growth is obtained anaerobically. The colonies on blood agar are either small, flat, grey-white and buttery, or are larger, heaped up and more granular. Both colony forms are β-haemolytic. Acid is produced from glucose, mannose, trehalose and fructose but not from lactose, maltose and sucrose. It is indole positive, nitratase positive, catalase positive and liquefies gelatin slowly. Slide agglutination is useful (Difco).

It is a commensal on the skin of man, in hair follicles and in sweat glands. Oral and systemic infections have been suspected.

Brevibacterium

This includes the organisms previously known as *Kurthia* and *Zopfius*. They grow on ordinary media, aerobically and anaerobically, are motile at 20–22 °C but not at 35 °C and fail to attack carbohydrates. They are non-haemolytic, are indole and VP negative, attack urea slowly and some strains liquefy gelatin. Some strains are thermoduric.

They are found in milk and dairy products, decomposing farmyard material and faeces and have been isolated from human material.

Erysipelothrix

This genus contains small (2×0.3 μm) Gram-positive rods that are found in and on healthy animals and on the scales of fish. Man may be infected through abrasions when handling such animals, particularly pigs and fish. The resulting skin lesions are called erysipeloid. Epizootic septicaemia of mice has also been reported. There is one species, *Erysipelothrix insidiosa*, also known as *E. rhusiopathiae*. Mouse strains have been called *E. muriseptica*.

Isolation and identification

Culture tissue or fluid from the edge of a lesion. Emulsified biopsy material is better than swabs of lesions. Plate on blood agar and on crystal violet or azide blood agar and incubate at 37 °C in 5–10% carbon dioxide for 24–48 h.

Colonies are either small (0.5–1.0 mm) and 'dew-drop' or larger and granular. Young cultures are Gram-variable; older cultures show filaments. Test for motility with a Craigie tube (Difco motility medium is recommended). Test for fermentation of glucose, maltose and mannitol using enriched medium, e.g. broth containing Fildes' extract. Inoculate aesculin broth. Inoculate three tubes of enriched agar and incubate one each at 4, 22 and 37 °C. Do a catalase test and test for neomycin sensitivity by the disc method.

Erysipelothrix insidiosa

Is non-motile, produces acid from glucose but not from maltose and mannitol, does not hydrolyse aesculin and is catalase negative and is resistant to neomycin. It grows at 22 °C but not at 4 °C (*see Table 33.3*).

Listeria

These small (2 × 0.5 μm) Gram-positive rods are found in sewage, soil and water. They cause various infections of mammals, including septicaemia (with monocytosis), encephalitis and abortion in farm animals and necrotic hepatitis in poultry. Human infections, e.g. meningo-encephalitis and granulomatis infantiseptica, occur. There is one species, *Listeria monocytogenes.*

Isolation and identification

Macerate suspected tissues in nutrient broth in a blender. Plate on blood agar, blood agar containing 0.004% nalidixic acid to discourage other organisms, and on tryptone agar. Incubate at 35–37 °C for 24–48 h. Keep some material at 4 °C if possible for up to two months, repeating cultures if negative results are obtained. *L. monocytogenes* grows at 4 °C; most other organisms do not. Incubate two tubes of nutrient broth containing 0.1% Tween 80 and 0.375% potassium thiocyanate. Incubate one tube at 37 °C and the other at 4 °C. Plate out the 37 °C culture at 24 and 48 h and the 4 °C culture daily for four days.

Colonies of *L. monocytogenes* are small, 'dew-drop'-like and on clear media by oblique light appear to be bluish green. β-Haemolysis is usually observed on blood agar. Test for motility in a Craigie tube at 22 °C (at 37 °C the organisms may be only feebly motile) (Difco motility medium is recommended) and test for fermentation of glucose, maltose and mannitol. Inoculate aesculin broth and a nutrient agar slope for the catalase test. Test for neomycin sensitivity by the disc method.

Listeria monocytogenes

Is motile, produces acid from glucose and maltose but not from mannitol, hydrolyses aesculin and is catalase positive and sensitive to neomycin (*see Table 33.3*).

Antisera for FA and agglutination tests are available.

TABLE 33.3 ***Erysipelothrix*** **and** ***Listeria***

	Growth at 4 °C	*Motility*	*Acid from*			*Aesculin hydrolysis*	*Catalase*	*Haemolysis*
			Glucose	*Maltose*	*Mannitol*			
Erysipelothrix insidiosa[a]	−	−	+	−	−	−	−	α
Listeria monocytogenes[b]	+	+[c]	+	+	+	+	+	β

[a]Resistant to neomycin
[b]Sensitive to neomycin
[c]At 22 °C

Lactobacillus

Lactobacilli are important to the food and dairy industry and are commensals in the human and animal bodies.

Isolation

Plate in duplicate on MRS, Rogosa and similar media at pH 5.0–5.8. LS Differential medium (*O*) is particularly useful in the examination of yoghurt. Many strains grow poorly, if at all at pH 7.0, but plate human pathological material on blood agar (*see below*). Incubate one set of plates aerobically and the other in a 5:95 carbon dioxide–hydrogen atmosphere. Incubate cultures from food at 28–30 °C and from human or animal material at 35–37 °C for 48–72 h. Examine for very small, white colonies. Few other organisms, apart from moulds, will grow on the acid media.

For provisional identification of *L. acidophilus*, which is the commonest species in human material, place on the blood agar plates

(1) a sulphonamide sensitivity disc which has been dipped in saturated sucrose solution, and
(2) a penicillin (5 unit) disc.

L. acidophilus gives a green haemolysis. Growth is stimulated by sucrose; it is resistant to sulphonamides and sensitive to penicillin.

Identification

Subculture if necessary to obtain pure growth and heavily inoculate the modified MRS broth containing the following sugars: glucose (plus Durham's tube), lactose, sucrose, salicin, mannitol, sorbose and xylose (p. 73); in MRS broth containing 2% glucose and in which the ammonium citrate has been replaced by 0.3% arginine (for arginine hydrolysis test), and in MRS broth containing 4% sodium chloride. Incubate at 28–30 °C for 3–4 days. Inoculate tubes of MRS broth and incubate at 15 °C and at 45 °C.

Culture on solid medium low in carbohydrate to test for catalase (high-carbohydrate medium may give false positives). The catalase test is negative (corynebacteria and listeria, which may be confused with lactobacilli, are catalase positive) (*see Table 33.4*). The API system is useful.

TABLE 33.4 Some species of lactobacilli

	Growth at		*Acid from*						*NH_3 Arginine*	*Growth in 4% NaCl broth*
	15 °C	45 °C	*la*	*su*	*sal*	*mn*	*so*	*xy*		
L. acidophilus	−	+	+	+	+	−	−	−	−	−
L. bulgaricus	−	+	+	−	−	−	−	−	−	−
L. casei	+	v	+/−	+/−	+	+	+	−	−	+
L. plantarum	+	v	+	+	+	+	−	−	−	+
L. delbruckii	−	+	−	+	−	−	−	−	−	−
L. leichmanii	−	+	+/−	+	+	−	−	−	+/−	−
L. brevis	+	−	+/−	+	+/−	+/−	−	+	+	+
L. fermenti	−	+	+	+/−	−	−	−	+/−	+	−

la, lactose; su, sucrose; sal, salicin; mn, mannitol; so, sorbose; xy, xylose

Identification of lactobacilli is not easy; reactions vary according to technique and the taxonomic position of some species is confused.

The glucose Durham tube method (above) may not always detect carbon dioxide produced by heterofermentative lactobacilli. The method of Williams and Leon-Campbell[7] is useful.

Inoculate a tube of medium, e.g. MRS broth, containing 2% glucose and, if necessary, 5% tomato juice. Place this tube inside a larger tube containing several millilitres of barium hydroxide solution and stopper the larger tube with a rubber bung. Incubate at 30 °C for 48–72 h. If carbon dioxide is produced, a precipitate of barium carbonate will be formed. It is advisable to set up a control with a known homofermentative strain.

To grow the heterofermentative lactobacilli which cause greening of cured meats, add 0.1 g of thiamine hydrochloride to each litre of medium, or use APT medium (*D*). For further information *see* Carr *et al.*[8].

Species of lactobacilli

Lactobacillus acidophilus

This is widely distributed and found in milk and dairy products and as a commensal in the alimentary tract of mammals. It is used to make 'acidophilus milk' and is thought to be asssociated with dental caries in man, when it may be described as *L. odontolyticus*. *L. acidophilus* is probably the organism described as the Boas–Oppler bacillus, observed in cases of carcinoma of the stomach in man, and also Doderlein's bacillus, found in the human vagina.

Colonies of this organism on agar media are described as either 'feathery' or 'crab-like'.

L. bulgaricus

Is also associated with milk and dairy products but is thermophilic and not found in the mammalian intestine. It is used in the manufacture of yoghurt, along with *S. thermophilus*, and cream cheese and to produce gas holes in hard cheese. The natural fermentation of silage is in part due to this organism. Colonies on agar medium are small and grey.

L. casei and L. plantarum

Are difficult to separate except by paper chromatography of the end-products of carbohydrate fermentation. *L. casei* is used in making whey and occurs naturally in milk and cheese. *L. plantarum* is widely distributed on plants, alive and dead, and is one of the organisms responsible for fermentation in pickles and manufacture of sauerkraut. Both organisms give very poor growth on agar.

L. delbruckii

Is found in vegetation and in grain, and is used as a 'starter' to initiate acid conditions in yeast fermentation of grains. This organism grows on agar medium in small, flat colonies with crenated edges.

L. leichmanii

Is found in most dairy products. It is used in the commercial production of lactic acid and grows on agar as small white colonies.

L. bifidus

See Bifidobacterium (*below*).

Counting lactobacilli

Enumeration of lactobacilli is a useful guide to hygiene and sterilization procedures in dairy industries. Suitable dilutions of bottle or churn rinsings or of swabbings steeped in 0.1% peptone water are prepared (*see* Plate counts, Chapter 9) in Rogosa medium or one of the other media recommended for lactobacilli. Plates are incubated at 37 °C for three days or at 30 °C for five days in an atmosphere of 5% carbon dioxide.

Dental surgeons and oral hygienists sometimes request lactobacillus counts on saliva. The patient chews paraffin wax to encourage salivation, then 10 ml of saliva are collected and diluted for plate counts. Usually dilutions of 1:10, 1:100 and 1:1000 are sufficient.

Counts of 10^5 lactobacilli/ml of saliva are thought to indicate the likelihood of caries. There is a simple colorimetric test in which 0.2 ml of saliva is added to medium, as a shake tube culture, and a colour change after incubation suggests a predisposition to caries: (Snyder Test agar (*D*)).

Bifidobacterium

Bifidobacteria are small Gram-positive rods resembling coryneforms which predominate in the faeces of breast-fed infants, but they are also commensal in the adult bowels, mouth and vagina. They are anaerobic and may need to be differentiated from anaerobic corynebacteria. Bifidobacteria are catalase negative, nitratase negative and do not produce gas from glucose. The anaerobic corynebacteria are catalase and nitratase positive and do produce gas from glucose.

References

1. BOUSFIELD, I.J. and CALLEY, A.G. (Editors) (1978) *Coryneform Bacteria.* London: Academic Press
2. COLLINS, C.H. and DULAKE, C. (1983) Diphtheria: the logistics of mass swabbing. *Journal of Infection*, **6,** 227
3. DAVIES, J.R. (1974) Identification of diphtheria bacilli. In *Laboratory Methods, 1.* Public Health Laboratory Service Monograph No. 4. London: HMSO
4. JAMIESON, J.E. (1965) *Monthly Bulletin of the Ministry of Health and Public Health Laboratory Service*, **24,** 55
5. GARDNER, G.A. (1966) *Journal of Applied Bacteriology*, **29,** 455
6. ROBERTS, T.A., HOBBS, G., CHRISTIAN, J.H.B. and SKOVGAARD, N. (Editors) (1981) *Psychrotrophic Micro-organisms in Spoilage and Pathogenicity.* London: Academic Press
7. WILLIAMS, O.B. and LEON-CAMPBELL, L. (1965) *Food Technology London*, **5,** 306
8. CARR, J.G., CUTTING, C.V. and WHITING, G.C. (Editors) (1975) *Lactic Acid Bacteria of Beverages and Foods.* London: Academic Press

Chapter 34

Bacillus

This is a large genus of Gram-positive spore-bearing bacilli that are aerobic, facultatively anaerobic and catalase positive. Many species are normally present in soil and in decaying animal and vegetable matter. One species, *Bacillus anthracis* is responsible for anthrax in man and animals. *B. cereus* may cause food poisoning. Some are responsible for food spoilage.

The bacilli are large (up to $10 \times 1\ \mu m$) and commonly adhere in chains. Most species are motile; some form motile colonies. Some species are capsulated, some produce a sticky levan on media that contain sucrose. Spores may be round or oval, central or subterminal.

Physiological characters vary. There are strict aerobes and species which are facultative anaerobes. A few are thermophiles. All are catalase positive. In old cultures, Gram-negative forms may be seen and some species are best described as 'Gram-indifferent'. Apart from *B. anthracis* and *B. cereus*, identification of other 'species' is not easy.

Isolation of *B. anthracis* from pathological material

Veterinarians often diagnose anthrax in animals on clinical grounds and by the examination of blood films stained with polychrome methylene blue. These show chains of large, square-ended bacilli with the remains of their capsules forming pink-coloured debris between the ends of adjacent organisms. This is MacFadyean's reaction.

Culture animal blood, spleen substance or other tissue ground in a Griffith's tube or macerator. In man, culture material from the cutaneous lesions, faeces, urine and sputum (pulmonary anthrax is now rare).

Plate on blood agar and on chloral hydrate blood agar and incubate overnight. Pick woolly or waxy 'medusa head' colonies of Gram-positive rods into Craigie tubes to test for motility and on chloral hydrate blood agar. It may be difficult to get pure growths without several subcultures. For identification, *see below*.

Isolation of *B. anthracis* from hairs, hides, feedingstuffs and fertilizers

The sample should be in a 200–300-ml screw-capped jar, and should occupy about 4 fluid ounces. Add sufficient warm 0.1% peptone solution to cover it. Shake, stand at 37 °C for 2 h, decant and heat the fluid at 70 °C for 10 min. Place 0.1, 1.0 and 2 ml of the uncentrifuged fluid and the deposit after centrifuging in petri dishes. Add to each plate either 0.5 ml of 5% egg albumen or 0.25 ml of 0.01% dibromopropamidine isethionate (May and Baker), pour on 15 ml of yeast extract agar at 50 °C, mix well, allow to set and leave on the bench until 5 p.m. Incubate at 37 °C until 9 a.m. the next day. Longer incubation produces atypical colonies. The lysozyme in the egg albumen and the dibromopropamidine compound reduce the numbers of other organisms. Polymyxin B (20 units/ml) in the agar is also useful for heavily contaminated material. Plate also on heart infusion agar containing polymyxin (30 units/ml), lysozyme (40 units/ml), EDTA (300 μg/ml) and thallous acetate (40 μg/ml).

Examine under a low-power binocular (plate) microscope. Ignore surface colonies and look for deep colonies that resemble dahlia tubers or Chinese artichokes. Pick into Craigie tubes and on chloral hydrate agar.

Identification of *B. anthracis*

Bacillus anthracis is non-motile. The Craigie tube is the best way of demonstrating this. *B. cereus*, which is commonly mistaken for *B. anthracis*, is motile.

The growth of *B. anthracis* on agar media has a characteristic woolly or waxy nature when touched with a wire (the 'tenacity test'). Inoculate ammonium salt sugar medium containing 1% salicin. Do not use media that contain peptones; sufficient ammonia may be produced by some species to mask acid production. Culture on nutrient agar containing 10 units/ml of penicillin, nutrient agar containing 0.3% 2-phenylethanol and in nutrient broth incubated at 45 °C.

B. anthracis fails to ferment salicin, is sensitive to penicillin, is inhibited by 2-phenylethanol and does not grow at 45 °C. *B. cereus* ferments salicin, is resistant to penicillin and 2-phenylethanol and grows at 45 °C.

TABLE 34.1 Properties of some mesophilic aerobic spore bearers

Species	*Acid from*		*VP*[a]	*Gelatin liquefaction*	*Citrate*	*Hydrolysis of starch*	*Anaerobic growth*
	Glucose	*Xylose*					
B. subtilis	A	A	+	+	+	+	–
B. pumilis	A	A	+	+	+	–	–
B. coagulans	A	v	v	–	–	+	+
B. megaterium	A	v	–	+	+	+	–
B. cereus	A	–	+	+	+	+	+
B. anthracis	A	–	+	+[b]	v	+	+
B. polymyxa	AG	AG	+	+	–	+	+
B. macerans	AG	AG	–	+	–	+	+
B. circulans	A	A	–	+	–	+	v
B. laterosporus	A	–	–	+	–	–	+
B. brevis	A	–	–	+	v	–	–

A, acid; AG, acid and gas; v, variable in ammonium salt sugar media
[a] In 1% Na glucose phosphate broth
[b] Slow

The biochemical properties are given in *Table 34.1* but the best way to identify *B. anthracis* is by immunofluorescence (FA). Sera are available commercially. Unfortunately the specific W phage, which we used successfully for many years at the Anthrax Reference Laboratory in London is no longer generally available.

B. anthracis is regarded by some bacteriologists as a non-motile variant of *B. cereus* (*B. cereus* var. *anthracis*). The 'medusa head' surface colonies of the two organisms are similar. It causes anthrax in animals, a fatal septicaemia. Anthrax is transmissible to man, producing a localized cutaneous necrosis ('malignant pustule') or pneumonia (woolsorters' disease) due to inhalation of spores.

Imported bone meal, meat meal and other fertilizers and sometimes feedingstuffs may be infected. These are often prepared from the remains of animals that died of anthrax. In most developed countries, an animal dying of anthrax must be buried in quicklime under the supervision of a veterinarian and below the depth at which earthworms are active.

There are regulations governing the importation of hairs and hides from foreign countries where anthrax is endemic.

For further information on *B. anthracis see* Feeley and Patton[1].

Food poisoning

Bacillus cereus

May be associated with food poisoning (Chapter 14). A common vehicle is fried rice.

Make 10% suspensions of the food in 0.1% peptone water, using a Stomacher. Inoculate glucose tryptone agar, chloral hydrate blood agar (to suppress some other organisms) and, if available, the Bacillus Cereus Selective agar (*O*) of Holbrook and Anderson[2]. Incubate at 30 °C for 24–48 h.

B. cereus produces large, flat, irregular, 'ground glass' colonies on glucose tryptone agar, showing acid reaction. On the selective agar the colonies are turquoise blue and are surrounded by a blue halo of precipitated egg.

Test colonies for the egg yolk reaction, acid production from glucose and xylose (use ammonium salt sugars), VP (use 1% NaCl glucose phosphate broth) gelatin liquefaction, citrate utilization, hydrolysis of starch and ability to grow anaerobically (*see Table 34.1*). The API 20B system is useful.

NB Food poisoning due to *B. megaterium* has been reported.

Rapid staining identification[2]

Place air-dried and fixed films made from young colonies over boiling water and stain with 5% malachite green for 2 min. Wash and blot dry. Stain with 0.3% sudan black in 70% alcohol for 15 min. Wash with xylol and blot dry. Counterstain with 0.5% safranin for 20 s. Wash and dry. *B. cereus* cells stain red and contain black-stained lipid granules. The spores do not swell the cells and stain green.

Food spoilage

Examine Gram-stained films for spore-bearing Gram-positive bacilli. Make 10% suspensions of food in a 0.1% peptone water, and pasteurize some of the suspension at 75–80 °C for 10 min to kill vegetative forms.

Inoculate the following media with both unheated and pasteurized material: glucose tryptone agar, 5% egg yolk agar.

Incubate at 25–30 °C overnight. If the food was canned, incubate replicate cultures at 60 °C. Identify colonies of Gram-positive bacilli.

Cold-tolerant spore-formers are important spoilage organisms. Incubate at 5 °C for seven days.

Test colonies of Gram-positive bacilli for egg yolk reaction, acid from glucose and xylose ammonium salt sugars, VP (use 1% NaCl glucose phosphate broth), gelatin liquefaction, citrate utilization, starch hydrolysis and ability to grow anaerobically (*see Table 34.1*). Or use the API 20B system.

Species of *Bacillus*

Bacillus cereus General properties are described above. Phospholipinase activity produces large haloes around colonies on egg yolk agar. With most other species, activity is usually limited to beneath the colony. Methylene blue milk is rapidly decolorized. This organism is responsible for 'bitty cream', particularly in warm weather, and causes milk and ice-cream to fail the methylene blue test. It may cause food poisoning (*see above*). Bacteraemia and pneumonia associated with *B. cereus* have been reported. It is a common contaminant on shell eggs. It does not ferment mannitol, unilke *B. megaterium*, which it may otherwise resemble.

B. subtilis Wrinkled or smooth and folded colonies, 4–5 mm across on glucose tryptone agar with yellow halo (acid production). Strongly alkaline reaction in Crossley medium with peptonization but no gas or blackening. No zone around colonies on egg yolk agar. Responsible for spoilage in dried milk and in some fruit and vegetable products.

B. mesentericus Now considered to be identical, in Europe, with *B. subtilis*, but American strains are thought to be *B. pumilis*. Both cause 'ropy bread'. Autoenzymes from *B. subtilis* are used in certain laundry products. These may be toxic to factory operatives.

B. stearothermophilus This and associated thermophiles form large colonies (4 mm in diameter) on glucose tryptone agar with a yellow halo due to acid production at 60 °C but grow poorly at 37 °C and not at 20 °C. This organism is responsible for 'flat-sour' (i.e. acid but no gas) spoilage in canned foods. For more detailed investigation *see* Chapter 20).

B. mycoides A variant of *B. cereus* with rhizoid colonial form.

B. anthracoides Another variant of *B. cereus*. Could be confused with *B. anthracis* because of surface colony appearance, but is motile.

B. megaterium A very large bacillus. Widely distributed. May cause *B. cereus*-like food poisoning.

B. pumilus Found on plants. Common contaminant of culture media.

B. coagulans Widely distributed. Found in canned foods.

B. anthracis *See* p. 357.

B. polymyxa Widely distributed in soil and decaying vegetables (possesses a pectinase). Source of antibiotic polymyxin.

B. macerans Common in soil. Found in rotting flax.

B. circulans Motile colonies with non-motile variants. Found in soil.

B. laterosporus Soil organism. Source of the antibiotics tyrothricin and gramicidin.

B. popilliae Insect pathogen.

B. lentimorphus Insect pathogen.

B. thuringensis Insect pathogen, used in America in pest control.

B. sphaericus Large spores, resembles *C. tetani* in appearance. Forms motile colonies. Soil organism.

B. rotans Probably synonym of *B. sphaericus*. Motile colonies.

B. pasteuri Found in decomposing urine. Grows only in media containing peptone and urea.

References

1. FEELEY, J.C. and PATTON, C.M. (1980) *Bacillus anthracis*. In *Manual of Clinical Microbiology* 3rd edn. Edited by E.H. Lennette, A. Bolous, W.J. Hausler and J.P. Truant. Washington: American Society for Microbiology
2. HOLBROOK, R. and ANDERSON, J.M. (1980) *Canadian Journal of Microbiology*, **26,** 753

Chapter 35

Clostridium

This is a large genus of Gram-positive spore-bearing anaerobes (a few are aerotolerant) that are catalase negative.

They are normally present in soil; some are responsible for human and animal disease; others are associated with food spoilage.

Clostridia are classified according to the shape and position of the spores (*see Table 35.1*) and by their physiological characteristics. They may be either predominantly saccharolytic or proteolytic in their energy-yielding activities.

Saccharolytic species decompose sugars to form butyric and acetic acids and alcohols. The meat in Robertson's medium is reddened and gas is produced.

Proteolytic species attack amino acids. Meat in Robertson's medium is blackened and decomposed, giving the culture a foul odour.

Most species are mesophiles; there are a few important thermophiles and some psychrophiles and psychrotrophs.

TABLE 35.1 Morphology and colonial appearances of some species of *Clostridium*

	Spores	*Bacilli*	*Haemolysis*	*Colony appearance on blood agar*
C. botulinus	OC or S	normal	+	large, fimbriate, transparent
C. perfringens	OC	large thick	+	flat, circular, regular
C. tetani	RT	normal	+	small, grey, fimbriate, translucent
C. novyi	OS	large	+	flat, spreading, transparent
C. septicum	OS	normal	+	irregular, transparent
C. fallax	OS	thick	–	large, irregular, opaque
C. sordelli	OC or S	large thick	+	small, crenated
C. bifermentans	OC or S	large thick	+	small, circular, transparent
C. histolyticum	OS	normal	–	small, regular, transparent
C. sporogenes	OS	thin	+	medusa head, fimbriate, opaque
C. tertium	OT	long thin	–	small, regular, transparent
C. cochlearium		thin	–	circular, transparent
C. butyricum	OC	normal	–	white, circular, irregular
C. nigrificans		normal		black
C. thermosaccharolyticum		normal		granular, feather edges

Spores: O, oval; R, round; S, subterminal; C, central; T, terminal

The clostridia are conveniently considered under four headings: food poisoning; tetanus and gas gangrene; pseudomembranous colitis; and food spoilage.

Food poisoning

Botulism

This is the least common but most fatal kind (Chapter 14) caused by *Clostridium botulinum*. This is a strict anaerobe and requires a neutral pH and absence of competition to grow in food (e.g. canned, underprocessed). It is a difficult organism to isolate in pure culture.

Emulsify the suspected material in 0.1% peptone water and inoculate several tubes of Robertson's cooked meat medium which has been heated to drive off air and then cooled to room temperature. Heat some of these tubes at 75–80 °C for 30 min and cool. Incubate both heated and unheated tubes at 35 °C for 3–5 days. Plate them on pre-reduced blood agar and egg yolk agar and incubate under strict anaerobic conditions at 35 °C for 3–5 days.

If the material is heavily contaminated, heat some of the emulsion as described above, make several serial dilutions of heated and unheated material and add 1-ml amounts to 15–20 ml of glucose nutrient agar melted and at 50 °C in 152 × 16 mm tubes (Burri tubes). Mix, cool and incubate. Cut the tube near the sites of suspected colonies, fish these with a pasteur pipette and plate on blood agar or on egg yolk agar containing (per ml): cycloserine, 250 μg; sulphamethoxazole, 76 μg; trimethoprim, 4 μg (Dezfullian *et al.*)[1].

Clostridium botulinum gives a positive egg yolk and Nagler (half-antitoxin plate) reaction (*see below*). Types A, B and F liquefy gelatin, blacken and digest cooked meat medium and produce hydrogen sulphide. Types C, D and E do none of these. All strains produce acid from glucose, fructose and maltose and are indole negative. *See* p. 369 for methods and *Tables 35.1* and *35.2*.

It is best to use fluorescent antibody methods and to send suspected cultures to a Reference laboratory. It may be necessary to inoculate guinea pigs with suspensions of food, vomit or faeces. Control animals should receive *C. botulinus* antitoxin and some animals should be protected by *C. perfringens* and polyvalent gas gangrene antisera.

Type A *C. botulinum* is usually associated with meat. Type E is found in fish and fish products and estuarine mud and is psychrotrophic.

Perfringens (welchii) food poisoning

Clostridium perfringens is a common cause of food poisoning (Chapter 14).

Faeces

Inoculate two plates of neomycin blood agar; incubate one aerobically and the other anaerobically at 37 °C overnight. Inoculate two tubes of Robertson's meat medium; heat one at 80 °C for 60 min and cool. Incubate both at 37 °C overnight. Plate out both on blood agar plates for aerobic and anaerobic incubation as above.

TABLE 35.2 Cultural and biochemical properties of some species of clostridia

	RCM				Purple milk	Acid from				Indole	Gelatin liquefaction	Lecithinase
	Colour	Digestion	Odour	Gas		Glucose	Sucrose	Lactose	Salicin			
C. botulinum	black	+	−	+	D	+	v	−	−	−	+	−
C. perfringens	black	+	+	+	CD	+	−	−	v	+	+	+
C. tetani	black	+	+	−	C	−	−	−	−	+	+	−
C. novyi	red	−	−	+	GC	+	−	−	−	v	+	v
C. septicum	red	−	−	+	AC	+	−	+	+	−	+	−
C. fallax	red	−	−	+	AC	+	+	+	+	−	−	−
C. sordelli[a]	black	+	+	+	CD	+	−	−	−	+	+	+
C. bifermentans[a]	black	+	+	+	CD	+	−	−	v	+	+	+
C. histolyticum	black	+	+	−	D	−	−	−	−	−	+	−
C. sporogenes	black	+	+	+	D	+	−	−	v	−	+	−
C. tertium		−	−	+	AC	+	+	+	−	−	−	−
C. cochlearium	red	−	−	+		−	−	−	−	−	−	−
C. butyricum		−	−	+	ACG	+	+	+	+	−	−	−
C. nigrificans		−	−	−		−	−	−	−	−	−	−
C. thermosaccharolyticum		−	−	+		+	+	+	+	−	−	−

RCM: Robertson's cooked meat medium
Purple milk: A, acid; C, clot; D, digestion; G, gas
[a] *C. sordelli* also splits urea; *C. bifermentans* does not.

The aerobic plates act as a control to enable anaerobic colonies to be distinguished.

Food

Examine Gram-stained films of the material (usually meat or meat dishes) for short, plump, swollen bacilli, which may be present in large numbers. Spores are rarely seen. Make 10% emulsions in 0.1% peptone water using a Stomacher. Plate on Willis and Hobbs medium, tryptose sulphite cycloserine agar (TSC) (*O*) and neomycin blood agar. Do pour plates in oleandomycin–polymyxin–sulphadiazine–perfringens agar (OPSP) (*O*). Incubate duplicate cultures aerobically and anaerobically at 37 °C overnight.

Add 2–3 ml of the emulsion to several tubes of Robertson's cooked meat medium. Heat some tubes at 80 °C for 1 h. Incubate overnight and plate out both heated cultures on TSC, neomycin blood agar and Willis and Hobbs medium.

Identification

Heat-resistant strains of *C. perfringens* yield very slightly haemolytic colonies (other strains usually show much more haemolysis) which are moist, raised and smooth, about 2 mm in diameter. Heat-sensitive strains do not grow from the pasteurized cultures. Spores are not usually seen in laboratory cultures of this organism. Heat-sensitive strains may also cause food poisoning.

It is usually sufficient to identify *C. perfringens* by the Nagler half antitoxin test (*see below*) but other properties are given in *Tables 35.1* and *35.2*. No great reliance should be placed on the 'stormy fermentation' of litmus milk. General cultural methods are given on p. 187.

Nagler half-antitoxin plate test

Dip a cotton wool swab in standard antitoxin and spread on one-half of a plate of Willis and Hobbs medium or on nutrient agar containing 10% egg yolk. Dry the plates and inoculate the organism across both halves of the plate. Incubate anaerobically overnight at 37 °C.

The lecithinases of *C. perfringens* and *C. novyi* produce white haloes round colonies on the untreated half of the plate. This activity is inhibited by the half treated with antitoxin. The lecithinases of *C. bifermentans* and *C. sordelli* are also inhibited by *C. perfringens* antitoxin, but they may be distinguished on Willis and Hobbs medium. A diffuse pink halo appears round colonies of *C. perfringens* due to fermentation of the lactose. *C. bifermentans* and *C. sordelli* give white haloes because they do not ferment lactose.

On Willis and Hobbs medium, some clostridia show a 'pearly layer' due to lipolysis[2]. This is a useful differential criterion.

Tetanus and gas gangrene

Tetanus is caused by *C. tetani*. Clostridia associated with gas gangrene include *C. novyi* (*oedematiens*), *C. perfringens*, *C. septicum*, *C. fallax* and *C. sordelli*.

C. histolyticum is a possible pathogen and some other species, such as *C. sporogenes*, *C. tertium* and *C. bifermentans* may also be found in wounds but are not known to be pathogenic.

Examine Gram-stained and FA films of wound exudates. Bacilli with swollen terminal spores ('drumstick') suggest *C. tetani* but are not diagnostic. Thick, rectangular, box-like bacilli suggest *C. perfringens*. Very large bacilli might be *C. novyi* and thin, almost filamentous bacilli may be seen.

Inoculate two plates of blood agar, one for anaerobic and the other for aerobic culture, and also Willis and Hobbs or Lowbury and Lilly medium. One of the commercial media may used as well as or instead of the latter. Inoculate two tubes of Robertson's cooked meat medium. Heat one at 80 °C for 30 min and cool. Incubate one blood agar and all the other media anaerobically at 37 °C for 24 h. Plate out the Robertson's medium on blood agar and other clostridial media and incubate anaerobically with 10% carbon dioxide.

Identification

Test suspected colonies (*see Table 35.1*) with half antitoxin plates (*see above*), inoculate Crossley milk medium, glucose, lactose, sucrose, maltose and salicin peptone waters and note nature of growth and appearance in the cooked meat medium (*see Tables 35.1* and *35.2*).

Pseudomembranous colitis

Recent concern about this condition, especially in neonates, and about colitis and diarrhoea associated with chemotherapy led to the incrimination of *C. difficile*.[3]

Plate material on blood agar, on reinforced clostridial agar containing 0.2% phenol, or better, on Clostridium Difficile agar (*O*)[4]. Incubate anaerobically with 10% carbon dioxide at 37 °C for 24–48 h.

Colonies are 2–5 mm, grey, irregular, raised and opaque. *C. difficile* is motile, produces hydrogen sulphide, acid from glucose, mannitol and salicin, but not from maltose and sucrose. There is no change in milk, gelatin is not liquefied, nitrates are not reduced and indole is not formed.

For useful information about this organism *see* the ACP Broadsheet[5].

Species of *Clostridium* of medical importance

C. botulinum

Strict anaerobe; requires also a neutral pH and absence of competition to grow in food (e.g. canned, underprocessed). Free spores may be seen. Cultural characteristics are variable within and between strains. There are six antigenic types (A–F). Types A, B and F are proteolytic, causing clearing on Willis and Hobbs medium; types C, D and E are not. Causative organism of botulism, one of the least common but most fatal kind of food poisoning. Type A is usually associated with meat and Type E (a psychrotroph) with fish products. Found in soil and marine mud.

C. perfringens (*C. welchii*)

Not strictly anaerobic, may grow in broth, e.g. in MacConkey broth inoculated with water. Spores are rarely seen in culture (a diagnostic feature) but can be obtained on Ellner's medium. There are six antigenic types (A–F). Type A is associated with gas gangrene and with food poisoning. Causes lamb dysentery, sheep 'struck', pulpy kidney in lambs. Found in soil, water and animal intestines.

C. tetani

Strict anaerobe, easily dies on exposure to air. Swarms over the medium, but this may be difficult to see. It is, however, an important diagnostic characteristic. Found in soil, especially animal manured, and in the animal intestine. Causative organism of tetanus.

C. novyi (C. oedematiens)

A strict anaerobe which dies rapidly in air. There are four antigenic types (A–D). Types A and B cause gas gangrene, type D bacillary haemoglobinuria in cattle and 'black disease' of sheep.

C. septicum and chauvoei

Strict anaerobes. Considered by some bacteriologists to be one species—*C. septicum* types A and B. *C. septicum* causes gas gangrene in man, braxy and blackleg in sheep. *C. chauvoei* is not pathogenic for man but causes quarter evil, blackleg and 'symptomatic anthrax' of cattle and sheep. Both can be identified by fluorescent antibody methods.

C. fallax

A strict anaerobe, associated with gas gangrene. Spores are not very resistant to heat and the organism may be killed in differential heating methods. Found in soil.

C. bifermentans and C. sordelli

Separate species, distinguished biochemically but conveniently considered together because both have a toxin (lecithinase) similar to that of *C. perfringens* and give the same reaction as that organism on half-antitoxin plates with *C. perfringens* type A antiserum. *C. bifermentans* is not pathogenic but *C. sordelli* is associated with wound infections. Both are found in soil.

C. difficile

Implicated in pseudomembranous colitis and antibiotic associated diarrhoea.[3]

Species of clostridia of doubtful or no significance in human pathological material

C. histolyticum Not a strict anaerobe. Filamentous forms may grow as surface colonies on blood agar aerobically. Associated with gas gangrene but usually in mixed infection with other clostridia. Found in soil and intestines of man and animals.

C. sporogenes Not pathogenic but is a common contaminant. Widely distributed; found in soil and animal intestines.

C. tertium Aerotolerant and non-pathogenic.

C. tetanomorphum Of note because its sporing forms resemble the drumsticks of *C. tetani*. It is not pathogenic.

C. cochlearium The only common clostridium that is biochemically inert. It is not pathogenic.

Food spoilage clostridia

Clostridium species are known to be responsible for spoilage of a wide variety of foods, including milk, meat products, fresh water fish and vegetables.

Prepare 10% emulsions of the product in 0.1% peptone water using a Stomacher or blender. Heat some of this emulsion at 75–80 °C for 30 min (pasteurized sample). Inoculate several tubes of Robertson's cooked meat medium and/or liquid reinforced clostridium medium with pasteurized and unpasteurized emulsion and incubate pairs at various temperatures, e.g. 5–7 °C for psychrophiles, 22 and 37 °C for mesophiles and 55 °C for thermophiles. Inoculated melted differential reinforced clostridial agar (DRCM) in deep tubes with dilutions of the emulsions. Allow to set. Incubate at the desired temperature and look for black colonies, or blackening of the media, which suggests clostridia. Inoculate several tubes of Crossley's Milk medium (*O*) each with 2 ml of suspension and incubate at the required temperatures. This gives a useful guide to the identity of the clostridias.

Appearance in Crossley's milk medium

C. putrificum, *C. sporogenes*, *C. oedematiens*, *C. histolyticum* are indicated by slightly alkaline reaction, gas, soft curd subsequently digested leaving clear brown liquid, a black sediment and a foul smell.
C. sphenoides gives a slightly acid reaction, soft curd, whey and some gas.
C. butyricum produces acid, firm clot and gas.
C. perfringens usually gives a stormy clot.
C. tertium usually gives a stormy clot.

Note that aerobic spore bearers may give similar reactions, so plate out and incubate both aerobically and anaerobically.

Thermophiles associated with canned food spoilage

Two kinds of clostridial spoilage occur due to underprocessing: 'hard swell' and 'sulphur stinkers'.

Examine Gram-stained films for Gram-positive spore-bearers. Inoculate reinforced clostridial medium or glucose tryptone agar and iron sulphite medium in deep tube cultures by adding about 1-ml amounts of dilutions of a 10% emulsion of the food material in peptone water diluent to 152 × 16 mm tubes containing 15–20 ml of medium melted and at 50 °C. Either solid or semisolid media may be used. Incubate duplicate sets of tubes for up to three days at 60 and 25 °C.

C. thermosaccharolyticum

Produces white lenticular colonies in all three media and changes the colour of glucose tryptone agar from purple to yellow. This organism is responsible for 'hard swell'.

C. nigrificans

Produces black colonies, particularly in media that contain iron. This causes 'sulphur stinkers'.

Neither organism grows appreciably at 25 °C.

Counting clostridia

Emulsify 10 g of the food in 90 ml of a 0.1% peptone water in a homogenizer. Divide into two aliquots and heat one at 75 °C for 30 min.

Do MPN tests (10-, 1- and 0.1-ml amounts) using the five or three tube method, on each aliquot in liquid differential reinforced clostridial medium (DRCM). Clostridia turn this medium black ('black tube method'). The unheated aliquot gives the total count, the heated aliquot the spore count.

For the pour plate method add 0.1-ml amounts of serial dilutions of the emulsion to melted OPSP agar (*O*). Incubate anaerobically for 24 h and count the large black colonies.

It may be necessary to dilute the inoculum 1:10 or 1:100 if many clostridia are present. If the load of other organisms is heavy, add 75 units/ml of polymyxin to the medium used for the unheated count.

To recover individual colonies, Burri tubes may be used. These are open at both ends and are closed with rubber bungs or Astell seals. Use solid DRCM and, after incubation and counting, remove both stoppers and extrude the agar cylinder. Cut it with a sterile knife and fish colonies with a wire. These tubes need not be incubated in an anaerobic jar. After inoculation, cover the medium with a layer of melted paraffin wax about 2 cm deep.

Miller–Prickett tubes (flattened test-tubes, Astell Laboratories) are also useful. Pipette the dilutions into the tubes and then add 15 ml of melted medium at 50 °C to each tube. Seal with paraffin wax as above.

Membrane filters can be used for liquid samples. Roll up the filters and place in test-tubes; cover with melted medium.

DRCM is suitable for most counts but for *C. nigrificans* iron sulphite medium is good.

Select incubation temperatures according to the species to be counted.

General identification procedure

Subculture each kind of colony of anaerobic Gram-positive bacillus in the following media: Robertson's cooked medium; purple milk; peptone water for indole production; gelatin medium; glucose, lactose, sucrose and salicin peptone waters. These media should be in cotton wool-plugged test-tubes, not in screw-capped bottles. Omit the indicator from these as it may be decolorized during anaerobiosis. Test for acid production after 24–48 h with bromocresol purple. It may be necessary to enrich some media with Fildes' extract and an iron nail in each tube assists anaerobiosis. Gas production in sugar media is not very helpful and Durham's tubes can be omitted. The API 20A system for anaerobes is very helpful.

Do half antitoxin (Nagler) plates as above using *C. perfringens* type A and *C. novyi* antitoxin *see Tables.*

Food spoilage species

(1) Thermophiles	
C. nigrificans	Sulphur stinkers
C. thermosaccharolyticum	Hard swell
(2) Mesophiles	
C. butyricum	Cheese disorders. Butter and milk products
C. sporogenes	Meat and dairy products
*C. sphenoides**	
C. novyi	
C. perfringens	
(3) Psychrophile	
*C. putrefaciens**	Bone taint, off-odours

For further information on clostridia *see* Willis[6] and Holdeman *et al.*[7].

* Not described in this book.

References

1. DEZFULLIAN, M., McCROSKEY, C.L., HATHEWAY, C.L. and DOWELL, V.R. (1981) *Journal of Clinical Microbiology*, **13**, 526
2. WILLIS, A.T. and HOBBS, G. (1959) *Journal of Bacteriology*, **77**, 511
3. KEIGHLEY, M.R.B., BURDON, D.W. and ALEXANDER-WILLIAMS, J. (1978) *Lancet*, **ii**, 1165
4. GEORGE, W.L., SUTTER, V.L., CITRON, D. and FINEGOLD, S.M. (1976) *Journal of Clinical Microbiology*, **9**, 214

5. ASSOCIATION OF CLINICAL PATHOLOGISTS (1982) *Clostridium difficile*, Broadsheet No. 102. London
6. WILLIS, A.T. (1977) *Anaerobic Bacteriology*. 3rd edn. London: Butterworths
7. HOLDEMAN, L.V., CATO, E.P. and MOORE, W.E.C. (Editors) (1977) *Anaerobic Laboratory Manual*. 4th edn. Blacksburg, VA: Virginia State University

Chapter 36

Mycobacterium

These organisms are acid-fast: if they are stained with a strong, phenolic solution of a dye, e.g. carbol fuchsin, they retain the stain when washed with dilute acid. Other organisms are decolorized. There are no grounds for the commonly held belief that some mycobacteria are acid- and alcohol-fast but that others are only acid-fast. These properties vary with the technique and with the physiological state of the organisms.

The most important members of this genus are obligate parasites. They include the three 'species' of tubercle bacilli, *Mycobacterium tuberculosis*, *M. bovis* and *M. africanum*; Johne's bacillus, *M. paratuberculosis* and the leprosy (Hansen's) bacillus *M. leprae*. Several species that appear to be free living are opportunist pathogens of man and animals. Others are not known to be associated with human or animal diseases.

Tubercle bacilli

Tubercle bacilli are Risk Group III pathogens, infectious by the airborne route. All manipulation which might produce aerosols should be done in microbiological safety cabinets in Containment laboratories.

Direct microscopic examination

Sputum

This is the most common material suspected of containing tubercle bacilli. It is often viscous and difficult to manipulate. To make direct films use disposable 10 μg plastic loops. These avoid using bunsens in safety cabinets. Discard them into disinfectant. Ordinary bacteriological loops are unsatisfactory and unsafe. Spread a small portion of the most purulent part of the material carefully on a slide, avoiding hard rubbing, which releases infected airborne particles. Allow the slides to dry in the safety cabinet then remove them for fixing and staining. This is quite safe; aerosols are released during the spreading, not from the dried film, although these should not be left unstained for any length of time. Fixing may not kill the organisms and dried material is easily detached.

For the direct microscopy of centrifuged deposits after homogenizing sputum for culture spread a 10-μl loopful over an area of about 1 cm^2. Dry slowly and handle with care. The material tends to float off the slide while it is being stained.

CSF

Make two parallel marks 2–3 mm apart and 10-mm long in the middle of a microscope slide. Place a loopful of the centrifuged deposit between the marks. Allow to dry and superimpose another loopful. Do this several times before drying, fixing and staining. Examine the whole area.

Urine

Direct microscopy of the centrifuged deposit is unreliable because urine frequently contains acid fast bacilli which may resemble tubercle bacilli but which have entered the specimen from the skin or the environment.

Aspirated fluids

Centrifuge if possible and make films from the deposit. Otherwise prepare thin films. Material containing much blood is difficult to examine and may give false positives.

Gastric lavage

Direct microscopy is unreliable because environmental mycobacteria are frequently present in food and hence in the stomach contents.

Laryngeal swabs

Direct microscopy is unrewarding.

Tissue

Cut the tissue into small pieces. Work in a safety cabinet. Remove caseous material with a scalpel and scrape the area between caseation and soft tissue. Spread this on a slide. It is more likely to contain the bacilli than other material.

Staining

Stain direct films of sputum or other material by the ZN method. The bacilli may be difficult to find and it is often necessary to spend several minutes examining each film. Fluorescence microscopy, using the auramine phenol (AP) method, is popular in some clinical laboratories. A lower-power objective may be used and more material examined in less time. False positives are not uncommon with this method and the presence of acid-fast bacilli should be confirmed by overstaining the film with ZN stain. These staining methods are described in Chapter 6.

Reporting

Report 'acid-fast bacilli seen/not seen'. If they are present report the number per high power field. *Table 36.1* gives a reporting system in common use in Europe for sputum films.

TABLE 36.1 Scale for reporting acid-fast bacilli in sputum smears examined microscopically

No. of bacilli observed	*Report*
0 per 300 fields	Negative for AFB
1–2 per 300 fields	(±) Repeat test
1–10 per 100 fields	+
1–10 per 10 fields	++
1–10 per field	+++
10 or more per field	++++

Note that this is a logarithmic scale; this facilitates plotting.

False positives

Single acid fast bacilli should be viewed with caution. They may be environmental contaminants, e.g. from water. They may be transferred from one slide to another during staining or if the same piece of blotting paper is used for more than one slide.

Isolation from pathological material

Mycobacteria are often scanty in pathological material and not uniformly distributed. Many other organisms may be present and the plating methods used in other bacteriological examinations are useless. Acid-fast bacilli usually grow at a very much slower rate than other bacteria and are overgrown on plate cultures. Isolation methods depend on homogenization of the suspected specimens with a reagent that is less lethal for mycobacteria than for other organisms and that reduces the viscosity of the preparation so that it may be centrifuged. No ideal reagent is known; there is a choice between 'hard' and 'soft' reagents.

Hard reagents are recommended for specimens that are heavily contaminated with other organisms. Exposure times are critical or the mycobacteria are also killed. Soft reagents are to be preferred when there are fewer other organisms, e.g. in freshly collected sputum. The reagent can be left in contact with the specimen for several hours without significantly reducing the numbers of mycobacteria. Soft reagents may permit the recovery of mycobacteria other than tubercle bacilli, which are often killed by hard reagents.

The methods described below are intended to reduce the number of manipulations, particularly in neutralization and centrifugation, so that the hazards to the operator are minimized.

Culture of the centrifuged deposit after treatment gives optimum results, but is avoided by some workers because of the dangers of centrifuging tuberculous liquids. Sealed centrifuge buckets (safety cups) overcome this problem (*see* Chapter 1). These reduce considerably the hazards of aerosol formation should a bottle leak or break in the centrifuge. They should be opened in a safety cabinet.

If material is not centrifuged, more tubes of culture media should be inoculated and a liquid medium should be included. This should contain an antibiotic mixture (p. 376).

Sputum culture

Encourage the sputum from the specimen container into the preparation bottle with a pipette made from a piece of sterile glass tubing (200 × 5 mm) with one end left rough to cut through sputum strands and the other smoothed in a flame to accept a rubber teat.

Hard method

Add about 1 ml of sputum to 2 ml of 4% NaOH solution in a 25-ml screw-capped bottle. Stopper securely, place in a self-sealing plastic bag to prevent accidental dispersal of aerosols and shake mechanically, but not vigorously, for not less than 15 min but not more than 30 min. Thin specimens require the shorter time. Incubation does not help.

Remove the bottle from the plastic bag and add 3 ml of 14% (approximately 1 M) dipotassium hydrogen orthophosphate (KH_2PO_4) solution containing enough phenol red to give a yellow colour. This neutralizing fluid should be dispensed ready for use and sterilized in small screw-capped bottles. It should not be pipetted from a stock bottle.

The colour change of the phenol red from yellow to orange pink indicates correct neutralization. Re-stopper, mix gently and centrifuge with the bottles in sealed centrifuge buckets ('safety cups') for 15 min at 3000 rev/min. Pour off the supernatant carefully into disinfectant. Wipe the neck of the bottle with a piece of filter-paper, which is then discarded into disinfectant. Culture the deposit.

Soft method

To 2–4 ml of sputum in a screw-capped bottle, add an equal volume of 23% trisodium orthophosphate solution. Mix gently and stand at room temperature or in a refrigerator for 18 h. Add the contents of a 10-ml bottle of distilled water to reduce the viscosity, centrifuge, decant and culture the deposit using the methods and precautions described above.

Laryngeal and other swabs

Place the swab in a tube containing 1 N sodium hydroxide solution for 5 min. Remove, drain and place in another tube containing 14% potassium dihydrogen orthophosphate (KH_2PO_4) solution for 5 min. Drain and inoculate culture media.

Urine culture

Use fresh, early morning mid-stream specimens. Do not bulk several specimens as this often leads to contamination of cultures. Centrifuge 50 ml and treat the deposit with 2 ml of 3% sulphuric acid for 15 min. Add 20–30 ml of sterile distilled water, centrifuge and culture the deposit.

Alternatively plate some of the urine on blood agar, incubate overnight and if this is sterile pass 50–100 ml of the urine through a membrane filter. Cut the filter into strips and place each strip on the surface of the culture media in screw-capped bottles.

Culture of CSF, pleural fluids, pus, etc.

Plate original specimen or centrifuged deposit on blood agar and incubate overnight. Keep the remainder of specimen in a deep-freeze. If the blood agar culture is sterile, inoculate media for mycobacteria without further treatment. Use as much material and inoculate as many tubes as possible. If the blood agar culture shows the presence of other organisms, proceed as for sputum and/or culture directly into media containing antibiotics (*see below*). Half fill the original container with Kirchner medium and incubate. Some tubercle bacilli may adhere to the glass or plastic.

Culture of tissue

Grind very small pieces of tissue, e.g. endometrial curettings, in Griffith's tubes or emulsify them in sterile water with glass beads on a Vortex mixer. Homogenize larger specimens with a Stomacher Lab-Blender or blender. Check the sterility and proceed as for CSF, etc. Keep some of the material in a deep freezer in case the cultures are contaminated and the tests need repeating.

Isolation from the environment

This may be done to trace the sources of opportunist and saprophytic mycobacteria which contaminate clinical material and laboratory reagents.

Water

Pass up to 2 litres of water from cold and hot taps through membrane filters (as for water bacteriology, p. 255). Drain the membranes and place them in 3% sulphuric or oxalic acid for 3 min and then in sterile water for 5 min. Cut them into strips and place each strip on the surface of the culture medium in screw-capped bottles. Use Lowenstein–Jensen medium and Middlebrook 7H11 medium containing antibiotics (*see below*).

Swab the insides of cold and hot water taps and treat them as laryngeal swabs.

Dust and soil

Place 2–5 g samples in 1 litre of sterile distilled water containing 0.5% Tween 80. Mix gently to avoid too much froth. Pass through a coarse filter to remove large particles and allow to settle in a refrigerator for 24 h. Decant the supernatant and pass it through membrane filters as described above for waters. Centrifuge the deposit and treat it with sodium hydroxide as for sputum. Neutralize, centrifuge and inoculate several tubes of medium.

Culture media

A wide variety of media is available (*see* p. 58). Egg media are most commonly used and there seems little to choose between those listed. In the UK, Lowenstein–Jensen medium is most popular, but in the USA bacteriologists seem to favour ATS or Piezer media. Both glycerol medium (which encourages the growth of the human tubercle bacillus) and pyruvate medium (which encourages the bovine bacillus) should be used.

Pyruvate medium should not be used alone, however: some opportunists grow very poorly on it.

Some workers prefer agar based media, Middlebrook 7H9 or 7H11. Kirchner liquid medium is also useful for fluid specimens that cannot be centrifuged or when it is desirable to culture a large amount of material.

The antibiotic media of Mitchison *et al.*[1] are useful for contaminated specimens, even after treatment with NaOH, etc. They offer a safety net for non-repeatable specimens such as biopsies. These media should be used as well as, not in place of egg media for tissues, fluids and urines. They are made as follows.

To 200 ml of complete Kirchner or Middlebrook 7H11 add 2 ml of each of the following: polymyxin, 20 000 units/ml; carbenicillin, 10 000 units/ml; trimethoprim lactate 1000 μg/ml; amphotericin, 100 μg/ml. Tube the Kirchner medium in 10-ml lots; tube the 7H11 medium in 7-ml lots and set in slopes.

Inoculating culture media

Use an inoculum of at least 0.2 ml (not a loopful) for each tube of medium. Inoculate two tubes each of egg medium, one containing glycerol and the other pyruvic acid and other media as indicated above.

Incubation

Incubate all cultures at 35–37 °C and cultures from superficial lesions also at 33 °C for at least six weeks. Prolonging the incubation for a further two to four weeks may result in a small increment of positives, especially from tissues. Examine weekly.

Identification of tubercle bacilli

Most mycobacteria cultured from pathological material will be tubercle bacilli. No growth will be evident on egg media until 10–14 days. Colonies of the human tubercle bacillus are cream coloured, dry, and look like breadcrumbs or cauliflowers. Those of the bovine tubercle bacillus are smaller, whiter and flat. Growth of this organism may be very poor; it grows much better on pyruvate medium than on glycerol medium.

On Middlebrook agar media, colonies of both types are flat and grey. In Kirchner fluid media, colonies are round, granular and may adhere to one another in strings or masses. They grow at the bottom of the tube and settle rapidly after the fluid has been shaken.

Make ZN films to check acid-fastness. Some yeasts and coryneform organisms grow on egg media and their colonies may resemble those of tubercle bacilli. Make the films in a drop of saturated mercuric chloride as a safety measure against dispersing live bacilli in aerosols, and spread the films gently. Note whether the organisms are difficult or easy to emulsify. Check the morphology microscopically. Tubercle bacilli may be arranged in serpentine cords, are usually uniformly stained and regularly 3–4 μm in length.

Make suspensions of the organisms as follows. Prepare small, glass screw-capped bottles containing two wire nails shorter than the diameter of the base of the bottles, a few glass beads 2–3 mm in diameter and 2 ml of phosphate buffer, pH 7.2–7.4 (the nails will rust if water is used; wash both nails and beads in dilute hydrochloric acid and then in water before use). Place several colonies of the organisms in a bottle and mix on a magnetic stirrer. Allow large particles to settle and use the supernatant.

Inoculate two tubes of egg medium and one tube of the same medium containing 500 μg/ml of *p*-nitrobenzoic acid (PNBA medium). Incubate one egg slope at 25 ± 0.5 °C, the other in an incubator with an internal light at 35–37 °C. Incubate the *p*-nitrobenzoic acid egg slope at 35–37 °C. To make the *p*-nitrobenzoic acid stock solution, dissolve 0.5 g of the compound in 50 ml of water to which a small volume of 1 N sodium hydroxide solution has been added. Carefully neutralize with hydrochloric acid and make up to 100 ml with water. This solution keeps for several months in a refrigerator. If an internally illuminated incubator is not available, grow the organisms for 14 days and then expose to the light of a 25-W lamp at a distance of 2–3 ft for several hours and re-incubate for one week.

Tubercle bacilli have the following characteristics:

(1) Not easily emulsified in water.
(2) Regular morphology, 3–4 μm in length and usually showing serpentine cords.
(3) Relatively slow growing, taking ten days or more to show visible growth.
(4) Produce no yellow, orange or red pigment.
(5) Fail to grow in the presence of 500 μg/ml of *p*-nitrobenzoic acid (the bovine organism may show a trace of growth).
(6) Fail to grow at 25 °C (the bovine organism may show a trace of growth).

Acid-fast organisms that grow in one week or less, or yield yellow or red colonies, or emulsify easily, or are morphologically small, coccoid, or long, thin and beaded or irregularly stained, are unlikely to be tubercle bacilli.

Variants of tubercle bacilli and BCG

Although the identification of human, bovine and 'African' variants of *M. tuberculosis* is of no clinical value it may be necessary to recognize them for epidemiological work[2]. It is also useful to identify BCG, particularly if it may have been responsible for disseminated disease as a result of vaccination of neonates or the treatment of malignancy.

Use suspensions prepared as described above and inoculate media for the following tests and incubate at 37 °C.

TCH sensitivity

Use Lowenstein–Jensen medium containing 5 μg/ml of thiophen-2-carboxylic acid hydrazide. Read at 18 days.

Pyruvate preference

Compare the growth on Lowenstein–Jensen medium containing glycerol with that containing pyruvate after 18 days' incubation.

Nitratase test

Use Middlebrook 7H10 broth and test after 18 days by method[3] described on p. 110.

Oxygen preference

Use semisolid medium (p. 80). Inoculate with about 0.02 ml of suspension. Mix gently to avoid air bubbles and incubate undisturbed for 18 days. Aerobic growth occurs at or near the surface; microaerophilic growth as a band 1–3 cm below the surface, sometimes extending upwards.

Pyrazinamide sensitivity

Use the method described on p. 391.

Cycloserine sensitivity

Use Lowenstein–Jensen medium containing 20 μg/ml cycloserine. Read after 18 days (*see Table 36.2*).

Identification of a strain as 'human', 'bovine', 'Asian' or 'African' does not indicate origin of strain or ethnic group of the patient. These variants are widely distributed.

TABLE 36.2 Variants of tubercle bacilli

	TCH	*Nitratase*	*Oxygen preference*	*Pyrazinamide*
Classical bovine	S	–	M	R
African I[a]	S	–	M	S
African II[a]	S	+	M	S
Asian human	S	+	A	S
Classical human	R	+	A	S
BCG[b]	S	–	A	R

TCH, thiophen-2-carboxylic acid hydrazide; S, sensitive; R, resistant
[a] both variants are known as *M. africanum*.
[b] Differs from others in being resistant to cycloserine.
M, microaerophilic.
A, aerobic.

The niacin test

It was originally claimed that this test distinguished between human tubercle bacilli, which synthesize niacin, and bovine tubercle bacilli, which do not. Unfortunately some human strains give negative niacin results and some other mycobacteria do synthesize niacin. It is an unreliable test and uses a hazardous chemical (cyanogen bromide). It has little to commend it.

Species of tubercle bacilli

M. tuberculosis

This species grows well (eugonic) on egg media containing glycerol or pyruvate. Colonies resemble breadcrumbs and are cream coloured. Films show clumping and cord formation—the bacilli are orientated in ropes or cords, especially on moist medium. There is no growth at 25 or 37 °C or on PNBA medium. It is usually resistant to TCH (the Asian variants are sensitive), is nitratase positive, aerobic and sensitive to pyrazinamide. It causes tuberculosis in humans and may infect domestic and wild animals (usually directly or indirectly from man). Simians are particularly likely to become infected.

M. bovis

Is also known as the bovine variant of *M. tuberculosis*. It grows poorly (dysgonic) on egg medium containing glycerol but growth is enhanced on pyruvate medium. Colonies are flat and grey or white; growth may be effuse. Cords are present in films. There is no or very poor growth at 25 °C and on PNBA medium compared with a control slope at 37 °C. It is sensitive to TCH, nitratase negative, microaerophilic and resistant to pyrazinamide. This is the classical bovine tubercle bacillus, common in milch cows before eradication schemes were introduced. It is still occasionally associated with human disease, usually extrapulmonary, but such infections are now rarely associated with the consumption of infected milk.

M. africanum

This is now regarded as a variant of the other tubercle bacilli as strains have the properties of both[2]. Growth is usually dysgonic, enhanced by pyruvate. It is sensitive to TCH, microaerophilic and sensitive to pyrazinamide. The nitratase test may be negative (most West African strains), or positive (most East African strains). It causes tuberculosis, clinically indistinguishable from that caused by the other tubercle bacilli.

BCG

This is the bacillus of Calmette and Guerin, an organism of attenuated virulence used in immunization against tuberculosis. It is eugonic; growth is not enhanced by pyruvate. It is sensitive to TCH, nitratase negative, aerobic, resistant to pyrazinamide and (unlike other strains in this group) is resistant to cycloserine.

Animal inoculation tests

Cultural methods have now largely superseded animal inoculation. They are generally more reliable and much less expensive. Tubercle bacilli that are resistant to isoniazid usually show limited or no virulence to guinea pigs, and so do some Afro-Asian strains. Reasons for ending the routine guinea pig test were given by Marks[3].

Animal inoculation may, however, be the method of choice for specimens that are grossly and consistently contaminated after treatment with sodium hydroxide, and for milk samples.

In the guinea pig, both *M. tuberculosis* and *M. bovis* give a progressive disease when inoculated into the thigh. *M. tuberculosis* gives minimal disease in rabbits, usually only a local lesion, but *M. bovis* gives a progressive disease. BCG is not virulent for guinea pigs.

Post-mortem examinations of tuberculous animals generate large numbers of infected airborne particles. Such examinations should be done with great care in a microbiological safety cabinet.

Opportunist mycobacteria

Some mycobacteria, normally present in the environment, may, under some circumstances, cause or be associated with human disease. Such diseases are correctly called mycobacteroses, not tuberculosis (Runyon[4]). Before specific names were given to these organisms and their taxonomic position was clarified, they were known as 'atypical' or 'anonymous' mycobacteria. Various temporary systems of classification were proposed[5,6,7]. Only that of Runyon[5] has survived and, although it is not now generally used by bacteriologists, it is still understood by physicians. It is shown in *Table 36.3*.

A number of other mycobacteria, not so far known to be associated with human disease, are frequently cultured in clinical laboratories. These are usually contaminants of the specimen or are introduced during collection and/or laboratory processing. Mostly they come from water but they may also be present in dust, soil and food. Occasionally they are seen in direct films but fail to grow on cultures incubated at 35–37 °C.

TABLE 36.3 Runyon classification of mycobacteria[5]

Group	*Members*	*Characteristics*
I	Photochromogens	Produce a yellow pigment only when exposed to light during the growth phase. No pigment when incubated in the dark. Now known to include the opportunist pathogen *M. kansasii*.
II	Scotochromogens	Produce a yellow pigment when incubated in light and in dark. Less often pathogenic than Group I. Now known to contain *M. scrofulaceum* as well as saprophytes.
III	Non-photochromogens	Produce no pigment or poorly developed pigment independent of light. Often pathogenic and now known to include the MAIS complex and *M. xenopi* as well as some saprophytes.
IV	Rapid growers	Rapidly growing mycobacteria. Not often pathogenic. Now known to contain *M. fortuitum* and *M. chelonei* as well as saprophytes.

Runyon[4] and Marks[8] suggested that these mycobacteria be reported by their specific names only if they are accepted opportunists or pathogens. The specific names of other mycobacteria need not be given nor time wasted on identifying them. While agreeing with this proposal, we describe most of the species likely to be encountered. There are many instances in bacteriology of saprophytic organisms assuming a pathogenic role, especially nowadays in immunosuppressed patients.

TABLE 36.4 Some species of mycobacteria

Species	*Pigment*	*TZ*	NO_3	*TW*	*Growth at*				*Sulphatase*		*Catalase*	*Tellurite reduced*	*Growth on N medium*	*Sensitive to*				*Rapid growth*	*Runyon group*
					20°C	25°C	42°C	44°C	*3 days*	*21 days*				*Sm*	*Et*	*Eb*	*R*		
M. kansasii	Ph	S	+	+	–	+	±	–		+	+ + +	–	–	R	S	S	S	–	I
M. marinum	Ph	R	–	+	±	+	–	–			+ +	–	±	R	S	S	S	+	I or IV
M. xenopi	poor or –	R	–	–	–	–	+	+		+ + +	–	–	–	v	S	v	v	–	III
MAIS	poor or Sc	v	–	–	–	+	v	v		v	–	+	–	R	v	R	R	–	III
			–	–	–	+	±	–		+	+ +	–	–	R	v	R	R	–	II or III
M. fortuitum	–	R	+	–	+	+	±	–	+ + +		+ + +	+	+	R	v	R	R	+	IV
M. chelonei	–	R	–	–	+	+	–	–	+ + +		+ + +	+	+	R	R	R	R	+	IV
M. szulgai	Sc	R	+	+	–	+	–	–		+ + +	+	–	–	R	S	S	S	–	II
M. gordonae	Sc	R	–	+	+	+	–	–		+	+ +	–	–	S	S	S	S	–	II
M. flavescens	Sc	R	+	+	+	+	–	–		+	+	–	+	v	S	S	R	+	II
M. gastri	–	S	–	+	±	+	±	–		+	+	–	–	S	S	S	S	–	III
M. terrae	–	R	+	+	±	+	±	–		+	+ + +	–	–	R	S	S	R	–	III
M. nonchromo-genicum	–	R	–	+	±	+	±	–		+	+ + +	–	–	v	S	S	v	–	III
M. smegmatis	–	R	+	+	+	+	+	+	–		+ +	+	+	S	S	S	R	+	IV
M. phlei	Sc	R	+	+	+	+	+	+	–		+ +	+	+	S	S	S	S	+	IV

TZ, Thiacetazone; Ph, Photochromogen; Sc, Scotochromogen: Sm, Streptomycin; Et, Ethionamide; Eb, Ethambutol; R, Rifampicin; S, Sensitive; R, Resistant; v, variable; TW, Tween hydrolysis; NO_3, Nitratase

It is also difficult to assess the significance of opportunists. Single isolations may be regarded with suspicion. Whenever possible, attempts should be made to recover the organisms from subsequent specimens.

Most of these mycobacteria may be identified with the aid of the following tests and with *Table 36.4*.

Inoculation

Unless otherwise stated, inoculate the media with one loopful or a 10-μl drop of a suspension prepared as described on p. 389.

Pigment production

Inoculate two egg medium slopes, Incubate both at 35–37 °C, one exposed to light and the other in a light-proof box. Examine at 14 days. Photochromogens show a yellow pigment only when exposed to light. Scotochromogens are pigmented in light and dark, but the culture exposed to light is usually deeper in colour. False photochromogenicity may arise if the inoculum is too heavy and pigment precursor is carried over. Continue to incubate the light slope and look for orange-coloured crystals of carotene pigment which may form in growth.

Temperature tests

Inoculate egg medium slopes with a single streak down the centre and incubate at the following (exact) temperatures: 20, 25, 42 and 44 °C. Examine for growth at three and seven days and thereafter weekly for three weeks.

Thiacetazone sensitivity

Inoculate egg medium slopes containing 20 μg/ml of thiacetazone (TZ). Make a stock 0.1% solution of this drug in dimethylformamide. This solution keeps well in a refrigerator. Incubate slopes of 18–21 days and observe the growth in comparison with that on an egg slope without the drug.

Nitratase test

Use method (3) on p. 110.

Aryl sulphate test

Use the method described on p. 112. If the organism is a rapid grower, add the ammonia after three days, or use sulphatase agar (**B**).

Catalase test

Prepare the egg medium in butts in screw-capped tubes (20 × 150 mm). Inoculate and incubate at 35–37 °C for 14 days with the cap loosened and then add 1 ml of a mixture of equal parts of Superoxol (30% hydrogen peroxide) and 10% Tween 80. Allow to stand for a few minutes and measure the height in millimetres of the column of bubbles. More than 40 mm of froth is a strong positive result (+ + +) (Wayne's test).

Tween hydrolysis

Add 0.5 ml of Tween 80 and 2.0 ml of a 0.1% aqueous solution of neutral red to 100 ml of M/15 phosphate buffer. Dispense this in 2-ml lots and autoclave at 115°C for 10 min. This solution keeps for about two weeks in the dark and should be a pale straw or amber colour. Add a large loopful of culture from solid medium and incubate at 35–37 °C. Examine at five and ten days for a colour change from amber to deep pink.

Tellurite reduction

Grow the organisms in Middlebrook 7H11 medium for 14 days or until there is a heavy growth. Add four drops of a sterile 0.02% aqueous solution of potassium tellurite and incubate for a further seven days. If tellurite is reduced there will be a black deposit of metallic tellurium. Ignore grey precipitates. This test is not much use for highly pigmented organisms.

Growth in N medium

Inoculate the medium (p. 78) with a straight wire and incubate. Rapid growers give a turbidity with or without a pellicle in three days. 'Frosting', i.e. adherence of a film of growth to the walls of the tube above the surface of the medium, is often seen. Ignore a very faint turbidity.

Resistance to antibiotics

Methods of doing these tests are described below. Some mycobacteria exhibit consistent patterns. All opportunists are resistant to PAS and pyrazinamide, and all but a few strains of *M. xenopi* are resistant to INAH.

Aerial hyphae

Some *Nocardia* spp. are partially acid-fast and resemble rapidly growing mycobacteria. Do a slide culture (p. 116) and look for aerial hyphae. In very young cultures of mycobacteria, mycelium may be seen, but it fragments early into bacilli. Aerial hyphae are not formed.

Emulsifiability and morphology

When making films for microscopic examination, note if the organisms emulsify easily. Note the morphology, whether very short, coccobacilli, long, poorly stained filaments or beaded forms.

Species of opportunist and other mycobacteria

M. kansasii

This is a photochromogen (but *see below*). Optimum pigment production is observed if the inoculum is not too heavy, if there is a plentiful supply of air (loosened cap) and if exposure to light is continuous. Young cultures grown in the dark and then exposed to light for 1 h will develop pigment, but old cultures that have reached the stationary phase may not show any pigment after exposure. Continuous incubation in the light results in the formation of orange-coloured crystals of carotene. The morphology is distinctive; the bacilli are long (5–6 μm) and are beaded. *M. kansasii* is nitratase positive, sensitive to 20 μg/ml of thiacetazone, hydrolyses Tween 80 rapidly and grows at 26 °C but not at 20 °C; some strains grow poorly at 42 °C but not at 44 °C. The sulphatase test is weakly positive and the catalase strongly positive (more than 40 mm in the Wayne test). It is resistant to streptomycin, isoniazid and PAS, but sensitive to ethionamide, ethambutol and rifampicin.

Some strains are sensitive to amikacin and erythromycin by the disc technique.

Occasional non-chromogenic or scotochromogenic strains have been reported but these may be recognized by their biochemical reactions and morphology.

This organism is an opportunist pathogen, associated with pulmonary infection. It is rarely significant when isolated from other sites. It has been found in water supplies[9, 10].

M. xenopi

Pigment, if any, is more obvious on cultures incubated at 42–44 °C, which become pale yellow. Morphologically it is easily recognized; it stains poorly and the bacilli are long (5–6 μm) and filamentous. It is nitratase negative, resistant to thiacetazone and does not hydrolyse Tween 80. It is a thermophile, growing well at 44 °C, slowly at 35 °C, when growth is effuse, and not at 25°C. (This is the only opportunist mycobacterium which fails to grow at 25 °C.) The sulphatase test is strongly positive, the catalase test negative. It does not reduce tellurite. It is more sensitive to INAH than other opportunist mycobacteria, usually giving a resistance ratio of 4. It is sensitive to ethionamide, but its sensitivity to other antituberculous drugs varies between strains. Some strains are sensitive to amikacin and erythromycin by the disc method.

It is an opportunist pathogen in human lung disease and is rarely significant in other sites. It is a frequent contaminant of pathological material, especially in urines, and has been found in hospital hot water supplies[9, 10]. 'Outbreaks' of laboratory contamination are not uncommon. This organism is very common in south-east England and north-west France.

M. avium–intracellulare–scrofulaceum

This is a complex of improperly defined species, usually described as the MAIS bacilli, after the initial letters of its components. A feeble yellow pigment, not influenced by light, is produced by some species, but *M. scrofulaceum* is usually regarded as a scotochromogen within the definition of Runyon[5]. The bacilli are small, almost coccoid. Some strains are sensitive to thiacetazone. All are nitratase negative, catalase negative or weakly positive, reduce tellurite and do not hydrolyse Tween 80. The sulphatase reaction varies from strongly positive to negative. There is growth on egg medium at 25 °C, growth at 20 °C and 42 °C is variable. Resistance to antituberculous drugs is usual, but some strains are sensitive to ethionamide.

MAIS bacilli are opportunist pathogens of man, associated with cervical adenitis (scrofula), especially in young children, but pulmonary infections also occur. They are also opportunist pathogens of pigs and birds (*M. avium* was once described as the avian tubercle bacillus). They have also been found in soil and water.

MAIS bacilli are Risk Group III pathogens, infectious by the air-borne route. All manipulations which might produce aerosols should be done in a microbiological safety cabinet in a Containment laboratory.

M. marinum (formerly M. balnei)

This is missed in clinical laboratories that do not culture material from superficial lesions at 30–33 °C. Primary cultures do not grow at 35–37 °C. It is a photochromogen. There is no distinctive morphology. It is nitratase negative, resistant to thiacetazone and hydrolyses Tween 80. After laboratory subculture its temperature range is modified. It will grow at 25 °C and at 37 °C but not at 44 °C. The catalase and sulphatase tests are weakly positive and growth may occur in N medium. Resistance to streptomycin, isoniazid and PAS is usual; resistance to other drugs is variable. Some strains are sensitive to cotrimoxazole, erythromycin and amikacin by the disc method. This is an opportunist pathogen responsible for superficial infections, e.g. swimming pool granulomas and fish tank 'sporotrichoid' disease on the hands and arms. It is found in sea-bathing pools and in tanks where tropical fish are kept, and is a pathogen of some fish.

Slowly-growing scotochromogens

The commonest species is probably *M. gordonae*, also known as the tap water scotochromogen, although other scotochromogens are found in water supplies. Organisms in this group grow slowly, produce a deep orange pigment in light and usually a yellow pigment in the dark. No crystals of carotene are formed. If such crystals are seen, the organisms may be a scotochromogenic strain of *M. kansasii*. Morphology is not distinctive. They are nitratase negative, resistant to thiosemicarbazone and hydrolyse Tween 80. Growth occurs at 20 °C but not at 44 °C. Some strains are psychrophilic. The sulphatase reaction is weak and the catalase test is usually strongly positive. Growth does not occur in N medium. They are usually resistant to isoniazid and PAS but sensitive to streptomycin and other antituberculous drugs.

So far as is known, the organisms in this group are not associated with human disease and may be reported as non-significant or environmental scotochromogens. They are not infrequent contaminants of pathological material but usually appear as single colonies on egg medium. They may be found in tap water, dust and soil.

Other slowly growing scotochromogens include *M. duvalii* and *M. gilvum*, which are difficult to distinguish from one another except by immunodiffusion techniques. They differ from *M. gordonae* in being nitratase positive and from *M. szulgai* by giving a weak sulphatase reaction. Both are rare; neither is considered to be associated with human disease.

M. szulgai

This scotochromogen differs from *M. gordonae* in being nitratase positive, strongly sulphatase positive and giving a weak catalase reaction. It is a rare human opportunist pathogen.

M. flavescens

This is a rapidly growing scotochromogen which differs from *M. gordonae* in being nitratase positive, growing in N medium and being highly resistant to rifampicin. It is not known to cause human disease but occurs in the environment and sometimes contaminates pathological material.

M. fortuitum and M. chelonei

These two non-pigmented rapid growers are conveniently considered together. There have been taxonomic problems: *M. fortuitum* has also been called *M. ranae* and *M. peregrinum*; *M. chelonei* was formerly named *M. abcessus*, *M. runyonii* and *M. borstelense*. Growth occurs in most media in three days. The bacilli tend to be rather fat and solidly stained. The nitratase test is positive (*M. fortuitum*) or negative (*M. chelonei*). Tween 80 is not hydrolysed. There is growth at 20 °C and in some strains at 42 °C. Psychrophilic strains, which fail to grow at 37 °C, are not uncommon. The sulphatase test is positive at three days and the catalase reaction is strongly positive. Tellurite is reduced rapidly. There is growth in N medium in three days. There is a general resistance to all antituberculous drugs, except that *M. fortuitum* is usually sensitive to ethionamide and *M. chelonei* is resistant.

These are opportunist pathogens, occurring in superficial infections (e.g. injection abcesses) and occasionally as secondary agents in pulmonary disease. They are common in the environment and frequently appear as laboratory contaminants.

M. smegmatis and M. phlei

These two rapidly growing saprophytes are more common in textbooks than in clinical laboratories. Growth occurs in three days. The bacilli have no distinguishing features. They are nitratase positive and hydrolyse Tween 80 in ten days. They grow at 20 °C and at 44 °C, *M. phlei* grows at 52 °C. The three-day sulphatase test is negative, but longer incubation gives positive reactions. The catalase test is strongly positive. Tellurite is reduced. There is growth in N medium.

They are not associated with human or animal disease.

Slowly-growing nonchromogens

There are at least four species, *M. terrae*, *M. nonchromogenicum*, *M. triviale* and *M. gastri*. All grow at 25 °C but poorly or not at all at 42 °C. All hydrolyse Tween 80 and none reduce tellurite, which permits differentiation from MAIS bacilli. *M. terrae* and *M. triviale* give a positive nitratase test but *M. nonchromogenicum* and *M. gastri* are negative. *M. gastri* is occasionally photochromogenic and is usually sensitive to thiacetazone and may therefore be confused with *M. kansasii*. Sensitivity to anti-tuberculous drugs is variable.

M. ulcerans

This is likely to be missed by clinical laboratories because it does not grow at 37 °C. It grows only between 31 and 34 °C. Growth is very slow, taking 10–12 weeks to give small colonies resembling those of *M. bovis*. It is biochemically inert and undistinguished. It causes Buruli ulcer, a serious superficial skin infection.

M. simiae

Originally isolated from monkeys, this is also known as *M. habana*. It is a photochromogen, is nitratase negative and does not hydrolyse Tween 80. Pulmonary and disseminated infections have been reported.

M. malmoense

This resembles the MAIS organisms, grows very slowly and is sometimes confused with *M. bovis* by inexperienced workers. This suggests that it is commoner than might be expected. It is difficult to identify except by lipid chromatography. It causes pulmonary disease in adults and cervical adenopathy in children.

M. haemophilum

Another rare species, possibly missed because it grows poorly or not at all on media that do not contain iron or haemin. It grows on 30 °C on LJ medium containing 2% ferric ammonium citrate and on Middlebrook 7H11 medium plus 60 μg/ml of haemin. All the usual tests are negative. It has been isolated from the skin and subcutaneous tissues of immunologically compromised patients.

Drug sensitivity tests

Most cases of tuberculosis yield tubercle bacilli that are sensitive to the commonly used anti-tuberculous drugs. Sensitivity tests are, however, frequently required by physicians at the commencement of and during treatment, particularly if the patient's condition does not improve. Sensitivity tests are also required in cases of infection with opportunist mycobacteria as these are usually resistant to some of these drugs.

It is customary for physicians to treat patients with combinations of three drugs. Treatment with one drug only usually results in the emergence of resistant organisms.

The drugs in general use are INAH (isoniazid), rifampicin, ethambutol, pyrazinamide and streptomycin. If resistance is found to more than one of these, ethionamide, capreomycin and cycloserine may be given. Thiacetazone is rarely used and PAS now appears to be unobtainable.

Sensitivity tests with mycobacteria are complex and are usually done in specialist and reference laboratories. There are three principal methods:

(1) The Absolute Concentration method which is popular in Europe;

(2) The Proportion Method, also used in Europe and popular in America;
(3) The Resistance Ratio method used in the UK.

Different techniques, are, however, used for pyrazinamide sensitivity tests.

The absolute concentration method

Carefully measured amounts of standardized inocula are placed on control media and on media containing varying amounts of drugs and the lowest concentration that will inhibit all, or nearly all, of the growth is reported. This method gives different results in different laboratories. It is not possible to use a medium that must be heated after the addition of the drug (e.g. egg medium) because some drugs are partially heat labile and heating times are rarely constant even in the same laboratory. Middlebrook 7H10 or 7H11 media are used but this method has not found much favour in the UK.

The proportion method

Several dilutions of the inoculum are made and media containing no drug and standard concentration of drugs are inoculated. The number of colonies growing on the control from suitable dilutions of the inoculum is counted and also the number growing on drug-containing medium receiving the same inoculum. Comparison of the two shows the proportion of organisms that are resistant. This is usually expressed as a percentage.

This method is popular in the USA and in Europe but it is technically very difficult and as it is usually done in petri dishes we regard it as highly hazardous. There are also more risks attached to standardizing the inocula than with the Resistance Ratio method.

The resistance ratio method

The minimal inhibitory concentrations of the drug that inhibit test strains are divided by those which inhibit control strains to give the resistance ratios. Ratios of 1 and 2 are considered sensitive, those of 4 or more resistant (*see below*). These results usually correlate with the clinical findings. This method is used extensively in the UK and is described below.

Each of the above methods may be used for 'direct' tests on sputum homogenates that contain enough tubercle bacilli to give positive direct films as well as for 'indirect' tests on cultures. The latter are more reliable and reproducible.

For information about these tests, *see* Canetti *et al.*[11] and Vestal[12]. For a discussion on the design of Resistance Ratio tests, *see* Marks[13].

Technique of resistance ratio method

Dilutions are incorporated in Lowenstein–Jensen medium, which is then inspissated. As some of these drugs are affected by heat, it is essential that the inspissation procedure is standardized. The inspissator must have a large circulating fan so that all tubes are raised to the same temperature in the same time. The load (i.e. number of tubes) and the time of exposure must be constant. To ensure that the drug-containing medium is not overheated, the machine should be raised to its correct temperature *before* it is loaded and racks can be devised that allow loading in a few seconds. A period of 45 min at 80 °C is usually sufficient to coagulate this medium and no further heat is necessary if it is prepared with a reasonable aseptic technique.

Control strains

The resistance ratio method compares the MIC of the unknown strain with that of control stains on the same batch of medium.

Some workers use the H37Rv strain of *M. tuberculosis* as the control strain, but the sensitivity of this to some drugs does not parallel that of wild tubercle bacilli and may give misleading ratios. It is better to use the modal resistance method of Marks[13, 15]. The unknown strains are compared with the modal resistance (i.e. that which occurs most often) of a number of known sensitive strains.

Drug concentrations

Each laboratory must determine its own range, as these will vary slightly according to local conditions. Initially use those suggested in *Table 36.5* and

TABLE 36.5 TB sensitivity tests: suggested concentrations of drugs in Lowenstein-Jensen medium

	Final concentration (μg/ml)						
INAH	0.007	0.015	0.03	0.06	0.125	0.25	0.5
Ethambutol	0.07	0.15	0.31	0.62	1.25	2.5	5.0
Rifampicin	0.53	1.06	3.12	6.25	12.5	25	50
Streptomycin	0.53	1.06	3.12	6.25	12.5	25	50
Capreomycin	1.06	3.12	6.25	12.5	25	50	100
Cycloserine	1.06	3.12	6.25	12.5	25	50	100
Ethionamide	1.06	3.12	6.25	12.5	25	50	100

These are for the preliminary titration. Choose six that give confluent or near confluent growth in the two lowest and no growth in the next four.

inoculate at least 12 sets with known sensitive organisms to arrive at a baseline for future work. A drug-free control slope must be included.

Stock solutions of the drugs are conveniently prepared as 1% solutions in water (except for rifampicin which should be dissolved in formdimethylamide. All stock solutions keep well at −4 °C.

Bacterial suspension

Smooth suspensions must be used. Large clumps or rafts of bacilli give irregular results and make readings difficult.

Sterilize 5 ml (bijou) screw-capped bottles containing a wire nail (shorter than the diameter of the bottle), a few glass beads and 2 ml of phosphate buffer (13.3 g of anhydrous Na_2HPO_4 and 3.5 g of KH_2PO_4 in 2 litres of water; pH 7.4). Into each bottle place a scrape of growth equal to about three or four large colonies and place the bottle on a magnetic stirrer for 3–4 min. Allow to stand for a further 5 min for any lumps to settle and use the supernatant to inoculate culture media.

Methods for inoculation

Place the bottle containing the suspension on a block of Plasticine at a suitable angle. Inoculate each tube with a 3-mm loopful of the suspension, withdrawn edgewise. Plastic disposable 10 μl loops are best.

A better and quicker method uses an automatic micropipette (the Jencon Micro-Repette is ideal) fitted with a plastic pipette tip. Deposit 10 μl of the suspension near to the top of each slope of medium. As it runs down it spreads out to give a suitable inoculum. The plastic tips are sterilized individually in small, capped test-tubes and after use are deposited in a jar containing glutaraldehyde, which is subsequently autoclaved. The tips may be re-used.

Incubation

Incubate at 37 °C for 18–21 days.

Reading results

Examine tubes with a hand lens and record as follows: confluent growth, CG; innumerable discrete colonies, IC; between 20 and 100 colonies, +; less than 20 colonies, 0.

The drug-free control slope must give growth equal to CG or IC. Less growth invalidates the test. The *modal* resistance, i.e. the MIC occurring most frequently in the sensitive control strains, should give readings of CG or IC on the first two tubes (*Table 36.6*). The range must therefore be adjusted to give this result during the preliminary exercises. Once the range is set it is seldom necessary to vary it by more than one dilution.

TABLE 36.6 'Modal resistance'

Tube no.	*Drug concentration*					
	1	*2*	*3*	*4*	*5*	*6*
Strain A	CG	CG	0	0	0	0
B	CG	IC	0	0	0	0
C	CG	CG	+	0	0	0
D	IC	IC	0	0	0	0
E	CG	CG	+	0	0	0
Mode	CG	CG	0	0	0	0

Interpretation

The Resistance Ratio is found by dividing the MIC of the test strain by the modal MIC of the control strains. When the readings are all CG or IC, this is easy, but when there are tubes showing +, care and experience may be required in interpretation. In general, a resistance ratio of two or less can be reported as *sensitive*, 4 as *resistant* and 8 as *highly resistant*. A ratio of 3 is *borderline* except for ethambutol, ethionamide and cycloserine, when it probably indicates resistance. *Mixed sensitive* and *resistant* strains occur. Examples of these findings are shown in *Table 36.7*.

TABLE 36.7 Interpretation of Tb drug sensitivity tests

Tube no.	*Drug concentration*						*Resistance ratio*	*Interpretation*
	1	*2*	*3*	*4*	*5*	*6*		
Mode	CG	IC	0	0	0	0		
Strain A	CG	0	0	0	0	0	0.5	Sensitive
B	CG	CG	0	0	0	0	1	Sensitive
C	CG	IC	+	0	0	0	1	Sensitive
D	CG	CG	CG	0	0	0	2	Sensitive
E	CG	CG	CG	+	0	0	3	Borderline[9]
F	CG	CG	CG	IC	0	0	4	Resistant
G	CG	CG	CG	CG	IC	0	8	Highly resistant
H	CG	CG	CG	CG	CG	CG	16	Highly resistant
I	CG	CG	CG	+	+	+	–	Mixed sensitive and resistant

[9] Probably resistant with ethambutol, cycloserine and ethionamide.

Pyrazinamide sensitivity tests

Pyrazinamide acts upon tubercle bacilli in the lysosomes, which have a pH of about 5.2. For reliable sensitivity tests the medium should therefore be at this pH. Tubercle bacilli do not grow well on egg medium at pH 5.2 and agar medium (e.g. Middlebrook 7H11) are frequently used. We have had good and consistent results with Yates (unpublished) modification of Marks[14] stepped pH method. This is a Kirchner semisolid medium layered on to butts of Lowenstein–Jensen medium.

Pyrazinamide stock solution

Dissolve 0.22 g of dry powder in 100 ml of distilled water and sterilize by filtration or steaming.

Solid medium

Use Lowenstein–Jensen medium containing only one half the usual concentration of malachite green (at an acid pH less dye is bound to the egg and the higher concentration of 'free' dye is inhibitory to tubercle bacilli). Adjust 600 ml of medium to pH 5.2 with N HC1. To 300 ml add 9 ml of water (control); to the other 300 ml add 9 ml of the stock pyrazinamide solution (test). Tube each batch in 1-ml amounts and inspissate upright, to make butts, at 87 °C for 1 hr.

Semisolid medium

Add 1 g of agar and 3 g of sodium pyruvate to 1 litre of Kirchner medium. Adjust pH to 5.2 with 5N HC1. To 500 ml add 15 ml of water (control); to the other 500 ml add 15 ml of the stock pyrazinamide solution. Steam both bottles to dissolve the agar. Cool to 40 °C and add to each bottle 40 ml of OADC supplement (*D*).

Final medium

Layer 2 ml of the control semisolid medium onto butts of the control Lowenstein-Jensen medium. Do the same with the test media. Store in a refrigerator and use within three weeks. (It may keep longer; we have never tried it.)

Bacterial suspensions

Two inocula are required. Use one as described above for other sensitivity tests and also a 1:10 dilution of it in sterile water.

Inoculation

With a suitable pipette (e.g. the Jencons MicroRepette and plastic tip) add 20 μl (approximately) of (a) the undiluted, and (b) the diluted suspension to pairs of test and control media.

Interpretation

Two concentrations of inoculum are used because

(1) some strains require a heavy inoculum to give growth at an acid pH, even in the control medium, and
(2) a heavy inoculum of some other strains may overcome the inhibitory action of pyrazinamide.

Colonies of tubercle bacilli should be distributed throughout the semisolid medium in one or both controls. If there is no growth in the test bottle from either the heavy or light inoculum report the strain as *sensitive*. If there is growth in the test bottle that received the heavy inoculum but not in that which received the diluted suspension, report the strain as *sensitive*. If there is growth in both test bottles, comparable with that in their respective controls report the strain as *resistant*.

NB Human strains of tubercle bacilli from untreated patients are invariably sensitive. 'Classical bovine' strains and all other species of mycobacteria found in clinical material are naturally resistant.

References

1. MITCHISON, D.A., ALLEN, B.W. and LAMBERT, R.A. (1973) *Journal of Clinical Pathology*, **26,** 250
2. COLLINS, C.H., YATES, M.D. and GRANGE, J.M. (1983) *Journal of Hygiene, Cambridge*, **89,** 235
3. MARKS, J. (1972) *Tubercle, London*, **53,** 31
4. RUNYON, E.H. (1974) *Tubercle, London*, **55,** 235
5. RUNYON, E.H. (1959) *Bulletin of the International Union Against Tuberculosis*, **29,** 65
6. MARKS, J. and RICHARDS, M. (1962) *Monthly Bulletin of the Ministry of Health and Public Health Laboratory Service*, **21,** 200
7. COLLINS, C.H. (1966) *Journal of Clinical Pathology*, **19,** 433
8. MARKS, J. (1972) *Tubercle, London*, **53,** 259
9. BULLIN, C.H., TANNER, E.I. and COLLINS, C.H. (1970) *Journal of Hygiene, Cambridge*, **68,** 97
10. McSWIGGAN, D.A. and COLLINS, C.H. (1974) *Tubercle, London*, **55,** 291

11. CANETTI, G. *et al.* (1963) *Bulletin of the World Health Organization*, **29,** 565
12. VESTAL, A.L. (1975) *Procedures for the Isolation and Identification of Mycobacteria.* US Department of Health Education and Welfare Publication (CDC) 75–8230. Washington: Government Printing Office
13. MARKS, J. (1961) *Tubercle, London*, **42,** 314
14. MARKS, J. (1965) *Tubercle, London*, **46,** 55
15. LEAT, J.L. and MARKS, J. (1970) *Tubercle, London*, **51,** 68

Chapter 37

Actinomyces, nocardia and streptomyces

Actinomyces, nocardias and streptomyces are Gram-positive mycelial bacteria. The mycelium may branch or fragment into bacillary forms. Actinomyces are anaerobic, microaerophilic or capnophilic. Nocardias and streptomyces are aerobic. Species of medical and veterinary importance occur in pus and discharges as colonies or granules.

Actinomyces

Isolation

Wash pus gently with saline in a sterile bottle or petri dish. Look for 'sulphur granules', about the size of a pinhead or smaller. Fish granules with a pasteur pipette and crush one between two slides and stain by the Gram method. Look for mycelium and large club forms radiating from the centre of the granule.

Crush granules with a sterile glass rod in a small sterile tube containing a drop of broth. Inoculate blood agar plates, brain heart infusion agar, phenylethyl alcohol broth (*D*) in order to prevent growth of Gram-negative rods, and enriched thioglycollate medium, all in triplicate and incubate

(1) anaerobically plus 5% carbon dioxide,
(2) in a 5% carbon dioxide atmosphere, and
(3) aerobically at 37 °C for 2–7 days.

Subculture any growth in the broth media to solid media and incubate under the same conditions. Look for colonies resembling 'spiders' or 'molar teeth'.

Identification

Do catalase tests on colonies from anaerobic or microaerophilic cultures that show colonies of Gram-positive coryneform or filamentous organisms. Actinomyces are catalase negative; corynebacteria and bifidobacteria are usually catalase positive.

Subculture for nitratase test and test for acid production from glucose,

TABLE 37.1 Species of *Actinomyces*

Species	*Nitrates reduced*	*Acid from*			*Starch hydrolysis*
		Glucose	*Mannitol*	*Xylose*	
A. bovis	–	+	–	v	+
A. israelii	v	+	+	+	–
A. naeslundii	+	+	–	–	–
A. eriksonii	–	+	+	+	+

mannitol and xylose and for starch utilization (*see Table 37.1*). The API ZYM system may be useful in distinguishing between actinomyces and related genera.

Species of *Actinomyces*

Actinomyces bovis

Colonies at 48 h are pinpoint and smooth, later becoming white and shining with an entire edge. Spider and molar tooth colonies are rare. In thioglycollate broth, growth is usually diffuse but occasionally there are colonies that resemble breadcrumbs. Microscopically the organisms are coryneform, rarely branching.

Nitrates are not reduced, acid is produced from glucose but not from mannitol or xylose. Starch is hydrolysed.

This is a pathogen of animals, usually of bovines, causing lumpy jaw.

A. israelii

Colonies at 48 h are microscopic, with a spider appearance, later becoming white and lobulated with the appearance of molar teeth. In thioglycollate broth, there are distinct colonies with a diffuse surface growth and a clear medium. The colonies do not break up when the medium is shaken. Microscopically the organisms are coryneform, with branching and filamentous forms.

Nitrates may be reduced, acid is produced from glucose, mannitol and xylose but starch is not hydrolysed.

This species causes human actinomycosis.

A. naeslundii

Colonies at 48 h are similar to those of *A. israelii*; spider forms are common, but not molar tooth forms. In thioglycollate broth, growth is diffuse and the medium is turbid. Microscopically the organisms are irregular and branched, with mycelial and diphtheroid forms.

Nitrates are reduced, acid is produced from glucose but not from mannitol or xylose and starch is not hydrolysed. This is a facultative aerobe.

It is not known to be a human pathogen but has been found in human material.

A. eriksonii

Colonies at 48 h are tiny and glistening, later dull white and soft. Growth in thioglycollate broth is diffuse.

Nitrates are not reduced, acid is formed from glucose, mannitol and xylose and starch is hydrolysed. Microscopically the organisms are coryneform.

This organism is said to cause actinomycosis without granules. It is considered by some bacteriologists to be a bifidobacterium.

Nocardia and *Streptomyces*

Isolation

Proceed as for actinomyces (p. 394).

Identification

Subculture aerobic growth on blood agar and incubate at 37 °C for three to ten days. Colonies vary in size and may be flat or wrinkled, sometimes 'star shaped'. They may be pigmented (pink, red, green, brown or yellow) and often covered with a whitish, downy or chalky aerial mycelium. Colonies of nocardias are usually on the surface; those of streptomyces may be embedded in the medium.

Examine Gram- and ZN-stained films (minimum decolorization with the latter). Do slide culture and test for the hydrolysis of casein, xanthine, hypoxanthine and tyrosine. Test also for acid production from arabinose and xylose (*see Table 37.2*). The API ZYM system may be used to distinguish between nocardias, streptomyces and related organisms.

TABLE 37.2 ***Nocardia*** **and** ***Streptomyces***

	Acid-fast	*Casein*	*Xanthine*	*Hypo-xanthine*	*Tyrosine*	*Acids from*	
						Arabinose	*Xylose*
Nocardia asteroides	+	−	−	−	−	−	−
Nocardia braziliensis	+	+	−	+	+	−	−
Nocardia caviae	+	−	+	+	−	−	−
Streptomyces madurae	−	+	−	+	+	+	+
Streptomyces pelletieri	−	+	−	+	+	−	−
Streptomyces somaliensis	−	+	−	−	+	−	−

Slide cultures

Melt a tube of malt agar, dilute with an equal volume of distilled water and cool to 45 °C. Suck about 1 ml of this mixture into a pasteur pipette, followed by a drop of a thin suspension of the organisms. Run this over the surface of a slide in a moist chamber. No cover-glass is needed. Examine daily with a low-power microscope until a mycelium is seen. Spores may be observed as powdery spots on the surface of the medium.

Partially acid-fast, Gram-positive mycelium which does not branch and which fragments early in slide cultures may be tentatively identified as nocardia. Non-acid fast Gram-positive mycelium which branches with aerial mycelium and conidiophores abstricted in chains suggest streptomyces.

Nocardias* and *Streptomyces

The species of medical importance are those associated with mycetomas but the organisms may occur in pathological material without being of clinical significance. The nomenclature is in dispute; the generic names of some species seem to be interchangeable.

Nocardia asteroides

Does not hydrolyse any of the substrates. This is the most common species.

N. braziliensis

Hydrolyses casein, hypoxanthine and tyrosine. It is uncommon in the northern hemisphere.

N. (Streptomyces) madurae

Hydrolyses casein, hypoxanthine and tyrosine but not xanthine and is the only one of these organisms to produce acid from arabinose and xylose.

N. (Streptomyces) pelletieri

Hydrolyses casein, hypoxanthine and tyrosine.

Streptomyces somaliensis

Hydrolyses casein and tyrosine.

Farmers' lung

Farmers' lung is caused by sensitization to the spores of *Micropolyspora faeni* and, less often, of *Thermoactinomyces vulgaris* or other thermophiles. These may be found in moulding hay. Antibodies to *M. faeni* or, in mushroom workers' lung, to *T. vulgaris*, may usually be demonstrated by double diffusion tests (*see* p. 121).

For more information about nocardias, etc. *see* Goodfellow and Schaal[1].

Reference

1 GOODFELLOW, M. and SCHALL, K.P. (1979) In *Identification Methods for Microbiologists*. Edited by Skinner, F.A. and Lovelock, D.W. Society for Applied Bacteriology Technical Series No. 14. London: Academic Press

Chapter 38

Spirochaetes

Three genera that are of medical importance are considered briefly here: *Borrelia, Treponema, Leptospira.*

Borrelia

These cause relapsing fever, transmitted by lice and ticks. Species are *Borrelia duttonii* and *B. recurrentis*.

Identification

Take blood during a febrile period. Examine wet preparations by dark field using a high power dry objective. Stain thick and thin films with Giemsa stain.

Method

Fix smear in 10% methyl alcohol for 30 s. Stain with 1 part Giemsa stock stain and 49% Sorensen's buffer at pH 7 in a Coplin jar for 45 min. Wash off Sorensen's buffer. Dry in air and examine.

Inoculate two laboratory rats with 1–2 ml of fresh or refrigerated defibrinated blood and examine the animal's blood daily between the second and seventh days.

Morphology

They are shallow, coarse, irregular, highly motile coils 0.25–0.5 × 8–16 μm.

Treponema

The species are *T. pallidum*, *T. pertinue*, *T. carateum* and *T. vincentii* (*Borrelia vincentii*).

Identification

In suspected syphilis

Examine exudate from lesion by dark field or fluorescence microscopy. This requires great care and skill. For a detailed description of the technique *see* Sonnenwirth[1].

T. pallidum

Is the causative organism of syphilis, transmitted by direct contact.

Morphology

They are tightly wound slender coils $0.1–0.2 \times 6–20\,\mu m$ with pointed ends each having three axial fibrils. They are sluggishly motile with drifting flexuous movements.

In suspected Vincent's angina

Stain smears with dilute carbol fuchsin. Large numbers of spirochetes are seen together with fusobacteria (*Fusiformis fusiformis*).

T. vincentii

Is the causative organism of Vincent's angina (ulcerative lesions of the mouth or genitals) and pulmonary infections. It can be spread by direct contact.

Culture

Inoculate peptone yeast extract medium with added serum or ascitic fluid under anaerobic conditions at 37 °C. Small, white colonies 12–15 mm in diameter having the appearance of a slight haze are visible after two weeks.

Morphology

It is a loosely wound single contoured spirochete $0.2–0.6\,\mu m \times 17–18\,\mu m$.

Other treponemes

T. pertinue

This organism causes yaws and is transmitted by direct contact. It is morphologically indistinguishable from *T. pallidum*. Laboratory tests are unhelpful.

T. carateum

Causes pinta. The mode of spread is erratic. It is morphologically indistinguishable from *T. pallidum*. Diagnosis is by silver impregnation of tissue (*see* Sonnenwirth[1]).

Leptospira

There are several species but identification is important only for epidemiological purposes. In man they all produce similar symptoms, e.g. an acute febrile illness with or without jaundice, conjunctivitis and usually meningitis. This is known as Weil's disease when it occurs in humans. They are, however, primarily animal parasites 0.25×6–$20\ \mu m$.

Isolation and identification

Collect blood during the first week of illness. Add 9 parts of blood to 1 part of 1% sodium oxalate in phosphate buffer at pH 8.1 and centrifuge 15 min at 1500 rev/min. Examine clear plasma under dark field microscopy using low power magnification. If this is negative, centrifuge the remainder of the plasma at 10 000 rev/min for 20 min and examine the sediment. Direct films rarely show leptospira either in dark field or Giemsa-stained preparations.

Morphology

They are short, fine, closely wound spirals, 0.25×6–$20\ \mu m$, resembling a string of beads. The ends are bent at right angles to the main body to form hooks.

Culture

Plate blood or CSF on Leptospira Medium (D) after 10–28 days illness. Incubate at 28–30 °C for several days. (Rapid growth can be achieved at 37 °C after one to two days but the leptospira tend to die out at this temperature.)

Membrane filtration is useful for retaining contaminating organisms. Leptospira pass into the filtrate.

Serological differentiation is possible using appropriate antiserum (*D*).

Reference

1. SONNENWIRTH, A.C. (1970) The spirochetes. In *Gradwohl's Clinical Laboratory Methods and Diagnosis, Vol. 2.* 7th edn. Edited by S. Frankel, S. Reitman and A.C. Sonnenwirth, pp. 1366–1382. St Louis: The C.V. Mosby Co.

Chapter 39

Yeasts

A.G.J. Proctor

Those yeasts which form ascospores are, of course, included in the *Ascomycotina*; the rest are *Deuteromycotina*. The position of *Sporobolomyces* is disputed: some consider it to be a basidiomycete and others an imperfect fungus; the latter view is adopted here.

The first step in the identification of yeasts is to study their morphology undisturbed on a plate of a medium such as corn meal agar, examining the plate directly with a low-power microscope.

Yeast morphology may be tabulated as follows.

	Morphology	*Genera*
Imperfect	Budding cells only	*Pityrosporon, Torulopsis, Cryptococcus, Rhodotorula, Sporobolomyces*
	Budding and pseudomycelium	*Candida*
	Budding pseudomycelium and arthrospores	*Trichosporon*
	Mycelium and arthrospores	*Geotrichum*
Perfect	Budding cells and ascospores	*Saccharomyces, Hansenula, Pichia, Debaromyces*
	Mycelium and ascospores	*Endomyces*

Ascospores are often difficult to demonstrate. If the yeast has been tested for fermentation, a film of the unfermented tube of lactose peptone water sometimes shows ascospores; otherwise, one of the special media in Chapter 4 should be used.

Identification of yeasts depends to a considerable extent on biochemical properties. *Table 39.1* gives the fermentation and assimilation patterns of some common yeasts.

Genera of yeasts

Torulopsis

Small, oval yeasts. Most of the 'wild yeasts' that cause trouble in breweries belong to this genus. They are very common contaminants. They may cause systemic infections in debilitated patients and are often resistant to 5-fluorocytosine.

TABLE 39.1 Biochemical differentiation of some common yeasts

	Fermentation of				*Assimilation of*								
	glu	*ma*	*su*	*la*	*glu*	*ma*	*su*	*la*	*mn*	*ra*	*ce*	*er*	*NO_3*
Candida albicans	+	+	−	−	+	+	+	−	+	−	−	−	−
C. tropicalis	+	+	+	−	+	+	+	−	+	−	+	−	−
C. pseudotropicalis	+	−	+	+	+	−	+	+	+	+	+	−	−
C. parapsilosis	+	−	−	−	+	+	+	−	+	−	−	−	−
C. guilliermondii	+	−	+	−	+	+	+	−	+	+	+	−	−
C. krusei	+	−	−	−	+	−	−	−	−	−	−	−	−
Torulopsis glabrata	+	−	−	−	+	−	−	−	−	−	−	−	−
T. candida	±	−	±	−	+	+	+	+	+	+	+	±	−
Cryptococcus neoformans	−	−	−	−	+	+	+	−	+	$+^{w}$	$+^{w}$	±	−
C. albidus	−	−	−	−	+	+	+	±	+	$+^{w}$	+	−	+
C. laurentii	−	−	−	−	+	+	+	+	−	−	−	−	−
Trichosporun cutaneum	−	−	−	−	+	+	+	+	±	±	±	±	−
T. capitatum	−	−	−	−	+	−	−	−	−	−	−	−	−
Saccharomyces cerevisiae	+	+	+	−	+	+	+	−	±	±	−	−	−

glu, glucose
ma, maltose
su, sucrose
la, lactose
mn, mannitol
ra, raffinose
ce, cellobiose
er, erythritol
NO_3, nitrate
+, gas produced or growth occurs
−, no gas produced or no growth occurs
w, week

Rhodotorula and Sporobolomyces

Red or pink yeasts. *Sporobolomyces* has characteristic ballistospores; if a plate culture is left on the bench, a mirror image of the inoculum is formed on the lid of the dish by a fine powdering of bean-shaped spores. Neither genus has any fermentative ability, thus they are pretty but useless. *Rhodotorula* is urease positive.

Cryptococcus

Has relatively large, round cells, with a demonstrable capsule. The growth on a malt slope flows down the slope like condensed milk. Cryptococci are found on fruits and in bird droppings: one species, *C. neoformans*, is pathogenic to man. A useful screening test is the formation of a dark brown colour on niger seed agar (p. 79). Cryptococci are also urease positive. Colonies of cryptococci are salmon pink in colour. Sugars are not fermented (*Table 39.1*) and species are identified by assimilation tests. *C. neoformans* may cause chest abscesses and, on occasions, a slow, progressive meningitis. In advanced cases, agglutination tests may be negative because the capsular material is poorly metabolized by the body, and the accumulation of it causes a high dose tolerance. This antigen can be detected in serum or in CSF by the agglutination of latex, sensitized with γ globulin from an immune rabbit serum as described on p. 120. For details, *see* Bloomfield, Gordon and Elmdorf[1] and Gordon and Vedder[2].

Candida

This genus is of medical importance, *C. albicans* especially so. All *Candida* species form creamy white, smooth colonies on malt or peptone agar, except *C. krusei*, which has a ground-glass appearance (it is, in fact, the imperfect form of *Pichia fermentans*). On corn meal agar, on which all yeasts should be grown to study their morphology, *C. albicans* produces typical chlamydospores (*see Figure 39.1c*).

Another rapid test for *C. albicans* is to inoculate the suspect yeast into about 1 ml of sterile horse serum in a bijou bottle, and incubate at 37 °C for three hours. At the end of this time almost all strains of *C. albicans* produce germ tubes, *but not if the inoculum is heavy*. High agglutinin titres and/or a positive precipitin test suggest deep-seated candida infection. For this method, *see* p. 119.

Candida utilis

This is the 'food yeast', hence its name. It is easily grown in bulk, using ammonia as the sole source of nitrogen and a cheap carbohydrate such as molasses. From these it produces a first-class protein supplement. Because of its flavour, which is noticeable except in low concentrations, its main use is in cattle food.

Trichosporon

T. cutaneum, whose appearance is very characteristic (*see Figure 39.1d*), is a common skin contaminant. On a malt slope it forms a firm, spreading colony, usually with a wet, yeasty, umbonate centre.

Geotrichum

Organisms of this genus, including the organism formerly called '*Oospora lactis*', are common in dairy products. They also cause spoilage of pickles.

Saccharomyces

This genus contains the brewing ('pitching') and baking yeasts. *S. cerevisiae* is the 'top yeast' used in making beer, and *S. carlsbergensis* the 'bottom yeast' used in making lager. There are many special strains and variants used for particular purposes.

S. pastorianus, S. rouxii and S. mellis

A bottom yeast with long sausage-shaped cells, which produces unpleasant flavours in beer. *S. rouxii* and *S. mellis* are 'osmiophilic yeasts' which will grow in concentrated sugar solution (*cf.* aspergilli of the glaucus group) and spoil honey and jam.

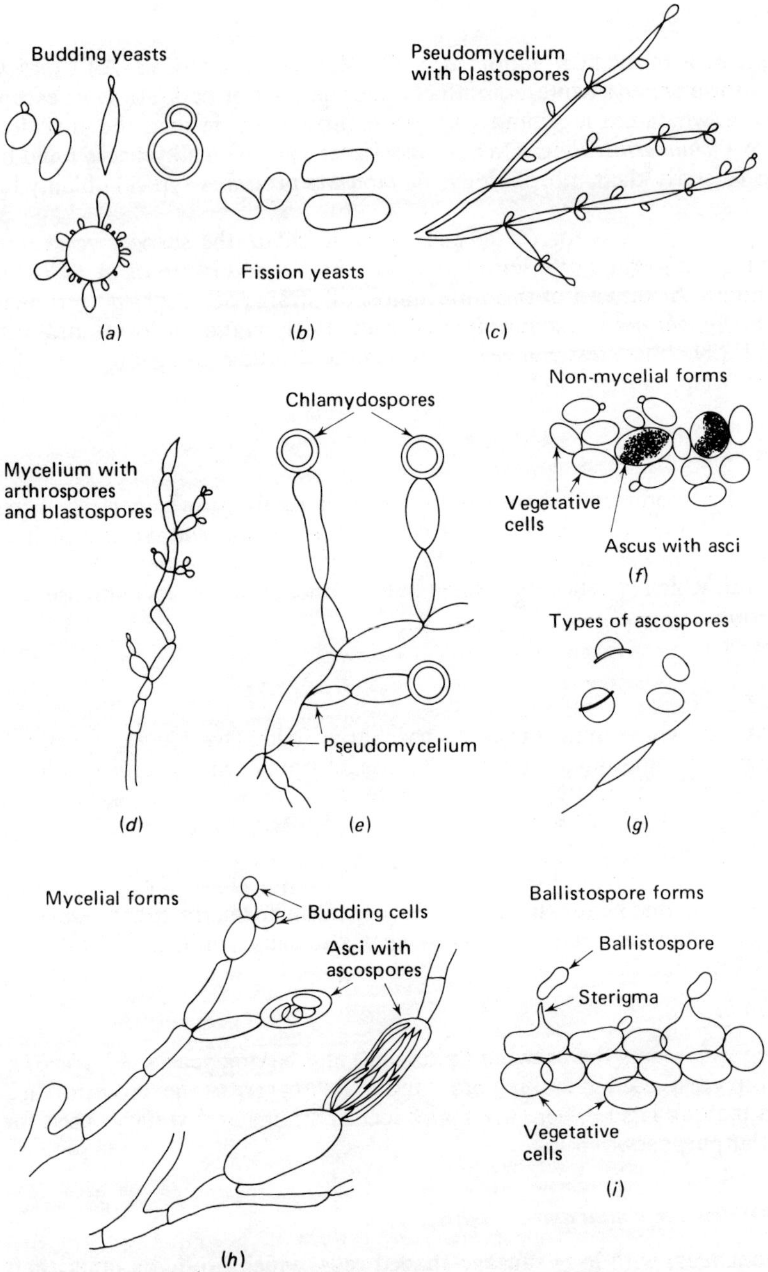

Figure 39.1 (a) Budding yeasts; (b) fission yeasts; (c) pseudomycelium with blastospores; (d) mycelium with arthrospores and blastospores; (e) pseudomycelium with chlamydospores; (f) yeasts with asci; (g) types of ascospores; (h) mycelial forms with asci; (i) ballistospore forms

Pichia and Hansenula

Occur as contaminants in alcoholic liquors. They can use alcohol as a source of carbon, and are a nuisance in the fermentation industry. They grow as a dry pellicle on the surface of the substrate. Some species occur as the 'flor' in sherry and in certain French wines, to which they give a distinctive flavour. They also cause pickle spoilage, particularly in low-salt (10–15%) brines.

Debaromyces

These are also 'pellicle forming' yeasts. They are found in animal products such as cheese, glue and rennet, and are also responsible for spoilage in high-salt (20–25%) pickle brines.

Endomyces

Strictly speaking, this is not a yeast but a filamentous fungus, a primitive form of ascomycete. It consists of mycelium with ascospores in thin-walled asci.

Apiculate yeasts

These lemon-shaped yeasts are common contaminants in food. Perfect forms are placed in the genus *Hanseniaspora* and imperfect forms in the genus *Kloeckera*.

Detection of contaminants in pitching yeasts

Wild yeasts utilize lysine but pitching yeasts do not. Wash the yeast three times by centrifuging with sterile water. Make a suspension containing approximately 10^7 cells/m (judged by the turbidity method or a counting chamber) and inoculate a plate of Lysine Agar (*O*) with 0.2 ml using a spreader. Incubate at 25 °C for four days and count the number of colonies. Calculate the number of wild yeast cells/10^6 cells of the original inoculum (*see* Fowell[3]).

References

1. BLOOMFIELD, N., GORDON, M.A. and ELMDORF, D.F. (1963) *Proceedings of the Society for Experimental Biology and Medicine*, **114**, 64
2. GORDON, M.A. and VEDDER, D.K. (1966) *Journal of the American Medical Association*, **197**, 961
3. FOWELL, R.R. (1965) *Journal of Applied Bacteriology*, **28**, 213

Chapter 40

Moulds

A.G.J. Proctor

The survey in this chapter is deliberately incomplete. The aim is to describe those moulds commonly encountered; lack of space precludes detailed treatment.

Zygomycotina

These are divided into two orders, the *Mucorales* and the *Entomophthorales*.

The *Mucorales* reproduce asexually by the production of sporangia. The majority fall into three genera: *Mucor*, *Rhizopus* and *Absidia*. These can be distinguished by the structure of the sporangiophore (*see Figure 40.1*).

In *Mucor* and *Rhizopus*, the sporangium makes a sharp angle with the sporangiophore, while the sporangium of *Absidia* swells out like the neck of a bottle. The columellae of *Absidia* and *Mucor* remain intact after rupture of the sporangium, but the columella of *Rhizopus* collapses to form a Chinese hat. In *Rhizopus*, root-like structures arise at the base of the sporangiophore.

All of these organisms produce an abundant, hairy growth, very like sheep's wool. They grow very rapidly on ordinary media at room temperature, probably on account of their non-septate mycelium.

The *Entomophthorales* reproduce asexually by means of ballistospores, borne singly on conidia. Two species, *Basidiobolus meristosporus* and *E. coronata*, have been described as causes of subcutaneous phycomycosis in hot countries (*Figure 40.1*). Unlike the *Mucorales*, their colony is mostly submerged, with little aerial mycelium.

Deuteromycetes

The two largest groups are the form genera *Aspergillus* and *Penicillium*. The fact that some species of these genera have been found to be *Ascomyctes* is interesting rather than useful.

In some cases, it is difficult to decide whether an organism is an *Aspergillus* or a *Penicillium*. In *Aspergillus* there is a well developed 'foot cell' at the base of the conidiphore and the phialides are produced simultaneously. In

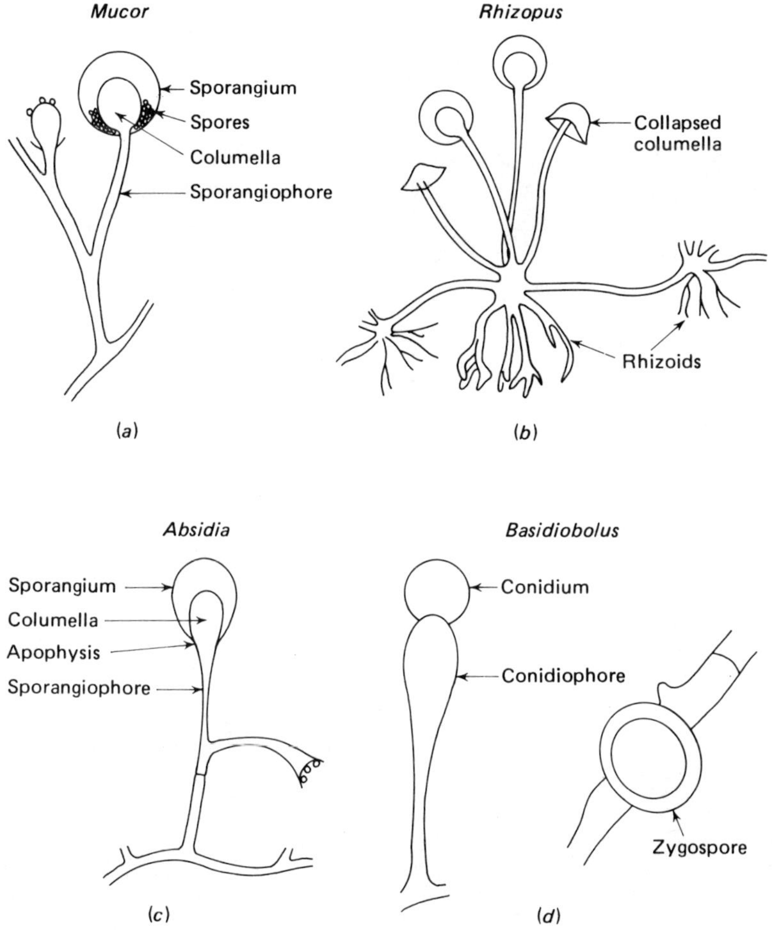

Figure 40.1 Zygomycetes: (a) *Mucor*; (b) *Rhizopus*; (c) *Absidia*; (d) *Basidiobolus*

Penicillium there is no 'foot cell' and the phialides are produced successively (*see Figure 40.2*).

Aspergillus

A colony consists of colourless mycelium from which conidiophores arise (*see Figure 40.2a*). The colour which develops is due entirely to the spores, which also give the surface of the colony a powdery appearance. The classification of the genus is not difficult. An excellent key may be found in Smith[2]. A description of some important groups is given below.

A. glaucus group

Members of this group flourish in conditions of physiological dryness, e.g. on the surface of jam, on textiles and on tobacco. They are very distinctive in

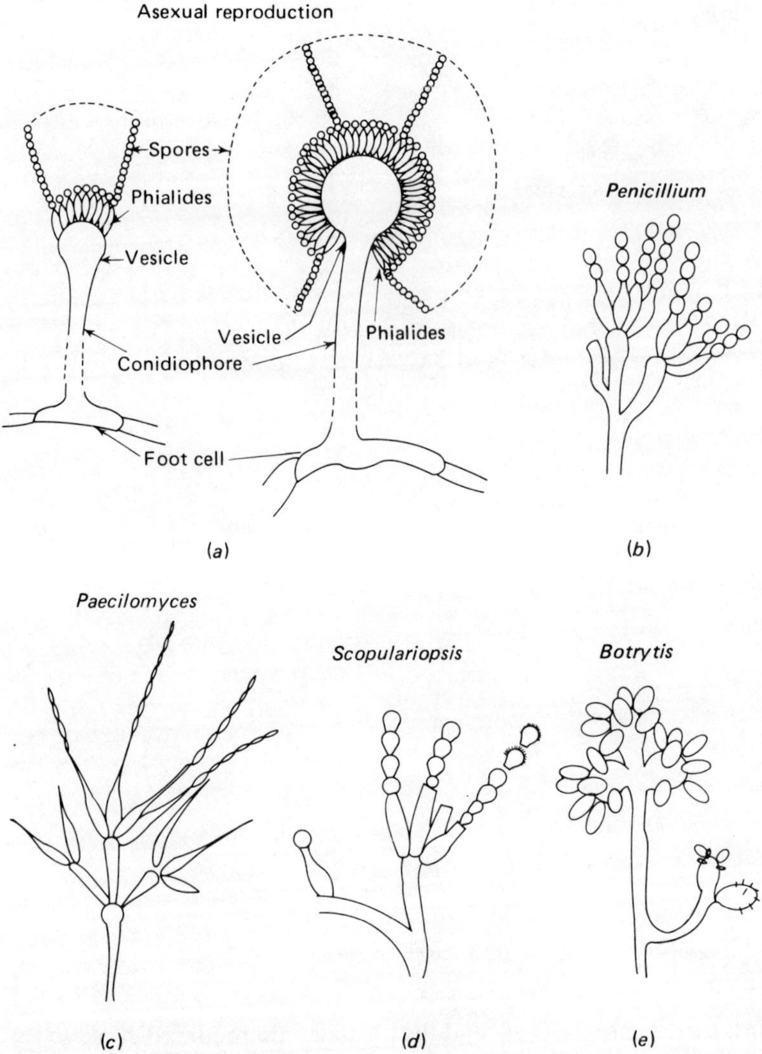

Figure 40.2 Aspergillus, Penicillium and *Penicillium*-like moulds: (a) *Aspergillus*; (b) *Penicillium*; (c) *Paecilomyces*; (d) *Scopulariopsis*; (e) *Botrytis*

appearance, with bluish green or grey-green conidia and large yellow perithicia.

A. restrictus group

Like *A. glaucus*, these will grow on dry materials, such as slightly dry textiles. As their name suggests, they are slow-growing organisms.

A. fumigatus

So-called from its smoky green colour. It is a common mould in compost, and can grow at temperatures above 40 °C and hence it is pathogenic for birds. It spores more readily at 37 °C than at 26 °C. It is now notorious as a human pathogen, underlying pulmonary eosinophilia. In this condition, the fungus may be found in the mucous plugs that are expectorated. It also invades old tuberculous cavities and gives rise to aspergilloma of the lung. When isolated from such cases, the organism often does not produce spores; incubation at 42 °C may encourage it to do so. Patients who harbour this organism develop precipitating antibody, which is rather weak in pulmonary eosinophilia but strong in cases of aspergilloma. For methods, *see* p. 117. If *A. fumigatus* is examined with a lens, the heads are seen to be columnar—the whole effect is like a test-tube brush.

A. niger

On the other hand, this organism has round heads which are large enough to appear discrete to the naked eye. This, together with their black colour, makes them easy to recognize.

The Flavus-Oryzae group

Members of this group are notorious for the production of aflatoxin in groundnuts and other foods such as millet when badly stored. They are also of value commercially, as a source of diastatic enzymes such as takadiastase. The specific name, which means yellow, is an error; the colony colour in this series is green. The heads are round, but smaller than those of *A. niger*.

Penicillium

This is an enormous and ill-defined genus. Like aspergilli, the penicillia form a felt of mycelium on and in the medium, and the conidiophores arise from this. A penicillus looks very like a hand with the fingers stretched out. The spores are always yellow or green, never brown or black. Various species of *Penicillium* will rot apples and citrus fruits, stunt the growth of tulips, parasitize toadstools, choke *A. niger*, damage fabrics, produce penicillin and glucose oxidase, and are used to ripen cheese (*Figure 40.2b*).

Fungi similar to *Penicillium*

Paecilomyces (*Figure 40.2c*)

Produces a rather woolly colony, sometimes with a pink tint, usually a dirty brown colour but never green. The phialides are long, slender and graceful. The perfect form of *Paecilomyces variotii* is *Byssochlamys fulva*, whose ascospores are relatively resistant to heat and able to withstand 87 °C for 30 min. They also germinate in the presence of low concentrations of oxygen. As they are common in soil, they are carried into canning factories on soft fruits such as strawberries and raspberries and spoil the finished product. The mycelium has usually disintegrated by the time the cans are examined.

Scopulariopsis

The common species is *S. brevicaulis*, which forms a brown, velvety colony on ordinary media. The sporiferous apparatus of this fungus is often a short, stumpy penicillus. The spores, however, are very characteristic (*Figure 40.2d*). It will grow on concentrated protein, and hence it infects human nail, and reduces Camembert cheese to an ammoniacal mess. It also liberates arsine from compounds that contain even a small trace of arsenic, hence the garlic-like smell of arsine when *Scopulariopsis* grows on ordinary gelatin. Arsine is also liberated when this mould grows on wallpaper coloured with Paris green.

Botrytis

This fungus grows as a loose, grey felt on the surface of the medium. The spores look like bunches of grapes, hence the name. Other characteristic features are dark brown sclerotia, which may reach 50 mm in diameter, and it has the tendency to form a dirty black line of spores along the wall of the tube. It occurs as a parasite of plants, notably tomatoes (*Figure 40.2e*).

Gliocladium

This is an example of a 'slimy spore' fungus. The colony is thin and bald-looking at first, and later greenish spots of spores appear. These are borne on penicillia, but after a time the spore chains stick together in a single slimy mass (*Figure 40.3a*).

Trichoderma

This is another 'slimy spore' species whose colonies are covered in dirty green patches of conidia (these also grow up the glass). The conidia are borne in balls on short conidiophores (*Figure 40.3b*).

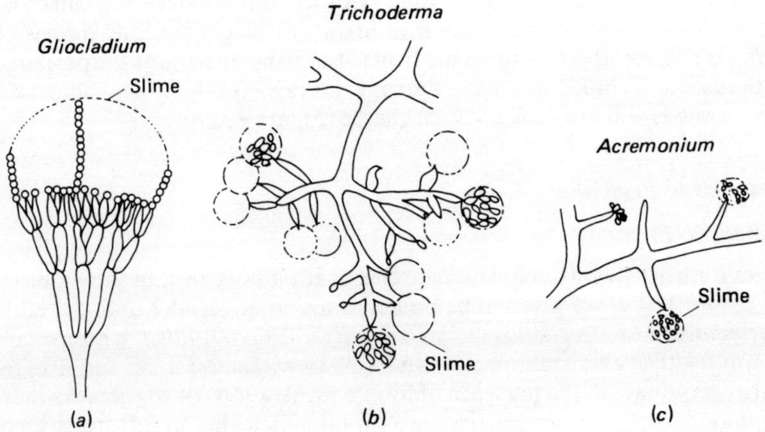

Figure 40.3 'Slimy spore fungi': (a) *Gliocladium*; (b) *Trichoderma*; (c) *Acremonium*

Acremonium

Produces a tough, white colony, turning pale pink. The 'slimy spores' are borne in small balls on the end of slender hyphae. These spore balls disintegrate when touched, making the organism a most awkward contaminant to eliminate (*Figure 40.3c*).

Neurospora sitophila

An ascomycete with resistant spores. Its ascospores need the stimulation of a fairly high temperature to germinate, and may survive in the centre of a loaf of bread. This leads to outbreaks of infection by the red bread mould, *M. sitophila*, the imperfect stage of *Neurospora*. As *Neurospora* is heterothallic, the perfect form is not found under these conditions.

Black spore types (excluding *Aspergillus niger*)

Alternaria

On first isolation, the colony has a thin, black aerial mycelium that consists almost entirely of spore chains. On subculture, aerial mycelium is more abundant and the spores become fewer, especially on rich media. It is found in damp situations, e.g. on damp wallpaper. The individual spores are septate, and irregular, often with a definite 'beak' (*Figure 40.4a*).

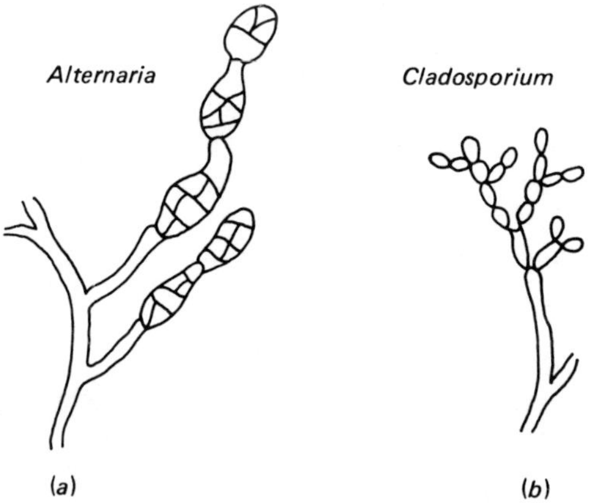

Figure 40.4 Black-spore fungi: (a) *Alternaria*; (b) *Cladosporium*

Helminthosporium

This fungus usually produces black, leathery colonies. The spores, which are long and septate, are produced at the ends of hyphae either singly or, more often, like bunches of bananas.

Cladosporium

A very common airborne mould. In plates exposed to the air, many of the colonies are of *Cladosporium*. They are a deep greenish black which is very characteristic. The spores occur in large, tree-like clusters, but they break up easily and are best studied in slide culture (*Figure 40.4b*). The species usually encountered is *C. herbarum*, which grows over a wide temperature range, on textiles, rubber, leather, paper and many foods, even on meat in cold storage.

Cladosporium wernickii

More yeast-like. Young cultures are moist but the yeast-like young culture becomes mycelial with the passage of time.

Pullularia (Aureobasidium) pullulans

This mould occurs on damp cellulosic materials and spoils painted surfaces. Colonies are at first white shiny and yeast-like. They soon darken and become black and leathery, through the formation of thick, dark hyphae that bear conidia as lateral buds (*see Figure 40.3d*).

Reference

ONIONS, A.H.S., ALSOPP, D. and HIGGINS, H.O.W. (Editors) (1981) *Smith's Introduction to Industrial Mycology*. London: Edward Arnold

Chapter 41

Pathogenic fungi

A.G.J. Proctor

The following fungi are considered in this chapter.

Dermatophytes, which grow only in the outer layers of the skin.

Those *dematiaceous fungi* which cause chromomycosis and the phycomycetes which cause subcutaneous phycomycosis.

Sporothrix and *Paracoccidioides*, which cause both superficial and systemic lesions.

Blastomyces, *Histoplasma* and *Coccidioides immitis*, where infection is by inhalation of spores and the primary lesion is usually pulmonary.

Dermatophytes

Appearance in morbid material

In skin and nail cleared with caustic soda, fungal hyphae can be seen, some of them breaking into arthrospores.

The appearance of infected hairs is as follows:

Organism	*Fluorescence*	*Spore arrangement*
Microsporum canis	+	Small spore ectothrix
M. audouinii	+	Small spore ectothrix
M. gypseum	–	Large spore ectothrix
Trichophyton mentagrophytes	–	Small spore ectothrix
T. verrucosum	–	Large spore ectothrix
T. tonsurans	–	Endothrix
T. violaceum	–	Endothrix
T. schoenleinii	+ or –	'Favic hairs'

Some dermatophyte-infected hairs fluoresce if examined under a Wood's lamp in a dark corner or darkroom. The various types of infected hair are shown in *Figure 41.1*, which also shows infected skin and nail.

The common dermatophytes are arranged in three genera, according to the type of macroconidia produced (*see Figure 41.2*). *Microsporum* and *Epidermophyton* species are easily distinguished by this means, but *Trichophyton* rarely produces macroconidia, and other features are used to separate it.

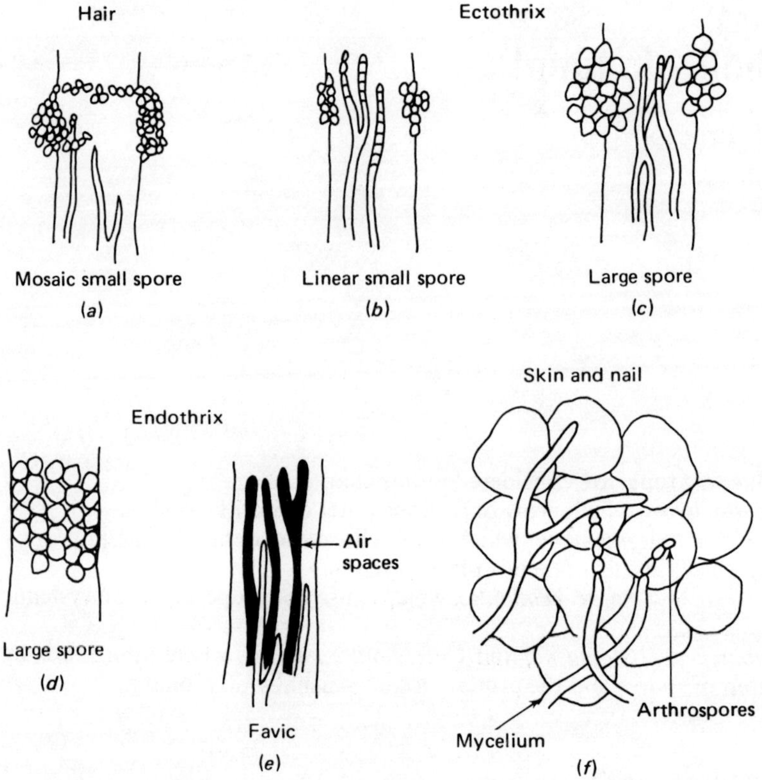

Figure 41.1 Infected hair, skin and nail: (a) mosaic, small-spore ectothrix; (b) linear small-spore ectothrix; (c) large-spore ectothrix; (d) large-spore endothrix; (e) favic-type endothrix; (f) skin and nail appearance

Microsporum audouinii

This is a scalp fungus with no animal or soil reservoir. It spreads from child to child in schools, and infection usually resolves at puberty. The organism is slow growing, forming a tenacious colony with little aerial mycelium, sometimes having a light pink colour on the back. Macroconidia are rare (*Figure 41.3a.*)

M. canis

As its name suggests, this is a common parasite of dogs and cats, where the infection is often difficult to detect. Children are commonly infected from pets. This organism, which is fast growing, usually produces a vivid yellow pigment and abundant macroconidia. Occasionally, strains are non-pigmented and slow growing. To differentiate them from *M. audouinii*, grow on rice (5 g of rice and 20 ml of water in a 100-ml flask, autoclaved at 115 °C for 20 min and cooled). *M. canis* produces aerial mycelium; *M. audouinii* does not (*Figure 41.3b*).

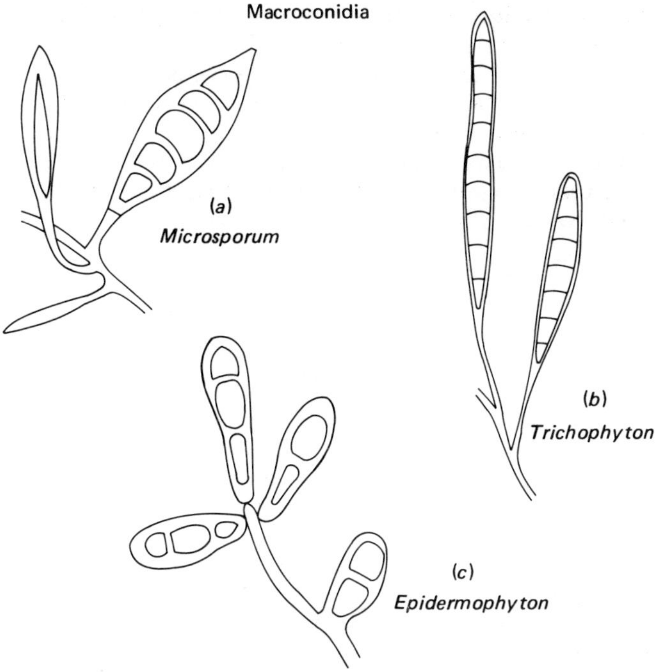

Figure 41.2 Macroconidia types: (a) microspores; (b) trichophyton; (c) epidermophyton

M. gypseum

This organism is very common in soil but not very virulent. The occasional case which occurs usually shows a solitary lesion, related to contact with the soil. The fungus grows fast, and is soon covered with a buff-coloured granular coating of spores. These tend to be arranged in radial strands on the surface, rather like a spider's web. Macroconidia are abundant (*Figure 41.3c*).

Trichophyton

A common cause of ringworm is one or another form of *T. mentagrophytes*. Any part of the body may be affected. The granular form produces a flat colony with a surface covered in spores, which give it a buff colour—the colour of official envelopes. On the back of the culture there is a brownish-red pigment. A slide mount of the mycelium shows abundant microconidia that resemble bunches of grapes and spiral hyphae. This granular form is the more virulent. The fluffy variety, *T. mentagrophytes var. interdigitale*, grows like cotton wool, with the same sort of pigment in some cases. The microconidia are round, and arranged along the sides of the hyphae.

T. rubrum, another very common trichophyton, usually grows as a fluffy colony, but more slowly than *T. interdigitale*. Its pigment, as its name suggests, is a more definite red. Microconidia are borne along the sides of the hyphae, but are rather elongated, resembling barbed wire under a low-power microscope. Both pigment production and spore formation may be slow: put the cultures back for a few days; the use of special media is not necessary. A useful differential test is to inoculate a urea slope. The medium differs from that of Christensen only in containing 1% of glucose and needing no pH adjustment (at pH 6.8, growth is slower). *T. mentagrophytes* will almost always produce a purple colour within eight days; *T. rubrum* never does. There is a granular form of *T. rubrum*, which is rather rare. Granular forms may show some macroconidia (*Figure 41.3d, e*).

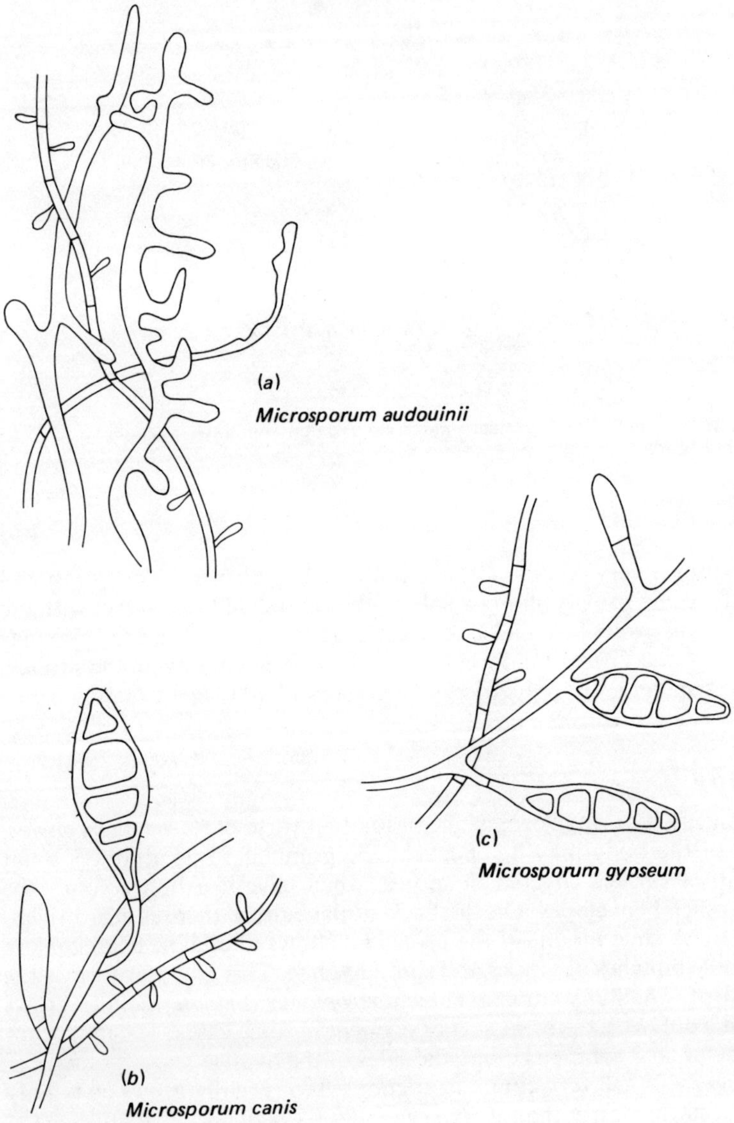

Figure 41.3 Microsporum and Trichophyton species: (a) *M. audouinii*; (b) *M. canis*; (c) *M. gypseum*

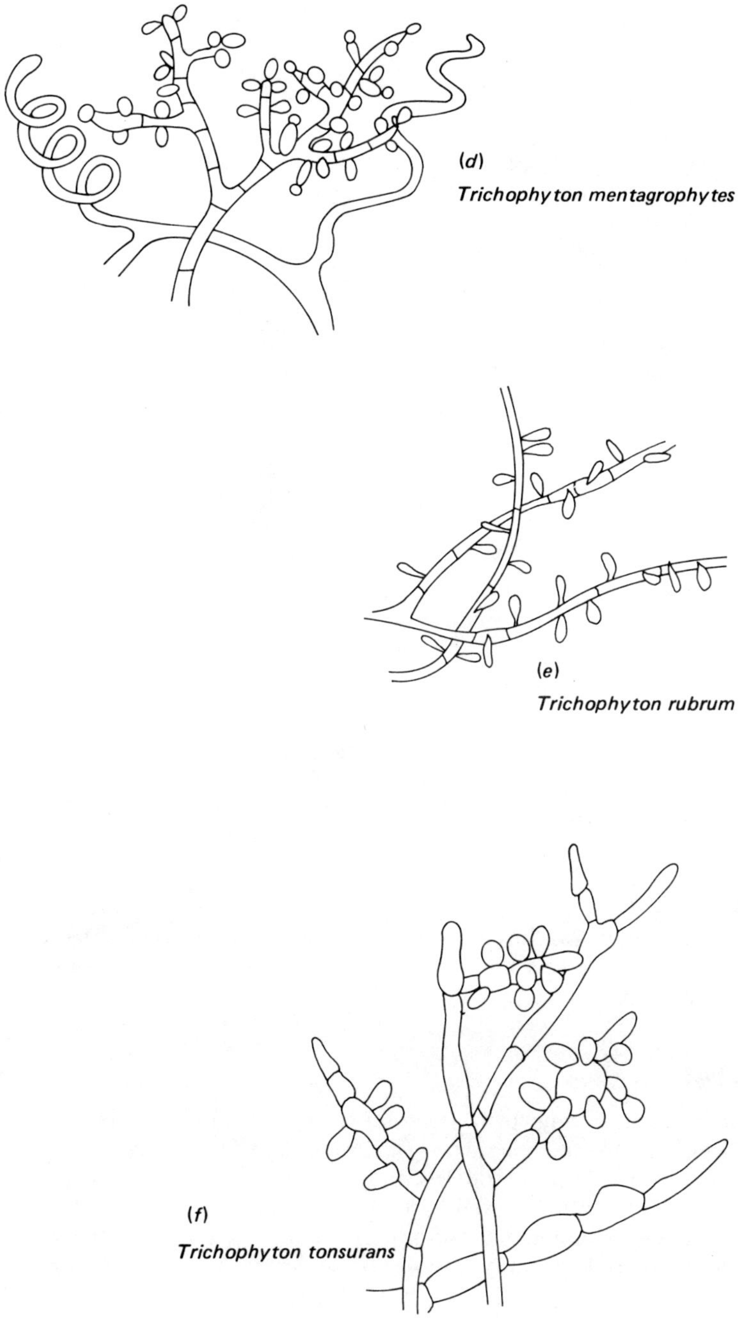

Figure 41.3 (*continued*) (d) *T. mentagrophytes*; (e) *T. rubrum*; (f) *T. tonsurans*

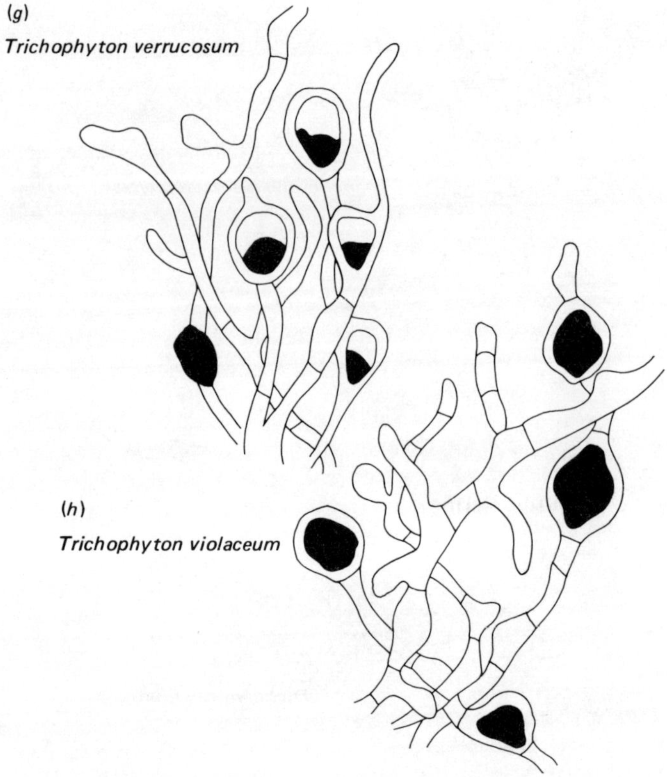

Figure 41.3 (*continued*) (g) *T. verrucosum*; (h) *T. violaceum*

T. tonsurans

A cause of scalp ringworm. Its colony is slow growing and velvety, often with a depressed centre. Microconidia are large and oval, arranged along the hyphae (*Figure 41.3f*).

T. verrucosum

The organism of cattle ringworm. It causes scalp, beard or nail infections in farm workers, and other people such as slaughterman and veterinary surgeons who come into contact with cattle. In culture it grows very poorly; suspected cultures should be examined carefully with a lens after 14 days' incubation, as colonies are often submerged and minute. This, together with the large, thin walled, balloon-like chlamydospores, is sufficient to identify the organism.

T. schoenleinii

The cause of favus. The colony is slow growing, hard and leathery, with a surface like white suede. The only microscopic features are thick, knobbly hyphae, like arthritic fingers—the so-called 'favic chandeliers'—and chlamydospores.

T. violaceum

Another cause of scalp ringworm, microscopically similar to *T. verrucosum*, although the hyphae are usually thinner. The characteristic feature is a deep red-purple pigment, which appears as a spot in the centre of a young colony, and gradually spreads to the edge (*Figure 41.3b*).

Epidermophyton

There is only one species, *E. floccosum*. This is the cause of 'dhobi itch' and 'jungle rot'. It also occasionally infects feet. Its growth on malt agar is khaki in colour, looking like a bit of old army blanket with a powdery surface due to masses of macronidia whose shape is very characteristic.

For further information about dermatophytes *see* Rebell and Taplin[1] and Mackenzie and Philpot[2].

Dematiaceous fungi

Phialophoras

These are associated with chromoblastomycosis or 'mossy foot', which responds to treatment with 5-fluorocytosine. This is presumed to be the result of skin injury by wood, on which some of the responsible fungi have been found. They all look like *Cladosporium*, forming black velvety colonies. Pathogens, however, grow at 37 °C and do not liquefy gelatin. They differ in the method of spore production (*see Figure 41.4*).

Fonsecaea pedrosoi

The most common, sporulates in *Cladosporium* or *Acrothera* form (*Figure 41.5*).

P. compacta

Slow growing, with *Cladosporium* heads and barrel-shaped conidia.

P. verrucosa

Produces bottle-shaped phialides, with wide, flaring mouths (*Phialophora* type). If pus or scales from the surface of a lesion are mounted in caustic soda, clusters of brown, thick-walled sclerotic cells are seen. The organisms can be cultivated on malt agar with chloramphenicol. They are inhibited by cycloheximide.

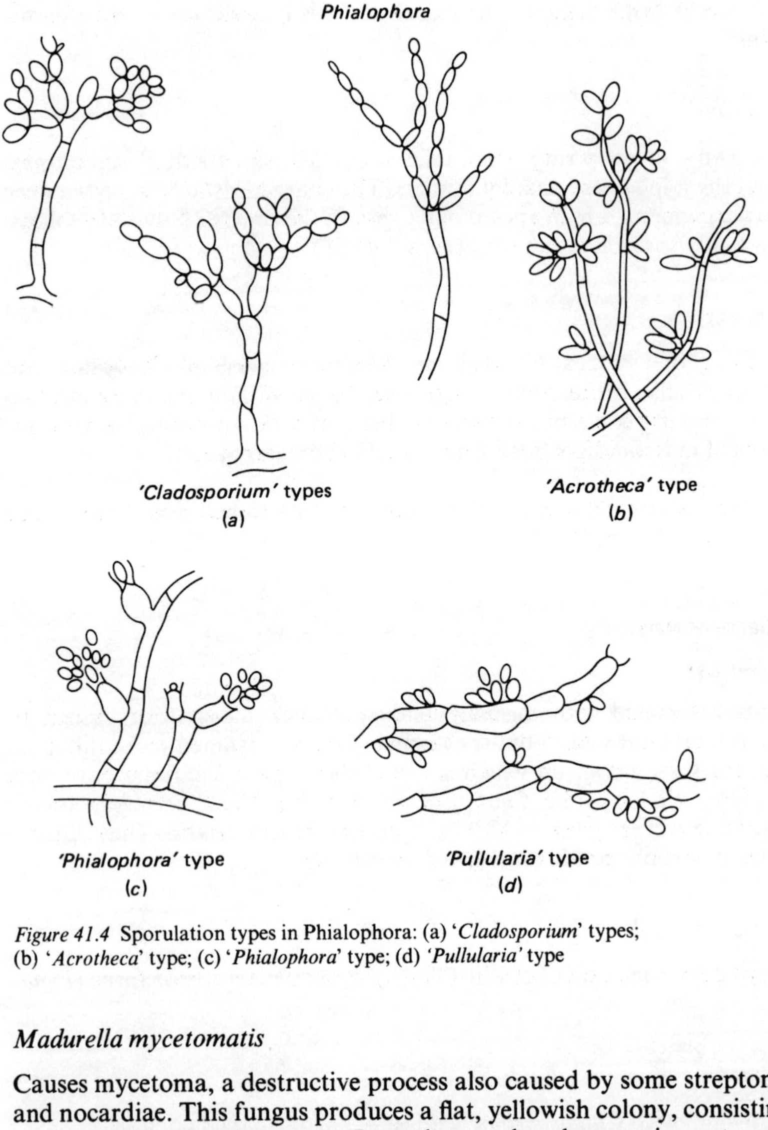

Figure 41.4 Sporulation types in Phialophora: (a) '*Cladosporium*' types; (b) '*Acrotheca*' type; (c) '*Phialophora*' type; (d) '*Pullularia*' type

Madurella mycetomatis

Causes mycetoma, a destructive process also caused by some streptomycetes and nocardiae. This fungus produces a flat, yellowish colony, consisting only of mycelium and arthrospores. It produces a deep brown pigment, especially on bacteriological dextrose nutrient agar. Many other fungi can cause mycetoma including the streptomycetes mentioned on p. 396. Although diagnosis is usually histological, double-diffusion tests are helpful.

Subcutaneous phycomycosis

This is a hard, extensive subcutaneous swelling, caused by *Basidiobolus* or *Entomophthora*.

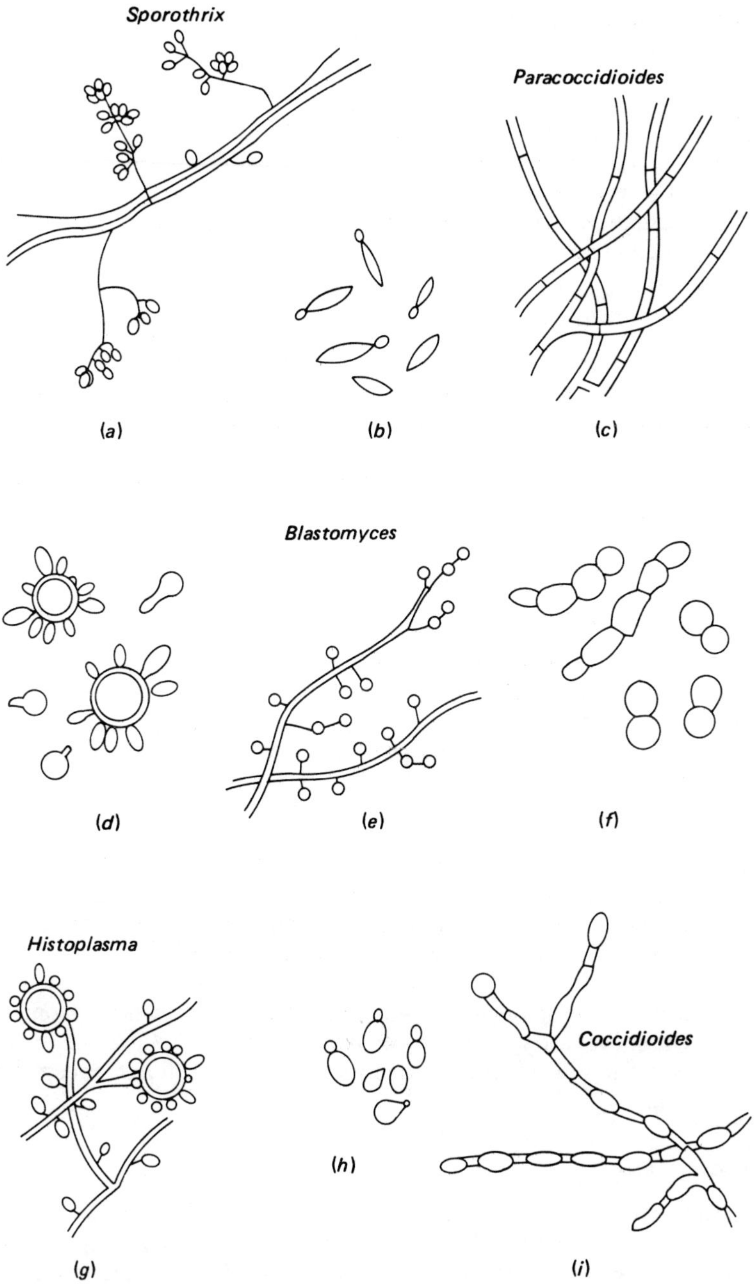

Figure 41.5 (a) (b) *Sporothrix*; (c) (d) *Paracoccidioides*; (e) (f) *Blastomyces*; (g) (h) *Histoplasma*; (i) *Coccidioides*. (a) (c) (e) (g) mycelial phases; (b) (d) (f) (h) yeast phases

Sporothrix* and *Paracoccidioides

Sporothrix schenckii

Causes sporotrichosis, a spreading, systemic infection caused by inoculation of wood splinters, rose thorns, etc. The condition begins as a local ulcer and spreads along the lymphatics. The fungus is silvery white when first isolated at 26°C, then the surface turns iron-grey to black with the production of spores. The yeast phase of this organism, found in tissue or in culture at 37 °C, is a spindle-shaped budding yeast (*Figure 41.5a, b*).

Paracoccidioides braziliensis

Causes South American blastomycosis. The primary lesion is almost always in the oral mucosa, possibly because of the habit of cleaning the teeth with wood splinters and chewing bark in the places where the disease is found. The spread, again, is through the lymphatics. The organism is very slow growing, white and featureless at 26 °C. At 37 °C on rich media, characteristic yeast cells are formed (*Figure 41.5c, d*).

Blastomyces, Histoplasma* and *Coccidioides

Blastomyces dermatitidis

Causes North American blastomycosis. Infection is almost always by inhalation and skin lesions are usually secondary to a pulmonary infection, even a mild one. The mycelial form, at 26 °C, is a moist colony at first, developing a fluffy aerial mycelium with microconidia. At 37 °C or in tissue, large oval multinucleate yeasts are seen (*Figure 41.5e, f*).

Histoplasma capsulatum

Causes histoplasmosis, which is very common in the southern half of the USA. Most people recover rapidly: only an occasional patient develops progressive disease. The route of infection is inhalation, and the pathology similar to tuberculosis (*Figure 41.5g, h*). *See* caution, *below*.

At 26 °C, the organism grows with a white aerial mycelium, which becomes light brown. The spores are striking and characteristic. At 37 °C, small, oval, budding yeasts are found. In tissue, these yeasts are intracellular, except in necrotic material.

Coccidioides immitis

Causes coccidiomycosis. This disease is limited to a few special areas in the south west of the USA and in Mexico and Venezuela. It is inhaled in desert dust. Again, most people suffer only a mild, transient, 'flu-like' illness. Occasionally, cases progress to generalized destructive lesions (*Figure 41.5*). *See* caution, *below*.

At 26 °C, the colony is at first moist, then develops an aerial mycelium, becoming brown with age and breaking up into arthrospores. There is no second form in artificial culture. In the body, however, large, thick-walled cells are seen. When mature, these are full of spores.

Serological diagnosis of the deeper mycoses

In chromomycosis and blastomycosis, lesions are usually accessible and the organism may be cultivated. Double-diffusion tests are a help in chromomycosis and in South American blastomycosis. In histoplasmosis, the double-diffusion test is positive in early cases, and the course of the disease can be followed by serial complement fixation tests. A combination of double-diffusion and complement fixation tests will detect 90% of cases of coccidioidomycosis (*see* p. 119 and p. 121).

Coccidioides and Histoplasma are Risk Group III organisms, infectious by the airborne route. They should be handled in microbiological safety cabinets in Containment laboratories.

References

1. REBELL, G. and TAPLIN, G. (1970) *Dermatophytes, their Recognition and Identification* (Revised). Miami: University of Miami Press.
2. MACKENZIE, D.W.R. and PHILPOT, C. (1981) *Isolation and Identification of Ringworm Fungi*. Public Health Laboratory Service Monograph No. 15. London: HMSO.

Appendix

Names and addresses of companies supplying apparatus mentioned in the text

Supplier	*UK*	*USA*
Air Flow Developments	Air Flow Developments Ltd, Lancaster Road, High Wycombe, Bucks.	
Ames	Ames Co., Stoke Court, Stoke Poges, Slough, Bucks, SL2 4LY.	
Amicon	Amicon Ltd, Upper Mill, Stonehouse, Glos, GL10 2 BJ.	Amicon Corporation, Lexington, Mass.
Anderman	Anderman & Co. Ltd, Central Avenue, East Molesey, Surrey, KT8 OQ2.	
API	API Laboratory Products Ltd, Grafton Way, Basingstoke, Hants, RG 22 6H9.	Analytab Products Inc., 514 Mineola Avenue, Carle Place, New York, 11514
Astell	Astell Hearson, Cray Road, Sidcup, Kent, DA14 5BZ.	
Azlon	Azlon Products Ltd, Glyn St, London, SE11 5JG	
Bactec	Laboratory Impex Ltd, Lion Road, Twickenham, Mdx	Becton Dickinson Co.

If no US agent is given for a British company or no UK agent for an American company, Arnold Horwell Ltd (*see* list) may be able to help.

Supplier	*UK*	*USA*
Bactomatic	Bactomatic Inc. PO Box 183, Marlow, Bucks, SL7 1SU.	Amicon Corporation, PO Box 3103, Princeton, NJ 08540.
Baker	Frost Instruments Ltd, Fishponds Road, Molly Millar's Lane, Wokingham, Berks, RG11 2QA.	Baker Co. Inc., Sanford Airport, Sanford, Maine, 04073.
Bassaire	John Bass Ltd, Bassaire Building, Duncan Rd, Swanwick, Southampton, SO3 7ZS.	
BBL Becton Dickinson Bioquest	Becton Dickinson (UK) Ltd, Between Towns Road, Cowley, Oxford, OX4 3LY.	BioQuest—East Division of Becton Dickinson Co., 250 Schilling Circle, Cockeysville, MD. 31030 BioQuest—West Division of Becton Dickinson Co., 1950 Williams Drive, Oxnard, Calif, 93030
BDH	British Drug Houses, (Laboratory Chemicals Division), Poole, Dorset.	
Beaumaris	Beaumaris Instrument Co. Ltd, Rosemary Lane, Beaumaris, Anglesey, LL58 8EB.	
Beckman	Beckman RIIC, Ltd, High Wycombe, Bucks.	Beckman Instruments Inc., PO Box 6100, Anaheim, Calif., 92806.
Bel Art	R.B. Radley & Co. Ltd, London Road, Sawbridgeworth, Herts, CM21 9JH.	Bel Art Products, Pequannock, NJ 07440.
Bennett (Thermalog)	Bennett & Co., Brimpton Common, Near Reading, Berks.	Biomedical Sciences Inc., 140 New Dutch Lane, NJ 07006.

Supplier	*UK*	*USA*
Bilbate	Bilbate Ltd, No. 16 Factory Unit, Middle March, Daventry, Northants.	
Blickman		S. Blickman Inc., 536 Gregory Ave, Weehawken, NJ 07087.
Boehringer	Boehringer Corporation, Mannhein House, Bell Lane, Lewes, East Sussex, BN7 1LG.	Oxford Laboratories, 1149 Chess Drive, Foster City, Calif. 94404.
Brown	Albert Brown Ltd, Chancery House, Abbey Gate, Leicester.	
Calbiochem	Calbiochem Ltd, 10 Wyndham Place, London, W1H 1AS.	Calbiochem Inc., 10933 North Torrey Pines Rd, La Jolla, Calif. 92037.
Camlab	Camlab Ltd, Nuffield Rd, Cambridge, CB4 1IT.	
Casella	Casella Ltd, Britannia Walk, London N1 7ND.	
Cherwell	Cherwell Laboratories Ltd, PO Box 3, 1 Murdock Rd, Bicester, Oxon, OX6 7XB.	
Clark	Clark Scientific Ltd, PO Box 22, Bognor Regis, West Sussex, PO21 3SX.	
Colab	*See* Inolex	
Comark	Comark Electronics Ltd, Dominion Way, Rustington, Littlehampton, West Sussex, BN16 3QZ.	
Coulter	Coulter Electronics Ltd, Northwell Drive, Luton, Beds, LU3 3RH.	
Denley	Denley Instruments, Daux Rd, Billingshurst, Sussex RH14 9SJ.	

Supplier	*UK*	*USA*
Diagnostic Products	Diagnostic Products Ltd, 62 St Mary's Street, Wallingford, Oxon, OX10 0EL.	
Diagnostic Research		Diagnostic Research Inc., Rosalyn, NY.
Difco	Difco Laboratories, PO Box 14B, Central Ave, West Molesey, Surrey, KT8 0SE.	Difco Laboratories Inc., Detroit, Mich 48201.
Disposable Products	LIP Ltd, 111 Dockfield Road, Shipley, West Yorks, BD 17 7AS.	
Don Whitley	Don Whitley Scientific Ltd., Green Lane, Baildon, Shipley, West Yorks, BD17 5JS.	
EDL	Engineering Developments Ltd, 11 Elmoor Rd, Farnborough, Hants, GU14 7NW.	
Edwards	Edwards High Vacuum, Manor Royal, Crawley, West Sussex, RH10 2LW.	
Elkay	Elkay Laboratory Products (UK) Ltd, Unit 2, Crockford Lane, Basingstoke, Hants, RG24 0HO.	
Envair	Envair (UK) Ltd, York Avenue, Broadway Industrial Estate, Haslingdean, Rossendale, Lancs, BB4 4HX.	
Exogen	Exogen Ltd, Beardmore Street, Clydebank Industrial Estate, Clydebank, G81 4SQ.	
Falcon	Becton Dickinson Co (qv)	Becton Dickinson Co (qv)

Supplier	*UK*	*USA*
Fisons	Fisons Scientific Apparatus Ltd, Bishops Meadow Rd, Loughborough, LE11 0RC.	
Flow	Flow Laboratories Ltd, PO Box 17, Second Avenue, Industrial Estate, Irvine, Ayrshire, KA12 8NB.	Flow Laboratories Inc., 7655 Old Springhouse Road, McLean, VA 22102.
Folex-Biotest	Folex-Biotest Ltd, 171 Alcester Road, Birmingham, B13 8JR.	Folex-Biotest-Schleussner Inc., 60 Commercial Avenue, Moonachie, NJ 07074.
Foramaflow	Peteric Instrumentation Ltd, Shepperton Studio Centre, Studios Road, Shepperton, Middx, TW17 0QD.	
Gallenkamp	A. Gallenkamp & Co. Ltd, PO Box 290, Technico House, Christopher St, London, EC2P ER.	
Gallay	Jean Galley (UK) Ltd, 13 Frogmore Road, Hemel Hempstead, Herts.	
Gelaire	Gelman Hawkesley Ltd, 10 Harrowden Road, Brackmills, Northampton, NN4 0EP.	
General Diagnostics	William R. Warner & Co. Ltd, Trafalgar Close, Chandlers Ford, Eastleigh, Hants, SO5 3WB.	Warner (qv)
Gibco	Gibco Europe Ltd, Unit 4, Cowley Mill Trading Estate, Longbridge Way, Uxbridge UB8 2YG.	
Gilson	Anachem Ltd, 15 Power Court, Luton, Beds, LU1 3JJ.	Gilson Medical Electronics Co., Box 27, Middleton, Wisc. 53562.

Supplier	*UK*	*USA*
Grant	Grant Instruments Ltd, Barrington, Cambridge.	
Gurr	British Drug Houses (qv)	
Heinicke		Heinicke Napco, 3000 Taft Street, Hollywood, FL, 33021.
Hepaire	Hepaire Ltd, 48/50 Fowler Rd, Hainault, Ilford, Essex, IG6 3XA.	
Hi-Tech	Hi-Tech Scientific Ltd, Brunel Road, Salisbury, Wilts, SP2 7PU.	
Horwell	Arnold Horwell Ltd, 73 Maygrove Rd, West Hampstead, London, NW6 2PB.	
How-Air	How-Air Ltd, Carne House, Markland Hill, Chorley New Road, Bolton, Lancs, BL1 5AP.	
Howorth	Howorth Air Engineering Ltd, Farnworth, Bolton, Lancs, BL4 7LZ.	
Hughes	Hughes & Hughes Ltd, Elms Industrial Estate, Church Rd, Harold Wood, Romford, Essex.	
Inolex		Inolex Biomedical Division, Wilson Pharmaceutical & Chemical Corporation, 3 Science Rd, Glenwood, Ill. 60425.
Jencons	Jencons (Scientific) Ltd, Cherry Court Way, Industrial Estate, Stanbridge Road, Leighton Buzzard, Beds, LU7 8UA.	

Supplier	*UK*	*USA*
Jennings	R.W. Jennings & Co. Ltd, Main St, East Bridgford, New Nottingham, NG13 BPQ.	
Jobling	Jobling Laboratory Division, Stone, Staffs., ST15 0BG.	Corning Glass Works, Scientific Products, Corning, NY 14830
Johnsons	Johnsons of Hendon Ltd, 335 Hendon Way, London, NW4	
Labco	Labco Ltd, 54 Mark Bottom Rd, Marlow, Bucks, SL7 3NF.	
Lab Line	HistoLab (Bethlehem Instruments) Ltd, PO Box 101, Leverstock Green Farm, Hemel Hempstead, Herts.	Lab Line Inc., Lab Line Plaza, Melrose Park, Ill 60160.
lab m	lab m Ltd, Ford Lane, Salford M6 6PB.	
Lamb	Raymond A. Lamb, 6 Sunbeam Rd., Acton, London, NW10 6JL.	
LEEC	Laboratory & Electrical Engineering Co., Private Road No. 7, Colwick Estates, Nottingham.	
Lewis Williams	Lewis Williams Ltd, 11 Dawlish Drive, Pinner, Middx, HA5 5LJ.	
Lovibond	Tintometer Ltd, Waterloo Road, Salisbury, SP1 2JY.	
LTE	Laboratory & Thermal Equipment Ltd, Private Rd, Colwick, Notttingham, NG4 2AJ.	
Luckhams	Luckhams Ltd, Victoria Gdns., Burgess Hill, West Sussex.	
Lumac	Comber & Son, 490 Manchester Road, Stockport, Cheshire, SK4 5DL.	Lumac Systems Inc., PO Box 2085, Titusville. FL, 32780.

Supplier	*UK*	*USA*
Mast	Mast Laboratories, Diamed Diagnostics Ltd, Mast House, Derby Rd, Bootle, Merseyside, L20 1EA.	
May & Baker	May & Baker Ltd, Dagenham, Essex, RM10 7XS.	
Medical Wire	Medical Wire & Equipment Co., (Bath), Potley, Corsham, Wilts.	Royal Scientific Inc., PO Box 1924, 3609 South Park Ave, Blasdell, New York, 14219.
Medikent	Medikent Ltd, 181 Verulam Road, St Albans, Herts.	
Mercia	Mercia Diagnostics Ltd, Paramount Estate, Sandown Rd, Watford, Herts, WD2 4XA.	
Merck	British Drug Houses (qv)	
Microflow	MDH Ltd, Walworth Road, Andover, Hants, SP10 5AA.	
Miele	Miele Co. Ltd, Fairacres, Marcham Road, Abingdon, Oxon, OX14 1TW.	
Millipore	Millipore (UK) Ltd, 11–15 Peterborough Road, Harrow, Middx, HA1 2YH.	Millipore Corporation, Ashley Rd, Bedford, Mass, 01730.
MSE	MSE Ltd, Manor Royal, Crawley, Sussex, RH10 2QQ.	
3M	3M United Kingdom Ltd, 3M House, Wigmore St, London, W1A 1ET.	3M Medical Products Division, 3M Center, Saint Paul, Minn, 55101.
National Instruments	Clandon Scientific, Lysons Avenue, Ash Vale, Aldershot, Hants, GU12 5QR.	National Instrument Co. Inc., 4119 Fordleigh Road, Baltimore, MD, 21215.

Supplier	*UK*	*USA*
Netsch	Netsch (UK) Ltd, Loomer Road Industrial Estate, Chesterton, Newcastle, Staffs, ST5 7PZ.	
New Brunswick	New Brunswick Scientific (UK), Ltd, 26–34 Emerald Street, London, WC1N 3QA.	
Nunc	Nunc UK Ltd, 16 Salter St, Stafford, ST16 2JU.	Vangard International Inc., Box 312, Red Bank, New Jersey, 07701.
Organon	Organon Teknika Ltd, Science Park, Milton Road, Cambridge, CB 4 4BH.	
Orion Diagnostica	Gibco (Europe) (qv)	
Partome	Bob Turner Marketing Ltd, 23 Wicks Lane, Formby, Liverpool.	
Pasteur Institute	Townson & Mercer Ltd, 93–96 Chadwick Rd, Astmoor, Runcorn, Cheshire, WA7 1 PR.	
Payne	Payne & Sons Ltd, 6 Ively Rd, London, SW4 0HS.	
PBI	Cherwell (qv)	
Pharmacia	Pharmacia Diagnostics, Pharmacia House, Midsummer Boulevard, Milton Keynes, Bucks, MK9 3HR.	Pharmacia Diagnostics, 800 Centennial Ave, Piscataway, New Jersey 08854
Prossor	Prossor Scientific Instruments Ltd, Lady Lane Industrial Estate, Hadleigh, Ipswich, IP7 6BQ.	
Raven	Raven Scientific Ltd, PO Box 2, The Mill, Station Road, Haverhill, Suffolk.	

Supplier	*UK*	*USA*
Reichert	C. Reichert Co., British American Optical Co. Ltd, 820 Yeovil Rd, Slough, Berks.	
Reynier		Reynier Inc., Chicago 13, Ill.
Richardsons	Richardsons of Leicester Ltd, Evington Valley Rd, Leicester, LE5 5LQ.	
Roche	Roche Products Ltd, Diagnostics Dept., PO Box 8, Welwyn Garden City, Herts, AL7 3AY.	Roche Diagnostics, Division of Hoffman La Roche Inc., 340 Kingsland St, Nutley, NJ 07110.
Roth	Roth Scientific Co. Ltd, Alpha House, Alexandra Road, Farnborough, Hants.	
Seward	A.J. Seward Ltd, UAC House, Blackfriars Road, London, SE1 9UG.	
Scientific Industries		Scientific Industries, 150 Herricks Rd, Mineola, NY 11501
Scientific Products		Scientific Products, 1430 Waukegan Rd, McGraw Park, Ill. 60085.
Sigma	Sigma London Chemical Co., Fancy Road, Poole, Dorset BH17 7NH	Sigma Chemical Co., PO Box 14508, St Louis, Mo. 63178
Spectra	Spectra-Physics Ltd, 17 Brick Knoll Park, St Albans, Herts.	
Sterilin	Sterilin Ltd, 43–45 Broad St, Teddington, Middx, TW11 8QZ.	York Scientific Inc., Bridge & Port Authority Building, New York, NY 13669
Syva	Syva Co., St Ives House, Maidenhead, Berks, SL5 1RD.	

Supplier	*UK*	*USA*
Techne	Techne (Cambridge) Ltd, Duxford, Cambridge, CB2 4PZ.	
Technilab		Technilab Instruments, 6 Industrial Rd, Penquannock, NJ, 07440.
Thomas	Arnold Horwell Ltd, (qv)	Arthur H. Thomas, Box 779, Philadelphia, PA, 19105.
Tillomed	Tillomed Laboratories, Unit 24, Henlow Trading Estate, Henlow, Beds.	
Tissue Culture	Tissue Culture Services Ltd, 10 Henry Road, Slough, Berks, SL1 2QL.	
Turner	James W. Turner (Liverpool) Ltd, Buckland Street, Aigburth, Liverpool, L17 7DS.	
VirTis	Techmation Ltd, 56 Edgware Way, London, HA8 8JP.	VirTis, Rte 208, Gardiner, NY 12525.
Vitek		Vitek Systems Inc., Hazelwood, MO.
Waring	Christison Ltd, Alabany Road, East Gateshead Industrial Estate, Gateshead, NE8 3AT.	Waring Products, Rte 44, New Hartford, Conn, 06057.
Warner	*See* General Diagnostics	
WCB Plastics	WCB Plastics, Stamford Works, Bayley St, Stalybridge, Cheshire, SK15 1QQ.	
Wellcome	Wellcome Diagnostics, Temple Hill, Dartford, Kent, DA1 5AH.	Burroughs Wellcome Co., Reagents Division, 3030 Cornwallis Rd, Research Triangle Park, NC 27709.
Whitley	*See* Don Whitley	

Index

Note: Numbers in italics refer to figures and tables

Absidia, 406, *407*
Access, to laboratory, 8
Accidents, laboratory, 5, 6
Acetic acid bacteria, 239, *269*
Acetobacter, *269*
Achromobacter, *261*, *266*, 267
Acid broth, 67
Acid foods, thermally processed, 67, 234
Acid-fast bacilli (*see also Mycobacterium*), 371
  fluorescent stain, 99
Acid-fast stain, 98–99
Acinetobacter, *261*, *265–266*
Acremonium, *410*, 411
Actinobacillus, *311–312*
Actinomyces, 394–396
  other species, 395–396
Actinomyces israellii, 152, 312, 395
Actinomycetes, thermophilic,
  antigen production, 117–119
  culture, 115
Actinomycosis, 395, 396
Additives, culture media, 59–60, 67
Aerobic spore bearers, media for, 64, 82
Aerococcus, 325, *328*, 331, 340
Aeromonas, *261*, *263*, *271*, *272*, *275*, 277–79
  media for, 57
Aerosols (infected airborne particles), 5, 16–17
Aesculin hydrolysis, 63, 105
Agar (*see also* Culture media, Individual media)
  contact plates, 241
  slopes, 24, 86
  sloppy, 80–81, 110
Agglutination inhibition, 179
Agglutination tests, 142–145
  Brucella, 143, 305–306
  Candida, 119
  Cryptococcus, 120–121
  Haemophilus, 307
  salmonellas, 143–144, 293–296, 298
  shigellas, 298
  streptococci, 333–335
Air,
  currents affecting safety cabinets, 18–19
  removal for autoclaving, 33–34, 35, 36
  sampling, 115, 136, 245
Airborne particles, infected (aerosols), 5, 16–17
Airflow testing, 17, *19–21*
Albert's stain, 99
Alcalescens-Dispar group, 281, 298, 299
  biochemical reactions, 292–293, *294*
Alcaligenes, *261*, *263*, *265*, *266–267*
Alcohol disinfectants, 40, *41*, 43
Alcoholic beverages (*see also* Beer spoilage), 239, 405
Aldehyde disinfectants (*see also* Formaldehyde, Glutaraldehyde), 40, 42–43
Alginate swabs, 242
Alternaria, *411*
Alteromonas, 265
Amino acid requirements, 94
Aminoglycosides, 167, 176
Ammonia test, 105
Ammonium salt basal synthetic medium, 64
Ammonium salt 'sugars', 72
Amphotericin B,
  assay, 126–127
  sensitivity testing, 68, 124–125
Ampicillin, 167
Anaerobe agar, 68
Anaerobic cabinets, 92
Anaerobic culture, 91–93
  additives and supplements, 59
  counting media, 136
  identification tests, 103, 104, 105
  indicators, 92
  media, 68, 72, 80, 82
Anaerobic jars, 91–92
Anaerobic sugars, 72

Andersen sampler, 115, 245
Andrade indicator, 83
Anemometers, 20
Animal by-products and feeds, 251–252, 358
Animal inoculation tests, 379–380
Anthrax bacilli, 70, 356–358
Anti-tuberculous drug sensitivity test, 387–392
Antibiotic(s) (*see also* Antifungal drugs)
  assay, 174–180
  discs, 103, *169*, 171–172, 173
  in milk and food, 213
  paper strips, 172
  sensitivity tests *see* Sensitivity tests
Antifungal drugs, *175*
  assay, 68, 123–124, 126–127
  sensitivity testing, 68, 123–126
Antigenic extracts of fungi, 117–119
Antigens,
  Aspergillus, 117
  Cryptococcus, 120
  enterobacteria, 283–284
  Salmonella, 287–289
  Shigella, 289–290
  streptococci, 333
Antisera, specificity, 164
API systems (identification), 103
Apiculate yeasts, 405
Arginine broth, 69
Arginine hydrolysis, 105
  media, 63, 69, 73
Arizona group, *280*, *292–293*, *294*, 297
Arizona hinshawii, food poisoning, 185, 188
Arnold steamer, 38
Arsine, 410
Aryl sulphatase test, 80
Ascospores, 116, 401, *404*
Aspergilloma, 409
Aspergillus, 406–409, *408*
  antigen production, 117–119
  sensitivity testing, 125
Assay methods *see* Antibiotics, Antifungal drugs
Astell culture bottles, 27
Auramine-phenol stain, 99
Aureobasidium, 412
Autoclaves, 33–38
  efficiency testing, 37–38
  gravity displacement type, 35–37
  pressure cooker type, 34–35
Autoclaving, *51*, *52*, 53
AutoMicrobic identification system, 105
Auxanograms, 94, 116–117
Auxotab identification system, 104

Baby food, 228
Bacillary dysentery, 299
Bacillus, 105, 357–360
Bacillus anthracis, 356–358, *357*
Bacillus cereus, 357, *357*, 358–359
  antibiotic assay, *174*
  food poisoning, 184, 187–188, 358
  media, 57, 59
Bacillus coagulans, 234
Bacillus megaterium, *357*, 358, 360
  food poisoning, 184, 358
Bacillus pumilis, *175*, *357*, 360
Bacillus stearothermophilus, 38, 234, 359
Bacillus subtilis, *174*, *176*, *357*, 359
Bacitracin Disc method, 335
Bacon, 199, 200
Bactec System, 148
Bactericidal activity of serum, 179
Bacteroides, 316–317, *318*
  culture, 152, 153, 317
  media, 57, 59
Bactiburner, 31
Baird-Parker media, 64, 72, 76
Baking yeasts, 403
Ballistospores, 406
Basal synthetic media, 69
Basidiobolus meristosporus, 406, *407*, 420
BCG, 7, 377–378, 379
Beer spoilage,
  media, 61
  organisms, 239, 269, 342, 403
Bench,
  disinfection, 6
  equipment, 31
Bethesda-Ballerup group, *280*, 281, 299–300
Bifidobacterium, 355
'Bijou' culture bottles, 27
Bile inhibition and solubility tests, 67
Biocides, 39–48
Biopsy specimens, 155
Bitty cream, 361
Black tube counts for *Clostridium*, 198, 258, 367
Blastomyces, 413, *421*, 422
Blastomycosis, 422, 423
Blenders, 15
Blood cultures, 147–148
Blood agar cultures, anaerobiosis in, 92
Blood agar media, 70–71, 81
Blue milk, 263
Blue slide test, 106
Boas-Oppler bacillus, 354
Bone meal, 251–252, 358
Bone taint, 195, 199, 369
Bordetella, 310–311
  media, 57, 59
Bordetella bronchiseptica, *174*, *261*, *265*, *310*, 311
Bordetella parapertussis, 311
Bordetella pertussis, 152, *310–311*
Borrelia, 398
Botrytis, *408*, 410
Bottles,
  counts on, 243–244
  culture, 26–27
Botulism, 184–185, 188, 362

Branhamella, *322*, 323
Bread spoilage, 227, 359, 411
Breed count, 130
Brevibacterium, 298, 351
Brewing (pitching) yeasts, 342, 403, 405
Brines, testing of, 201–202
British Association of Chemical Specialities, 47–48
Brochothrix, 61, 75, 350
Bromocresol purple, 83
Bromothymol blue, 83
Bronchopneumonia, 311
Broths *see* Culture media
Browne's tubes, 38
Brucella, 303–306
  media, 57, 59
  serological diagnosis, 142–143, 305–306
Brucellosis, 142–143, 305
Bunsen burners, 31, 32
Burns, 158–159
Buruli ulcer, 387
Butter spoilage organisms, 81, 216–217
Buttermilk, 216, 338
Byssochlamys, 234

Cabinets *see* Anaerobic cabinets, Safety cabinets
Calgon Ringer solution, 82
Campylobacter, 317–*319*
  food poisoning, 185, 188
  media, 57, 59
Candida, 401, *402*, 403
  culture, 116, 117, 158
  identification, 104
  serological methods, 117–119, 125
Candida albicans, 104, *175*, *402*, 403
Candida utilis, 403
Canned food,
  leaker spoilage, 233–234
  pathogens, 238
  spoilage organisms, 234–237, 359, 368, 369
Cans, 230
  defects, 232–237
  sampling, 235–236
  types, 231–232
Cap-O-Test closures, 27
Capnophiles, incubation, 12, 92–93
Capreomycin, 387, *389*
Capsular swelling test, 307
Capsules, staining, 100
Carbenicillin, *169*, *174*, *176*
Carbohydrate additives, 67
Carbohydrate fermentation and oxidation tests (*see also* Hugh and Leifson test), 72–73, 105–106, 116
Carbohydrate utilization tests, media, 64, 69–70
Carbol-fuchsin stain, 97, 98–99, 100
Carbon assimilation media, for yeasts, 73
Carbon dioxide for anaerobic culture, 12, 92–93, 147
Cardiobacterium hominis, *311*, 312
Casamino acids, 124
Casein digestion, 64, 106
Castenada blood culture bottles, 147
Catalase test, 106
Centrifuges, 6, 13–15
Cephaloridine, *169*, *174*
Cereals, 226–228
Cerebrospinal fluid (CSF), 148–149
Cheese, 217–218, 338, 350, 354
Chick-Martin test, 44
Chlamydia trachomatis, 150
Chloral hydrate agar, 70
Chloramphenicol, *169*, 174
Chocolate, 226
Chocolate agar, 70–71
Chocolate cystine agar, 71
Cholera, 151, 274
Chopping blocks, counts, 244
Chromobacterium, *263*, 268–*269*
Chromoblastomycosis, 419
Chromomycosis, 413, 423
Churchman's method of capsule staining, 100
Citrate utilization, 64, 106
Citrobacter, *280*, 281, *292–294*
  heat resistance, 192
Citrobacter ballerupensis, 299–300
Citrobacter freundii, *282*, 284
Citrobacter–Bethesda–Ballerup group, 299–300
Cladosporium, 411–412
Clean air cabinets, laminar flow, 24, 86
Clindamycin, *169*
Clinical material, investigation methods, 146–159
Clostridium, 361–369
  black tubes, 198, 258, 367
  counting, 194, 368–369
  food spoilage, 198, 235, 367–369
  media, 57, 82
  in poultry, 198
  in water, 258
  other species, 365–368
Clostridium botulinum, *361*, *363*, 365
  food poisoning, 184, 200, 362
Clostridium difficile, 59, 365, 366
Clostridium perfringens, *361*, *363*, 366
  food poisoning, 151, 183–184, 187, 362–364
  lecithinase test, 108–109, 110
  media, 74, 75
  Nagler half-antitoxin plate test, 110, 364
  puerperal infections, 157
  in water, 258, 259
Clostridium tetani, *361*, *363*, 364–365
Clothing, protective, 7
Cloths and towels, counts on, 244
Clue cell, 157
Coagglutination, 333–334
Coagulase test, 106, 326–327

Coccidioides immitis, 413, *421*, 422
  serological diagnosis, 119, 423
Coccidiomycosis, 119, 422–423
Cockles, 204
Codes of Practice,
  disinfectant testing, 47–48
  incineration, 54
  prevention of laboratory-acquired infection, 5–9
Cold shock, 191
Coliform bacilli (*see also* individual organisms), 280, 281–286
  in food, 193
  hygiene indicators, 190, 192
  media, 58, 63
  in milk, 210–212
  Most Probable Number estimates, *137–141*
  in water, 254–257
Colistin-nalidixic acid agar, 71
Colony,
  appearances, *93*
  count test for UHT milk, 210
  counting methods, 132–136, 141
Colworth Droplette apparatus, 133–134
Complement fixation tests, 119, 423
Condensed milk, 215
Congo red, 83
Contagious abortion, 305
Containers,
  for contaminated material, 48–51
  food preparation, counts on, 242–244
  for processed food, 230–231
  specimen, 27–28, 191
Containment Laboratory, 2, 6, 8
Coomassie Brilliant Blue R, 123
Corynebacterium, 344–350
  carbohydrate media for, 105
  media, 59, 64, 81
  species,
    animal, 349–350
    human, 347–348
    staining, 99
Corynebacterium diphtheriae (diphtheria bacillus), 344–348
  biotypes, 347–348
  media, 58, 74–75, 81
  plate toxigenicity test, *345–347*
Corynebacterium vaginalis, 157, 308
Corynebacterium xerosis, *174*, 348
Coryneforms, 350
Cosmetics, 247–249
  media, 68, 77, 81
Cotton wool,
  impurities, 84–85
  pipette plugs, 28, 53
Counterimmunoelectrophoresis, 123, 149
Counting methods, 128–141
  accuracy, 128
  automated, 132, 141
  for food, 191–192
  media, 61, 62
  Most Probable Number (MPN) estimates, *136–141*
  total counts, 128–130
  viable counts, 128–129, 130–136
Cover-glasses, 30
Craigie tubes, 80–81, 110
Cream,
  bitty, 359
  natural and imitation, 214
Crockery, counts, 242
Crustaceans, raw and frozen, 204, 205
Cryptococcus, 401, *402*
  cultural methods, 114
  latex agglutination tests, 120–121, 402
  media, 79
  sensitivity testing, 124, 125
Cryptococcus neoformans, 104, *402*
Crystal violet blood agar, 71
Cultural methods, 89–96
Culture bottles, 26–27, 30
Culture media, 56–88
  additives and supplements, 59–60, 67
  antibiotic assay, 175
  antiboiotic sensitivity tests, 63, 167
  automatic distributors, 25, 85
  dermatophytes, 60
  diluents, 82–83
  disposal, 52
  fastidious organisms, 57
  food microbiology, 60–62, 191–192
  general purpose, 56, 80
  identification, 63–67, 87
  indicators, 83–84
  ingredients affecting zones of inhibition, 166–167
  isolation, 86–87
  labelling, 85
  laboratory prepared, 67–82
  manufacturers, 56
  membrane filtration, 62–63, 135
  pathogenic yeasts, 60
  plate pouring, 86
  preparation, 84–86
  quality control, 86–87
  selective, 57–59, 86–87
  slopes, preparation, 24, 86
  sterility test, 63
  storage, 27, 85
  transport, 60
  water microbiology, 62
Cultures (*see also* Subcultures)
  anaerobic, 91–93
  appearance, *93*
  drying, 94
  freeze-drying, 96
  freezing, 94
  incubation, 93
  moulds, 114–116
  shake tube, 91
  slide, 116
  stab, 91

stock, 74, 86–87, 94–96
Cured meat, 199–202, 233
Cycloserine, 387, *389*
Cytochrome oxidase test, 111

Dairy equipment, 213, 244, 355
Dairy products (*see also* Milk), 214–222
Leuconostoc, 341
media, 62, 332
spoilage organisms, 109, 350, 351, 354, 369, 403
Debaromyces, 401, 405
Decarboxylase tests, 106
media, 65, 73–74
Decontamination of safety cabinets, 21–23
Deionized water, 53, 84
Dematiaceous fungi, 413, 419–421
Dental caries, 339, 354, 355
Dental specimens, 149, 331
Dermatophytes, 413–419
cultural methods, 114
culture media, 60, 77–78
Deuteromycotina, 401
Dextran formation, 81
Dhobi itch, 419
Diffusion techniques (*see also* Double diffusion tests)
antibiotic assay, 175–178
sensitivity tests, 166–173
Dilution methods, for counting, 130–131
Dimethyl sulphoxide, 101
Dip slides, 136
Diphosphopyridine nucleotide, 306
Diphtheria bacilli, 344
Discard bins, 6, 48, 49, 52
Discard jars, 6, 48–50, 53
Discs,
antibiotic, 103, *169*, 171–172, 173
identification, 102, 103
Disinfectants, 6, 39–48
for discarded infected material, 48–50
hospital use, 48
precautions, 44, 47–48
properties, *41*
for safety cabinets, 22
testing efficiency, 44–48
types, 40–44
Disinfection, definition, 32
Disposable laboratory equipment, 25–26
treatment procedures, 50, *51*
Disposal of contaminated material, 51–54
Distilled water, 53, 84
DNAse test, 65, 107, 327
Doderlein's bacillus, 354
Donovan's medium, 74
Donovan's test for lactose fermenters, *283*
Dorset's egg medium, 74
Double diffusion tests, 119, 121–*122*, 123, 423
Drains, sampling, 291
Dried milk, 214–215
Drinks, soft, 239, 243
Drop count method, 133
Droplette method of counting, 133–134
Drugs *see* Antibiotics, Antifungal drugs
Drying cabinets, glassware, 25
Drying cultures, 95
Ducrey's bacillus, 308
Durham's (fermentation) tubes, 30, 72, 105, 116
Dysentery, 299

Ear discharges, 150
Efficiency of plating (EOP), 86–87
Egg media, 74
Egg yolk agar (Lowbury and Lilly), 74
Eggs, 218–219, 359
Eijkman test, 107
Eikenella, 316–317, *318*
Electrophoresis, 119, *122–123*, 149
Elek plate toxigenicity test, 74–75, 345–346
Ellner's medium, 75
Emulsifiers, tissue, 15
Endocarditis, 179, 312
Endomyces, 405
Enterobacter, 285–286
biochemical reactions, *280*, *282*, *294*
decarboxylase test, 106–107
Enterobacteria, 280–301
biochemical reactions, *261*, *271*, *280*
in food, 190, 192, 193–194
identification, 103, 104
lactose fermenters, 280–286
media, 58, 64, 75–76
non-lactose fermenters, 281, 286–300
Enterococci, 335, 337–308
food poisoning, 185, 188
as sanitary indicators, 194, 257–258
Enteroset identification system, 104
Enterotubes identification system, 103
Entomophthora, 420
Entomophthorales, 406
Enzyme multiplied immunoassay technique (EMIT), 180
Epidermophyton, 413, *415*, 419
Erwinia herbicola, 286
Erysipelotherix, 351, *352*
Erythromycin, *169*, *174*
Escherichia, *280*
Escherichia coli, 284
biochemical reactions, 282, *283*
in brine, 202
drug assay, 174
food poisoning, 183
hygiene standards, 190, 192, 193
sensitivity testing, 170
Esculin (Aesculin), 63, 105
Ethambutol, 387, *389*
Ethanol disinfection, 43
Ethionamide, 387, *389*
Ethylene oxide, 40

Evaporated milk, 215
Eye discharges, 150
Eye infections, 307, 309

Faecal coli *see Escherichia coli*
Faecal streptococci (*see also* Enterococci), 257-258
Faeces, examination of, 151-152, 186-188, 291
Falkow's decarboxylase method, 106
  medium for, 73-74
Farmers' lung, 399
Favus, 418-419
Fermentation (Durham's) tubes, 30, 72, 105, 116
Fermentation and oxidation tests, 72-73, 105-106, 116
Fertilizers, 251-252, 291, 358
Filters,
  bacterial, 39
  membrane, 39
  safety cabinet, 21, 23-24
Fish, 203-206
  food poisoning, 183, 185
  furunculosis, 278, 308
Fish tank 'sporotrichoid' disease, 385
Flagella,
  identification, 109
  staining, 100
Flaming, sterilization by, 32
Flat sour organisms, 234, 235, 359
  media for, 61, 63
Flavobacterium, *261*, 267-*268*
Flavus-Oryzae group, *Aspergillus*, 409
Flemming's method of spore staining, 100
Floors, counts, 244
Flour spoilage, 227
Flow Laboratories identification system, 104
Fluorescein isothiocyanate (FITC), 160
Fluorescence,
  incident light, 164-166
  of infected hairs, 413
  microscope, 160-162
Fluorescent antibody tests, 160, 162-164
  Bacillus anthracis, 358
  streptococci, 336
Fluorescent dyes, 160
Fluorescent immunoassay, 180
Fluorescent stain, 99, 161
5-Fluorocytosine, 68, 123-125, 126-127
Fonsecaea pedrosoi, 419
Food (*see also* Individual products),
  antibiotics in, 213
  culture media, 60-62, 67
  cultures from, 93, 115
  frozen *see* Frozen food
  general methods, 190-194
  preparation utensils, sanitation control, 241-245
  streptococci, 332
  thermally processed, 67, 192, 230-238
  yeasts, 403
Food poisoning, 182-188
  Arizona winshawii, 185
  Bacillus cereus, 184, 187-188, 358
  botulism, 184-185, 188, 362
  campylobacter, 185
  chemical, 182
  Clostridium perfringens, 151, 183-184, 187, 362-364
  detection of organisms, 190, 194
  enterococci, 185
  Escherichia coli, 183
  examination methods, 186-188
  salmonella, 182-183, 186-187, 297
  scomboid fish, 203-204
  staphylococci, 183, 186-187
  Staphylococcus aureus, 187, 218, 235, 329
  types, 182-186
  Vibrio parahaemolyticus, 151, 183, 186
  Yersinia enterocolytica, 185
Food spoilage *see* individual products
Formaldehyde, 22-23, *41*, 42-43
Formalin, 22, 42
Francisella tularensis, 312, *314*, 316
Freeze-drying cultures, 96
Freezing cultures, 95
Friedlander's pneumobacillus, 285
Frozen food,
  convenience, 223-224
  fruit and vegetables, 225, 341
  ice-cream, 220-222
  meat, 195-197
  sampling, 191
  sea-food, 205
Fruit, 224-225, 411
Fruit juices, 239, 341
Fuchsin stain, 97
Fucidin, *169*, *174*
Fungi (*see also* Antifungal drugs, Moulds, Yeasts)
  black spore, 411-412
  clearing tissue for staining, 101
  direct examination, 114-147
  pathogenic, 413-423
  Risk Groups, *4-5*
  serological identification methods, 117-123
  slimy spore, 410, 411
  in water, 258-259
Fungizone, 124
Furunculosis of fish, 278, 308
Fusidic acid, 167
Fusobacteria, 399
Fusobacterium, 316-317, *318*

Gaffkya homari, 340
Gardnerella vaginalis, 157, 308
Gardner's STAA medium, 75
Gas gangrene, 364-365, 366
Gelatin,
  agar, 75
  food processing, 228

liquefaction test, 65, 107
nutrient, 79
Gemella, *322*, 323
Gentamicin,
assay, *174*, *176*, 179-180
blood agar, 71
sensitivity testing, 167, *169*
Geotrichum, 401, 403
Gillies medium, 64
Gliocladium, *410*
Glanders, 264
Glassware, 25-30
cleaning, 52-53
drying cabinets, 25
washing machines, 24, 52-53
Globulin production, anti-cryptococcal, 120
Gluconate oxidation test, 107
media for, 65, 75-76
Gluconobacter, *269*
Glucose, sterilization of, 84
Glucose phosphate medium, 76
Glutaraldehyde, 22, *41*, 42-43
Gonococcal arthritis, 153
Gonococcus, 321
Gonorrhoea, 155, 321
Gram stain, 98
Grass, 252
Griffin Hollow Shaft Stirrer, 15
Griffith's egg medium, 74
Griffith's tubes, 6, 15
Grinders,
counts, 244
tissue, 15
Growth promoters and inhibitors, 59-60

H antigens, 283, 287-288, 295-296
Haematin, 306
Haemin solution, stock, 68
Haemolytic organisms, media for, 70
Haemolytic streptococci, 332-333
Haemophilus, 306-308
media, 58, 60, 71
Haemophilus vaginalis, 157, 308
Haemorrhagic septicaemia in animals, 315
Hairs, infected, 114, 413, *414*
Half-antitoxin plate test,
Ham, 201, 228, 233
Hand washing, 7
Hanging drop culture, 109-110
Hanseniaspora, 405
Hansen's bacillus, 371
Hansenula, 401, 405
Hard swell, 234, 235, 368, 369
Heaters, metal block, 13
Helber counting chamber, *129*
Helminthosporium, 411
High pressure liquid chromatography (HPLC), 179-180
Histoplasma, 413, *421*, 422, 423
Histoplasmata, sensitivity testing, 125
Histoplasmosis, 119, 422, 423
Hoffman's bacillus, 348
Hospital use of disinfectants, 48
Hot air sterilizers, 32-33
Hot boning of meat, 195
Hot-rooms, 12-13
Howie Code of Practice, 47-48
Hucker Gram stain, 98
Hugh and Leifson test, 107-108
media for, 65, 76
Hydrogen culture for anaerobes, 92, 93
Hydrogen sulphide production, 108, 339
media for, 64, 65, 67
Hydrogen swell organisms, 61, 232
Hygiene standards,
indicators of, 190, 192, 355
testing methods, 241-245
Hypochlorites, 22, 40-42, 43

Ice-cream, 220-221, 222, 228, 359
Ice-lollies, 220, 221
Identification,
media, 63-67
methods (*see also* Individual methods), 102-113
mycological methods, 114-117
Illnesses, reporting, 7
Imidazole, 68
Immunoassay of antibiotics, 180
Immunoelectrophoresis, 119, *122-123*
Immunofluorescence *see* Fluorescent antibody tests
Immunosuppressive drugs, 147, 148, 380
'In-use' test for disinfectants, 45, 46-47
INAH (Isoniazid), 387, *389*
Incident light fluorescence, 164-*166*
Incineration, 32, *51*, *52*, 54
Incubator rooms, 12-13
Incubators, 11-12, 92-93
Indian ink, 100
Indicators,
of anaerobiosis, 92
culture media, 83-84
Indole formation, 65, 108
Infant formulae, 228
Infected airborne particles (aerosols), 5, 16-17
Infected material, treatment and disposal, *48-54*
Infections, laboratory-associated, safety procedures, 1-9
Inoculation methods, 89-91
Inspissated serum, 77
Inspissators, 24
International biohazard warning sign, *8*
International Collaboration Study, 166, 172
Intravenous infusion fluids, 249
Iodophors, 40, *41*, 43
Isoniazid (INAH), 387, *389*

Jam spoilage, 403
Jameson's plate toxigenicity test, 346-*347*
Jars, 235, 236
  anaerobic, 91-92
  counts in, 243
  for processed food, 230, 232
  sample, 28
Jencon Accuramatic, 25
Jensen's Gram stain, 98, 100
Johne's bacillus, 371
Jungle rot, 419

K antigen, 144, 283-284
Kampff Micro-bunsen burner, 31
Kanamycin, *169*, *174*, *176*
KCN *see* Potassium cyanide
Kelsey-Sykes capacity test, 45-46
Ketchups, 226
Ketoconazole, 124-125, 126
Kinyoun's acid-fast stain, 99
Kirby-Bauer method, 166, 172-173
Kirchner's medium, 76
Kit identification systems, 102, 103-105
Kitchens, sanitation control, 241-245
Klebsiella, 280, 282-283, 285
  antibiotic assay, *176*, 177
  hygiene indicators, 192
  identification, 106-107
  media, 74
Klebsiella aerogenes, *282*, 285
Klebsiella pneumoniae, *174*, *282*, 285
Kloeckera, 405
Koch steamer, 38
Koch-Weekes bacillus, 307
Kohn two-tube medium, 64
Krumwiede double sugar agar, 64
Kurthia, 351

Laboratory,
  Basic, *2*, 7-8
  Containment, *2*, 6, 8
  equipment, 10-31
  infections, prevention, 1-9
  Maximum Containment, *2*, 8
Laboratory Pasteurization Count (LPC), 211
Lactobacillus, 352-355
  identification, 105
  media, 58, 60, 64, 69, 73
Lactophenol stain, 101
Lactophenol-cotton blue stain, 101
Laminar flow cabinet, 17
Laminar flow clean air work stations, 24
Latex agglutination tests,
  Cryptococcus, 120-121
  streptococci, 333-334
Laybourn's modification of Albert's stain, 99
Leaker spoilage, 233-234, 235
Lecithinase activity, 108-109, 110
  media, 65, 67
Legionella, 321-322
  media for, 58, 60, 154
Leprosy, 371
Leptospira, 400
Lethheen agar, 77
Lethheen broth, 77
Leuconostoc, 234, 235, *340-343*
Levan production, 81, 109
Lincomycin, *174*
Lipolytic activity test, 109
  media for, 61, 81
Lissamine rhodamine B (RB200), 160
Listeris monocytogenes, 158, *352*
Loeffler medium, 77
Looping-out method, *90*
Loops, bacteriological, 6, 30
Lowbury and Lilly medium, 74
Lowenstein-Jensen medium, 77
Lumpy jaw of animals, 395
Lyophilization of cultures, 96
Lysine agar, 67

'McCartney' culture bottles, 27
MacFadyean's reaction, 356
Macro-Vue card test, 179
Macroconidia, types, *415*
Madurella mycetomatis, 420
Malonate utilization test, 65, 109
Malt agar, modifications, 77-78
Malta fever, 305
Mastitis, bovine, 213, 332, 336, 337
Mead's medium, 78
Meat, 195-202
  boned out, 195-196
  botulism, 184, 185
  Brochothrix, 75, 350
  carcase, 195
  Clostridium perfringens, 184
  comminuted, 196
  cured, 199-202, 233
  media, 62
  pies, 197-198
  pre-packed fresh, 197
  spoilage organisms, 196, 197, 199, 200, 369
  vacuum-packed, 200-201
Mediterranean fever, 307
Melioidosis, 265
Membrane filters, 39
  bottle counts, 243
  counting methods, 134-136
  media for, 62-63, 135
  for water examination, 253-257, 258
Meningitis, 149, 307, 321
Meningococcus (*see also Neisseria meningitidis*) 322-323
Mercurial compounds, 40, 44
Metal block heaters, 13
Methicillin resistance, 168-*169*
Methyl blue stain, 115

Methyl red, 83
Methyl red tests *see* MR tests
Methyl violet, 98, 100
Methylene blue reduction tests,
  for cream, 214
  for ice-cream, 220–221
  for milk, 208–209, 211
Methylene blue stain, 97, 136
Miconazole, 124–126
Micro ID system, 104
Micro-incinerators, 32
Micro-slides, 29–30
Microaerophiles, culture of, 92
Microbact identifications system, 104
Microbacterium, 350
Microbiological Safety cabinets (*see also* Safety cabinets) 16–24
Micrococci (*see also Aerococcus, Pediococcus*), 325–328
  media for, 64, 72, 76
Micrococcus, *328*, 330
Microfungi in water, 258–259
Micropolyspora faeni, 397
Microscopes,
  fluorescence, 160–162
  incident light, 164–166
  low power, 11
Microsporum, 413–*415*
Miles and Misra drop counts, 86–87, 133
Milk, 208–213
  acidophilus, 354
  antibiotics in, 213
  bottles, 243
  Brucella, 304, 305–306
  churns, counts in, 242–243
  dried, 359
  human, 216
  Lactobacillus, 354
  Leuconostoc, 341
  media for, 65, 78
  microbial content, 212–213
  processed, 214–215
  UK statutory tests, 207–210
  US standard methods, 210–211
Milk salt agar, 78
Milk (Special Designation) Regulations, 207, 208–209
Mincers, counts in, 244
Minerals Modified Glutamate broth, 254–255
Minimum Inhibitory Concentration (MIC), 172
Minitek systems, 104
Mitcheson and Spicer method of antibiotic assay, 176–177
Modal resistance, *390*
Moeller's decarboxylase method, 73, 106
Molasses, 226
Morax-Axenfeld bacillus, 309
Moraxella, *261*, 308–*310*
Mossy foot, 419
Most Probable Number method,
  tables, *136–141*
  water examination, 254–255, 257, 258
Motility tests,
  media, 66, 80–81
  methods, 109–110
Moulds (*see also* Individual organisms), 406–412
  counting, 136, 194
  examination methods, 114–116
  media, 61, 62, 63, 69, 78
  spoilage, 196
Mouth, infection via, 6, 29
MR (Methyl red) tests, 109
  media for, 66, 76
MRS medium, 64, 78
MSE Homogenizer, 15
Mucor, 406, *407*
Muller-Chermack acid-fast stain, 99
Mushroom workers' lung, 397
Mycetoma, 397, 420
Mycetoma grains, 114
Mycobacterium, 371–392
  anonymous, 380
  media, 58, 60, 76, 77, 78–79, 80
  opportunist, 371, 375, 380–387
  Runyon's classification, *380*
  saprophytic, 375, 380
  sensitivity tests, 387
  tubercle bacilli, 379
Mycobacteroses, 380
Mycological methods, 114–127

N medium, 78–79
Nagler-half-antitoxin plate test, 110
  Clostridium perfringens, 364
  media, 66
Nagler test, 110
Nails, mycological examination, 114, 410, 413, *414*
Nalidixic acid, *169*
Nasal swabs, 152
National Collection of Type Cultures, 94
Neisseria, 321–323
  carbohydrate media for, 105
  fluorescent antibody tests, 162–163
  media, 58, 60, 71, 105
Neisseria gonorrhoeae, 150, 321–322
Neisseria meingitidis, 152, 321, 322–*323*
Neomycin, 82
Neomycin blood agar, 71
Neurospora sitophila, 411
Niacin test, 378
Niger seed agar, 79
Nitratase test, 110
  media, 66, 67
Nitrofurantoin, 167, *169*
Nitrogen for anaerobic culture, 92, 93
Nitrogen assimilation media, 79
Nocardia, 383, 396–*397*
Nutrient gelatin, 79
Nutritional requirements, testing, 94
Nystatin, 124–125

O antigens, 283, 287–288
o-Nitrophenyl-D-galactopyranoside, 110
OF test, 107–108
Oily food, 229
Onion powder, 228–229
ONPG test, 110
  media for, 66, 79
Oospora lactis, 403
Opacity tube method, 130
Opportunist mycobacteria, 371, 375, 380–387
Optochin test, 110–111
Organic acid utilization, 66, 70
Ovens, sterilizing, 32–33
Oxferm test, 107–108
Oxford Staphylococcus, 170
Oxi-Ferm identification system, 104
Oxidase test, 111
Oxidation fermentation test, 107–108
Oxygen preference test, media for, 66, 80
Oysters, 204

p-Aminobenzoic acid, 167
p-Aminobenzoic acid (PABA) blood agar, 71
Paecilomyces, *408*, 409
Paper discs *see* Discs
Paper strips,
  antibiotic, 172
  identification tests, 102, 103
Paracoccidioides, 413
Paraoccidioides braziliensis, *421*, 422
Paracolon bacilli, 300
Paraformaldehyde, 22, 42
Pasteur pipettes, 6, 28
Pasteurella, 312, 315
  additives for, 60
Pasteurized milk, 207, 212
Pastry, 227
Pathotec system, 103
Pediococcus, 342–343
  beer spoilage, 239, 342
  media for, 58
Penicillin, *169*, *174*, *176*
Penicillinase-producing organisms, 168
Penicillium, 406–407, *408*, 409
Peptococcus, 325, 329
Peptone water,
  alkaline, 67
  diluent, 83
  sugars, 72
Peptostreptococcus, 339
Petri dishes, 25, 26
  incubation, 93
  pouring, 25, 86
pH,
  antibiotic assay, 174–175
  culture media, 85
  sensitivity tests, 167
Pharmaceutical products, 247–249
Phenol coefficient tests, 44
Phenol red, 84
Phenolic disinfectants, 22, 40, *41*, 44
Phenolphthalein, 83
Phenolphthalein phosphate agar, 79–80
Phenolphthalein sulphate agar, 80
Phenylalanine (PPA or PPD) test, 66, 111
Phialophoras, 419, *420*
Phosphatase test, 111
  media for, 66, 79
  milk, 209–210
  staphylococci, 328
Photobacterium, 279
Phycomycosis, subcutaneous, 406, 413, 420
Pichia, 401, 405
Pickles, 225–226
  bacteria in, 341, 342, 354
  yeasts in, 403, 405
Pimaricin, 124–125
Pinta, 399
Pipette(s),
  graduated, 28–29
  jars, 48, 50
  pasteur, 6, 28
  washing, 53
Pipetting, hazards of, 6, 29
Pipetting aids, 29
Pipetting machines, automatic, 25
Pipetting methods, for dilutions, 130–131
'Pipettors', 29
Pitching (brewing) yeasts, 342, 403, 405
Plague, 312–313
Plague bacillus (*Yersinia pestis*), 80, 312–313, *314*
Plastic bags, 28, 49, 51, 52
Plastic laboratory equipment, 25–31
Plate counts, 131–132
  preparation methods, 89–90
  water examination, 254
Plate diffusion, antibiotic assay, 177–178
Plate test, half-antitoxin *see* Nagler half-antitoxin plate test
Plate toxigenicity tests, 74–75, *345–347*
Plates, petri *see* Petri dishes
Plesiomonas, *271*, *272*, *275*
Pneumococci (*Streptococcus pneumoniae*), 339–340
  countercurrent immuno-electrophoresis, 149
  optochin test, 103, 110, 339
Polymyxins, *169*, *174*
Porton impinger, 115
Post-mortem material, 155
Posting specimens, 6–7, 28
Potassium cyanide (KCN) test, 65, 108
Potassium permanganate oxidation, 22–23
Poultry, 184, 198–199
Pour plate cultures, 89
PPA test, 66, 111
Prawns, 204, 205
Precipitin tests, 334, *335*
Pregnancy, 7
Preisz-Nocard bacillus, 349

Preparation room,
design features, 51-*52*
equipment, 24-25
Propanol, 43
Propiolactone, 40
Propionibacterium, 218, 350-351
Proteolysis test, 111
Proteus, 280, 281, 300-*301*
in pus, 152
swarming, 168, 300
Providence, *280*, 300-*301*
Pseudomembranous colitis, 365, 366
Pseudomonas, *262*-265, *271*
media, 58, 64, 72
other species, 263-265
Pseudomonas aeruginosa, 170, *174*, *176*, *263*
Pseudotuberculosis in animals, 315, 349
Psychrophiles, incubation of, 93
Puerperal infections, 157
Pullularia pullulans, 412
Pulmonary eosinophilia, 409
Pus, 152-153
Pyrazinamide, 387, 391-392

Quality control of culture media, 86-87
Quaternary ammonium compounds (QAC), 40, *41*, 43

Racks, test-tube, 30
Radio immunoassay (RIA), 180
Rayson's method of drying bacteria, 95
Re-usable laboratory equipment, 25-26, *51*
Rectal swabs, 151-152, 291
Redigel media, 56
Reference laboratories, 297-298
Rennet, 350, 405
Resistance ratio method, 388-391
Rhizopus, 406, *407*
Rhodotorula, 401, 402
Rice, fried, 184, 358
Rideal-Walker test, 44
Rifampicin, 387, *389*
Ring test, 305
Ringer solutions, 82-83
Ringworm, 415, 418, 419
Rinse method, 242-243
Risk Group classifications, *1-5*
Robertson's cooked meat medium, 56, 80, 147-148
Robinson's serum water sugars, 64, 73
Roll-tube counts, 132-133
bottles, 243-244
shellfish, 204
Rope organisms, 227
Ropy beer, 269
Ropy bread, 227, 359
Rotary plating, *90*, 171
Rubber teats, 29
Runyon's classification of mycobacteria, *380*
Russell's double sugar agar, 64

Saccharomyces, 401, 403
Saccharomyces cerevisiae, 127, *175*
Safety (*see also* Codes of Practice), 1-9
Safety cabinets, 16-24
airflow testing, *19-21*
Class I, 16-*17*, *20-21*
Class II, *17*, *20-21*
Class III, 17-*18*, 20-*21*
decontamination, 21-23
maintenance, 23-24
siting, 18-*19*
Salad creams, 226
Saline solutions, 82-83
Saliva, 154, 355
Salmonella,
in animal by-products, 251-252
antigens, 287-*289*
biochemical reactions, *280*, 292-*293*, *294*
culture, 290-292
food poisoning, 182-183, 186, 297
media for, 58-59, 60, 63
serological identification, *293-296*
species and serotypes, *288*, 297-298
in water, 259
Salmonella typhi, 297
agglutination tests, 143-144
disinfectant testing, 45, 47
media, 63
Salt agar, 80
Salt beef, 199
Salt meat broth, 80
Sample jars and containers, 28
Sampling,
air, 115, 136, 245
clinical material, 146-147
food processing factory, 191
methods in sanitation control, 241-245
water, 253
Sandwich technique, 162
Sanitation control, 241-245
Sarcina, 325, 329
Sarcina sickness, 342
Satellitism, 168, 306
Sauces, 226
Sauerkraut, 338, 341, 342, 354
Sausages, 197, 350
Scarlet fever, 336
Schleifer and Kloos test, 66
Scopulariopsis, *408*, 410
Scomboid fish poisoning, 203-204
Sea food, 203-206
Sea water, 260
Semisolid oxygen preference medium, 80
Sensitivity tests, 166-173
controls, 170
errors, 168, 173
media for, 63, 71
methods, 169-173
mycobacteria, 387-392
Septicaemia, 147, 148
Serous fluids, 153

Serratia, 280, 281, 286, 294, 329
Serratia marcescens, 300
Serum,
  antibiotic assay, 174–180
  bactericidal activity, 179
Settle plates, 245
Sewers, 291
Shake tube cultures, 91
Shaking machines, 16
Shellfish, 204–205
Shelving, for incubator rooms, 12
Shigella, *280*, *281*, *292–293*, *294*, 299
  culture, 290–292
  media for, 59
  serotypes, 289–290
Shrimps, 204, 205
Silage, 252, 341, 354
Silver nitrate stain, 100–101
Skim milk agar, 80
Skin (*see also* Dermatophytes)
  cultural examination, 114
  erysipeloid lesions, 351
  infected, 413, *414*
Slide agglutination, 145
Slide culture, 116
Slopes of culture medium, preparation, 24, 86
'Sloppy agar' Craigie tubes, 80–81, 110
Sodium bicarbonate activator, 42
Soft drinks, 239, 243
Soil, cultures from, 115
Specimens,
  collection of clinical, 146–147, 174
  containers, 7, 27–28, 146
Spicer–Edwards Rapid Salmonella Diagnostic agglutination
  sera, *295*
Spices, 228–229
Spiral plating, *90*, 134
Spirochaetes, 398–400
Sporangiophores, 406, *407*
Spore bearers, media, 64, 72, 82
Spores,
  examination, 115
  staining, 100
Sporobolomyces, 401, 402
Sporothrix, 413, *421*, 422
Sporotrichosis, 422
Spreaders, 30
Spreading plates, 89–*90*
Sputum, 147, 153–154
  mycobacteria, 371–372, *373*, 374
Stability test, 45, 46
Staining methods, 97–101
  counting, 136
  double diffusion tests, 123
  fungi, 115
Stamp's method of drying cultures, 95
Standard solutions, antibiotic assay, 175, *176*
Staphylococcal enterocolitis, 151
Staphylococcus, 325–*329*
  burns, 158–159
  food poisoning (*see also S. aureus*), 183, 186–187
  identification tests, 103, 326–328
  media, 59, 63, 64, 72, 76
  nasal swabs, 152
  resistance, 168–169
Staphylococcus aureus, 328–329
  antibiotic sensitivity testing, 170
  drug assay, *174*, *176*
  food poisoning, 187, 218, 235, 329
  identification tests, 106, 326–328
  media for, 78, 80, 151
  V factor, 306
Starch agar, 81
Starch hydrolysis, 66, 111
Steam sterilization (*see also* Autoclaves, Tyndallization), 33–34
Steamers, 24, 38
Sterility test media, 63
Sterilization,
  definition, 32
  indicators, 38
  methods, 32–39
Sterilized milk, 208, 210
Stock cultures, 74, 86–87, 94–96
Stokes method of sensitivity testing, 166, 170–*172*
Stomacher Lab-Blender, 15
Streptococcus, 331–340
  burns, 158–159
  food processing, 192
  grouping methods, 333–335
  groups and species, 336–340
  media, 59, 60, 63, 69
  water, 257–258
Streptococcus faecalis, 78, 337
Streptococcus pneumoniae, 339–340
Streptococcus pyogenes, 154, 177, *336*
Streptomyces, 396–397
Streptomycetes, in water, 258–259
Streptomycin,
  assay, *174*, *176*
  sensitivity testing, *169*, 387, *389*
Subcultures,
  blood cultures, 148
  of mixed cultures, 91
  stock cultures, 95
Sucrose agar, 81
Sugar sickness, 341
Sugars, spoilage of, 226
Sulphafurazole, *169*
Sulphatase test, 67, 112
Sulphonamides, 71, 167
Sulphur granules, 152
Sulphur stinkers, 235, 368, 369
Supplements, culture media, 59–60, 67
Surface count method, 134
Surfaces, counts on, 241, 242
Swab counts, 242
Swab rinse method, 242
Swab sticks, 31

Swabs, clinical material, 146
Swarming, 168, 300
Swimming pools, 259, 385
Swine influenza, 307
Syphilis, 399
Syrups, 226

Teats, rubber, 29
Teeth, 149
Tellurite blood agar media, 81
Tellurite reduction, 112
Temperatures, incubation, 11–13, 93, 191–192
Test-tubes, 26–27
  baskets and racks, 30–31
  stoppers, 26–27
Tetanus, 364–365, 366
Tetracycline, 167, *169*, *174*
Thermoactinomyces, 397
Thermocouples, 33, 37
Thermophiles, incubation, 93
Thermostats, for incubator rooms, 12
Thiacetazone, 387
Thioglycollate medium, 81
Thiosulphate Ringer solution, 83
Throat swabs, 154–155
Thymidine, 167
Thymol blue, 84
Tissue grinders, 15
Tissue homogenizers, 6
Tissue specimens, 155
Titanium tetrachloride, 20
Tobramycin, *176*
Tonsillitis, 336
Torulopsis, 401, *402*
Total counts, 128–130
Towels, counts on, 244
Toxigenicity tests, diphtheria, 74–75, *345–347*
Training, safety, 8–9
Transport media, 60
Treponema, 398–399
Treponema pallidum, 163–164, 399
Treponema vincentii, 399
Tributyrin agar, 81
Trichoderma, *410*
Trichomonas vaginalis, 158
Trichophyton, 413, 415–419
Triphosporon, 117, 401, *402*, 403
Trimethoprim, *169*, *174*
Triple sugar iron agar, 64
Tubercle bacilli, 371–380
  animal inoculation tests, 379–380
  examination, 371–376
  identification, 376–378
  sensitivity tests, 387–392
  varieties, 379–380
Tuberculosis, 149, 379, 387
Tuberculous materials, safety measures, 7, 380
Tularaemia, 312, 316
Turbidity test, milk, 210
Tween hydrolysis, 112
Tyndallization, 38
Type Culture Collections, 94–95
Typhoid carriers, agglutination tests, 143–144
Tyrosine agar, 82
Tyrosine decomposition, 67, 78, 112

Ultra-heat-treated (UHT) milk, 207, 210
Ultraviolet light, 160
Undulant fever, 305
'Universal Container', 27
Urease test, 67, 112
Urethral discharges, 155–156
Urinary tract infections, 156
Urine,
  24-h, disposal of, 53
  antibiotic assay, 175
  countercurrent immunoelectrophoresis, 149
  mid-stream, 156
  samples, 146–147, 156–157, 291
Use-dilution tests, 45–47

V factor, 306–307
Vacuum blood collection systems, 147
Vacuum packaging, 200–201
Vaginal discharges, 157,–158
Vancomycin, *174*, *176*
Vegetables,
  botulism, 184, 185
  spoilage, 224–225, 341
Veillonella, 323–324
Ventilation, laboratory, 8, 19
Vertical diffusion method of antibiotic assay, 176–177
Vi agglutination tests, 143–144
Vi antigens, 144, 283–284, 288, 296
Viable counts, 128–129, 130–136
  food, 190, 191–192
Vibrio, *261*, *263*, *271–277*
  media for, 59
  in water, 259
  other species, 276–277
Vibrio cholerae, 67, *272*, *274–276*
Vibrio parahaemolyticus, *272*, *275*, 276
  food poisoning, 151, 183, 186
Vincent's angina, 399
Vitamin K solution, stock, 68
Vitamin requirements, 94
Voges Proskauer Reaction (VP) tests, 66, 76, 112
Vortex Mixer, 16
VP (Voges Proskauer Reaction) tests, 66, 76, 112

Walls, counts on, 244
Washing,
  glassware, 52–53
  hand, 7

Washing up, hygiene, 244
Water, 253-260
  Clostridium, 258
  deionized and distilled, 53, 84
  faecal streptococci, 257-258
  media for, 62
  mycobacteria, 375, 384, 385
  pathogens, 259
  Streptococcus, 332
Water-baths, 13
Weil's disease, 400
Whelks, 204
Whey, 354
Whey agglutination test, 305-306
Whooping cough, 152, 310-311
Willis and Hobbs medium, 82
Winkles, 204
Winter Vomiting Disease, 182
Wires, inoculating, 30
Wood splinters, infection from, 422
Woolsorter's disease, 358
Work surfaces,
  cleaning, 6
  counts, 242
World Health Organization,
  laboratory design, 7-8
  Risk Groups, *1-5*
Wort, 342
Wounds, deep and superficial, 157-159, 363
    365
Wright's stain, 100

X factor, 306-307
Xanthine agar, 82
Xanthine decomposition, 67

Yaws, 399
Yeast agar, buffered, 69
Yeasts, 401-405
  acidophilic, media for, 61, 77
  antigen production, 117-119
  auxanograms, 116-117
  baking, 403
  brewers' (pitching), 342, 403, 405
  counting, 136, 194
  cultural methods, 114, 115, 116
  food, 403
  identification, 103, 104, 401, *402*
  media, 61, 62, 69, 73, 79
  morphology, 401, *404*
  pathogenic, media for, 60, 77-78
  pitching (brewing), 342, 403, 405
  Risk Groups, 4-*5*
  sensitivity testing, 125
Yersinia, *261*, 312-315
  media for, 60
Yersinia enterocolitica, 313-315
  food poisoning, 185, 188
  media for, 59, 151
Yersinia pestis (Plague bacillus), 80, 312-313,
    314
Yoghurt, 215, 339, 354

Ziehl-Neelsen acid-fast stain, 98-99, 100
Zipettes, 25
Zones of inhibition,
  antibiotic assay, 175
  factors affecting, 166-169
  measurement, 172, 177
Zopfius, 351
Zygomycetes, 115, *407*
Zygomycotina, 406